Microplastics
Transport, Impacts, Monitoring and Mitigation

Microplastics
Transport, Impacts, Monitoring and Mitigation

Natalie Welden

*School of Social and Environmental Sustainability,
University of Glasgow, Dumfries, United Kingdom*

ELSEVIER

Elsevier
Radarweg 29, PO Box 211, 1000 AE Amsterdam, Netherlands
125 London Wall, London EC2Y 5AS, United Kingdom
50 Hampshire Street, 5th Floor, Cambridge, MA 02139, United States

ISBN: 978-0-443-13324-4

For Information on all Elsevier publications
visit our website at https://www.elsevier.com/books-and-journals

Publisher: Candice Janco
Acquisitions Editor: Jessica Mack
Editorial Project Manager: Teddy A. Lewis
Production Project Manager: Paul Prasad Chandramohan
Cover Designer: Vicky Pearson Esser

Typeset by MPS Limited, Chennai, India

Dedication

*This book is dedicated to my ever-patient parents,
Sandie and Paul, even though Jack had to do all the housework
while I was writing it.*

Contents

About the author

Dr. Natalie Welden began her work on microplastics in 2011, undertaking her PhD at the University Marine Biological Station in Millport at the onset of the current explosion of microplastic literature. As a result of the paucity of information at the time, Natalie's initial research was unavoidably holistic, exploring microplastic sources, formation, transport, uptake and impacts. Building on these themes, her subsequent work has focused on solutions to the plastic problem, including the use of alternative materials and improved filtration to reduce our plastic output. Over this period, she has worked closely with policymakers, industry stakeholders and the public to increase our understanding and encourage behavioural change. Natalie now lives in Dumfries and Galloway where, in addition to writing, walking, wild swimming and caring for her chickens, she is working as a lecturer in environmental science and sustainability at the University of Glasgow.

It is much easier to prevent plastics from getting into the environment than to take them back out again, especially when trying to pick them up with enormous tweezers.

Preface

Microplastic pollution has become one of the most significant environmental stories of the past two decades, initially noted as much for its apparent inconspicuousness as for its scale and scope. Despite their apparent abundance, microplastics were 'lost at sea', present but unperceived, hiding in plain sight. The same was quickly seen to be the case when it came to their impacts. Plastics are inert (not very chemically active) but are apparently able to affect both organisms and environments in a variety of previously unexpected ways.

Once conscious of the existence of microplastics, they seemed to be everywhere, and academic circles quickly came to acknowledge the fact that microplastics are all-pervasive. Subsequently, I do not believe that any researcher in the field was surprised when microplastics turned up in deep ocean sediments in 2014, on Mount Everest in 2020 or in Antarctic snow in 2022. Indeed, the identification of microplastics in novel settings is decidedly an issue of *when* (when funding is available, when new methods are developed, when the opportunity arises), rather than *if*. The impacts may not be limited to our own atmosphere either, there is plenty of plastic in space junk. Even the flag we left on the Moon is Nylon.

Nevertheless, the apparent contamination of remote and supposedly pristine habitats — along with a slew of charismatic or commercially important animals — continues to make the news on a semi-regular basis. Fortunately, this consistent narrative has resulted in a global movement to manage and mitigate microplastic pollution, often by innovative or unexpected means.

It is understood that the key factor driving the proliferation of microplastics is our uncontrolled use of plastics and the mishandling of plastic wastes. Combined, these factors have resulted in a huge standing stock of plastic material from which microplastics may form. However, the efficacy of our efforts to manage plastic wastes varies significantly depending on both their origin and the capabilities of the region in which they are generated.

In addition to managing plastic at the end of their useful life, behavioural change is key to reducing our total consumption. Unlike many pollutants, whose complex names and obscure origins have little to do with our day-to-day lives, plastics are a problem which most are able to understand and regularly interact with. This familiarity has given the wider public significant agency in identifying and addressing various sources of plastic and microplastic pollution, from cosmetics and clothing to face scrubs and straws. While voluntary behavioural change has led the way, changes in regional, national and international policy have sought to minimise the use of the most damaging or wasteful products, by banning their use, introducing levies or enabling more sustainable management.

Nevertheless, it is important to remember that, despite the scale of recent interest, microplastics are only a small part of the widespread negative impacts on our environment and the biological communities and people that depend upon them. The global levels of microplastic contamination and the impacts we have thus far observed are set against the dual foils of climate change and biodiversity loss (influenced not only by climate but by a host of primary and secondary anthropogenic effects).

Additionally, the steps that we take to address the issue of microplastics are seldom considered 'in the round', with insufficient comparison made between products and services and their varied impacts on people and the environment. Thus, as we strive to limit our output of plastic to the environment, we must ensure that our actions are not at the expense of other concerns, such as competition for land, water and food. Our response to this issue must be carefully considered. Successful microplastic management requires a holistic overview of the issue: of key sources, of the most damaging forms and of sustainable management approaches. To achieve these aims, we must bring together practitioners from a range of industries and backgrounds and ensure that the outcomes of these collaborations are accessible to all.

A problem on the scale of plastic production requires a collaborative response. Predominantly, it is public pressure that has led to the current changes in both policy and corporate behaviour, public demand that has driven the growth in low- and zero-waste industries and public participation that improved conditions in local rivers, beaches, parks and roadsides through litter picks and plastic surveys. There is already widespread will to tackle microplastic pollution if we can only identify the appropriate means.

Natalie Welden

Acknowledgements

I would like to acknowledge the willingness of so many friends and colleagues for their thoughts, insights, and opinions throughout the writing of this book. In particular, I would thank Dr. Amy Lusher, Dr. Bryce Stewart and Dr. Matthew Cole for the use of their wonderful images.

Finally, I would like to express my deepest gratitude to the Scottish Cancer Network, particularly Drs. Dutton and Ghaoui, Mses. Dawson, Chitambo and Killen, and the staff of both Dumfries and Galloway and Edinburgh Royal Infirmaries.

List of abbreviations

AAS	atomic absorption spectrometry
ABS	acrylonitrile-butadiene-styrene
ALDFG	abandoned, lost and discarded fishing gear
BPA	bisphenol A
CE	cellulose
CGC-ECD-MD	capillary gas chromatograph equipped with electron capture detector and mass detector
CP	chlorinated parafin
CP/MAS NMR	cross-polarisation magic-angle-spinning nuclear magnetic resonance spectroscopy
DDE	dichlorodiphenyldichloroethylene
DDT	dichlorodiphenyltrichloroethane
DRS	deposit returns scheme
EFW	energy from waste
EPRS	extended producer responsibility scheme
F-gas	fluorinated hydrocarbons
EPS	extracellular polymeric substances
FPA	focal plane array
FTIR	fourier-transform infrared spectroscopy
GC-ECD	gas chromatography-electron capture detector
GC-IMS	gas chromatography with ion mobility spectrometry
GCMS	gas chromatography mass spectrometry
GESAMP	group of experts on the scientific aspects of marine environmental protection
HCB	hexachlorobenzene
HCH	hexachlorocyclohexane
HDPE	high-density polyethylene
HPLC−MS	high-pressure liquid chromatography−mass spectrometry
ICP-MS	inductively coupled plasma mass spectrometry
ICP-OES	inductively coupled plasma optical emission spectroscopy
FAAS	flame atomic absorption spectroscopy
FP-XRF	field portable x-ray fluorescence
LCA	lifecycle assessment
LDPE	low-density polyethylene
LOD	limit of detection
LOI	loss on ignition
LOQ	limit of quantitation
LSC	liquid scintillation counting
MP	microplastic
MSFD	marine strategy framework directive
NGO	nongovernmental organisation
NP	nanoplastic
NR	natural rubber
OECD	Organisation for Economic Cooperation and Development

OSPAR	Oslo and Paris Conventions
PA	polyamide = nylon
PAH	polycylic aromatic hydrocarbons
PHA	polyhydroxyalkanoates
PB	polybutylene
PBDE	polybrominated diphenyl ethers
PBAT	polybutylene adipate terephthalate
PBT	polybutylene terephthalate = ptmt
PC	polycarbonate
PCB	polychlorinated biphenyls
PE	polyethylene
PET	polyethylene terephthalate
PFAS	perfluoroalkylated substances
PLA	polylactic acid
POP	persistent organic pollutant
PP	polypropylene
PRS	producer responsibility scheme
PS	polystyrene
PUR	polyurethane
PVA	polyvinyl acetate
PVC	polyvinyl chloride
PVOH	polyvinyl alcohol
REACH	registration, evaluation, authorisation and restriction of chemicals
ROV	remotely operated vehicle
SA:V	surface-area-to-volume ratio
SDG	sustainable development goals
TG-FTIR	thermogravimetric interfaced with fourier-transform infrared spectroscopy
TPA	terephthalic acid
WEEE	waste electrical and electronic equipment
WTE	waste to energy
WWT	wastewater treatment

The proliferation of plastics

1.1 Introduction

Observations of plastic debris in the environment have been common for many decades, with the composition, size and apparent sources of this litter being both spatially and temporally diverse. Scientific reports of plastic in marine settings arise as early as the 1970s (Jewett, 1976), and include the global movements of tsunami debris and lost material from shipping and fishing activities, in addition to the occurrence of smaller plastic objects and plastic fragments. Records of plastic litter encompass some of the world's most remote locations, including deep seas and uninhabited islands. Similarly, plastic litter is regularly observed in studies of freshwater and terrestrial habitats, a result of the mass of mishandled plastics, currently estimated at approximately 22% of that produced each year. Over the intervening decades following the first observations of widespread plastic litter, our monitoring and reporting of plastics and other debris in the environment has become increasingly standardised and widespread. Subsequently, certain environments, such as the deep sea, have been identified as long-term plastic sinks (Davies, 1987).

1.1.1 New polymers solve old problems

Plastics are synthetic polymers (molecules made of chains of repeating elements called monomers) used in many industrial and domestic products. In order to understand the widespread distribution of these man-made polymers in the environment, we must explore the drivers of plastic proliferation, beginning in the mid-19th century, with the synthesis of Parkesine from nitrocellulose and camphor by Birmingham-based chemist Alexander Parkes (in collaboration with the inventor Daniel Spill). Although unable to successfully market and produce the material at an industrial scale, the subsequent sale of the patented process to American John Wesley Hyatt resulted in the development of celluloid, the first — notably flammable and sometimes explosive — thermoplastic. Other early developments include the production of Shellac in 1894, Polyethylene (PE) in 1898, Bakelite in 1907 and Cellophane in 1912, followed by Polystyrene (PS) in the 1930s.

As with other modern environmental concerns, the synthesis and development of these novel materials was received with widespread appreciation at the time. Plastics represented a potential answer to the limited availability of numerous

Microplastics. DOI: https://doi.org/10.1016/B978-0-443-13324-4.00001-7

natural resources, such as of ivory, silk and rubber. Shortages of ivory had previously been driven by widespread hunting (driving manufacturers Phelan and Collender to offer a $10,000 reward in 1863 for an ivory alternative suitable for use in billiard balls), whereas those of silk and rubber were caused by multiple factors, not least, international conflict. For example, in 1941 the supply of Japanese silk to America was interrupted by increasing political tensions. In response, the US Office of Production Management acquired the entire national silk supply for the construction of parachutes and powder bags. Particularly affected was the hosiery industry, which responded by first raising the price of their products followed by turning entirely to the use of Nylon (polyamide), which had been introduced to the industry just two years before. Nevertheless, the availability of American legwear was dealt a further blow just a year later, when the renamed War Production Board laid claim to DuPont's entire Nylon supply for the manufacture of tyres, ropes and parachutes. Similarly, the development and widespread uptake of a replacement for natural rubber were also linked to the shortages of the Second World War, which drove German researchers, led by Professor Otto Bayer, to develop the first polyurethanes (PU). This new material was used as a chemical-resistant coating, of particular importance to soldiers exposed to mustard gas (Seymour & Kauffman, 1992).

Over the intervening century, research and development by a variety of commercial chemical companies has led to rapid growth in the number and diversity of plastic polymers. Unlike the earliest plastics, which employed nitrocellulose (a plant-based organic polymer treated with nitric and sulphuric acid) as a key component, the majority of these modern plastics utilise naphtha and other products of the fractionation processes in the oil industry. These processes provide a variety of light-weight monomers which may be subjected to either addition or condensation polymerisation, the products of which form the feedstock for the manufacturing industry (Fig. 1.1). The precise polymer formed depends on the mix of monomers and the bonds that are formed between them, for example ester

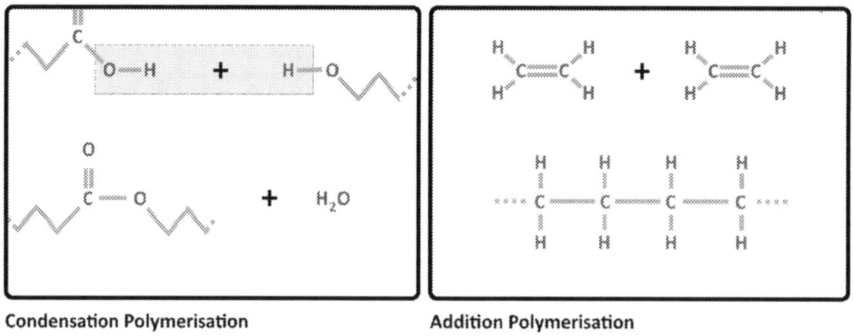

Condensation Polymerisation   **Addition Polymerisation**

FIGURE 1.1 Polymerisation steps.

Comparison of condensation and addition polymerisation steps.

bonds are found in polyethylene terephthalate, polycaprolactone and poly-lactic acid among others. We will discuss the significance of this chemical structure in later chapters.

The benefits of these new materials were seen to be numerous when compared to both traditional materials of the time (and, indeed, the earlier, less stable plastics). Plastic polymers have a low density and weight, low degradation rates, mechanical properties which could be easily adjusted to suit the requirements of the manufacturer and user, and, unlike early plastics, were less likely to explode in the event of exposure to heat or being struck. These positive properties are conferred by the chemical structure of the polymer. For example, the length of polymer chains results in a high molecular weight which limits degradation by the action of biota. The mechanical properties of a plastic may also be affected by alterations to the polymer structure or the inclusion of additives and fillers

Table 1.1 Examples of polymer fillers and their applications.

Additive	Examples
Antimicrobials/biostabilisers	l0′,l0′-Oxybisphenox arsine
	2-n-Octyl-4-isothiazolin-3-one (OIT)
	Dichloro-2-noctyl-4-isothiazolin-3-one (DCOIT)
	Tributyl tin (TBT) silver and silver compounds
Antistatics	Migrating antistatic agents
	Long-chain alkyl phenols
	Ethoxylated amines
	Glycerol esters
	Permanent antistatics
Biodegraders	Starches
	Prooxidants
Blowing agents	Ammonium bicarbonate
	Sodium bicarbonate
	Sodium borohydrate
Fillers	Chalk
	Talc
	Barium sulphate
	Glass
	Carbon
Flame retardants	Halogenated flame retardants
	Organo-phosphorous flame retardants
Impact modifiers	Acrylonitrile butadiene styrene (ABS)
	Acrylate resin (ACR)
	Chlorinated polyethylene (CPE)
	Ethylene vinyl acetate (EVA)
	Methyl methacrylate-butadiene-styrene copolymer (MBS)
	Ternary Ethylene propylene rubber (EPT)

(Continued)

Table 1.1 Examples of polymer fillers and their applications. *Continued*

Additive	Examples
UV stabilisers	Carbon black
	Rutile titanium oxide
	Hydroxybenzophenone
	Hydroxyphenyl benzotriazole

during the manufacturing process. Table 1.1 outlines a selection of additives and their associated effects on the polymer, we will discuss some of the issues related to various fillers in more detail in Chapter 6.

1.1.2 The development of an industry

In addition to the mechanical benefits of plastics to the manufacturing industry, the rapid growth in plastic production has been, in part, driven by marketing and consumer behaviour. As indicated above, the second world war had greatly impacted the availability of resources due to, among other factors, ongoing rationing. The post-war era was one of depression, with reduced manufacturing and subsequent effects on associated industries. As numerous businesses sought ways in which to increase demand for their products, as well as increasing the frequency at which they were replaced, a greater emphasis was placed on items as being disposable rather than durable. Disposable goods were marketed as time-saving and emancipating options to enable the consumer to spend less time on household chores. Subsequent repurchasing would then increase both national and international demand for new goods and the scale of the production process.

Since this post-war era, plastics have made their way into an astounding array of products. This proliferation of plastic goods, as well as increases in the scale of the consumer base (driven by population growth), have resulted in massive increases in the volume of plastics produced and used annually. Indeed, at time of writing we produce 400.3 million tonnes of plastic resins and pellets in a single year, approximately 270 times that produced in the 1950s, with this production exhibiting high spatial variation (Table 1.2).

1.1.3 From home to habitat

As with the production of polymers, the generation of plastic waste varies in relation to geographic differences production and consumption. We can observe these differences in the waste production per capita of the top ten waste-producing countries, as indicated in Table 1.3.

Of course, there are many ways by which this waste may be managed, many of which will be discussed in Chapter 7, and the fate of plastics is complicated to map in any one country let alone at a global scale. Additionally, the estimates of waste generation above fail to consider the mass of mishandled plastics that

Table 1.2 The per capita production of plastic by region (2022).

Region	Annual plastic production (Mt)	Percentage of global plastic production
China	128.096	32
Rest of Asia	76.057	19
North America	68.051	17
EU	56.042	14
Middle East/ Africa	36.027	9
South America	16.012	4
Japan	12.009	3
CIS	8.006	2

Table 1.3 The per capita production of plastic waste by country (2019).

Region	Waste production (million Mt)	Population (million)	Per capita production (Mt)
China	25.4	1408	0.01804
US	17.4	328.3	0.05300
India	5.6	1383	0.00404
Japan	4.7	126.6	0.03712
UK	2.9	66.84	0.04339
Brazil	2.8	211.8	0.01322
S. Korea	2.3	51.76	0.04444
France	2.3	67.39	0.03413
Russia	2.3	144.4	0.01593
Indonesia	2.3	269.6	0.00853
Mexico	2.2	125.2	0.01757

becomes litter before it can be enumerated. Despite the variety of waste management measures in place to handle plastics at the end of their useful life, almost one-quarter of the annually produced plastics are believed to be mishandled in some way. These plastics may eventually reach the environment by numerous routes, depending on their form, origin, use and mode of disposal (Fig. 1.2).

1.2 Microplastics

In the early 2000s, observations of plastics in the marine environment highlighted an apparently increasing abundance of small plastic litter in environmental samples (Fig. 1.3). Whilst small plastic litter was often noted in previous studies,

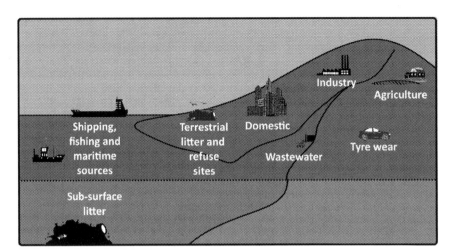

FIGURE 1.2 Sources of plastic and microplastic pollution.

Point and diffuse sources of both microplastics and plastic parent material.

FIGURE 1.3 Microplastics and plastic fragments.

A mixture of mesoplastics, primary and secondary microplastics.

Image Credit: Matthew Cole.

these new records showed that this size class was coming to represent an alarming proportion of wider debris, leading to significant scientific interest in 'microplastics' as a distinct subcategory of plastic pollution. More recently, advanced analytical techniques have demonstrated the presence of smaller 'nanoplastics', although this fraction is less routinely monitored. Subsequently, microplastics have been recognised to be a ubiquitous component of the wider environment; found at both poles, on our highest mountains and in the deep sea (Shahul Hamid et al., 2018).

A proportion of the recovered microplastic observed may be explained by the release of small plastic material directly introduced into the environment from both *point* and *diffuse* sources. Point sources include wastewater treatment facilities, textile mills and refuse sites at which plastics are produced or aggregated for management. Diffuse sources may include airborne fibres released during the wear of clothing, windblown litter and other spatially distributed, heterogeneous origins. A proportion of the microplastics observed in the environment are manyfactured at this small scale, reffered to as *primary* microplastics. However, a large proportion of the microplastics observed are the result of the breakdown of larger plastic debris. Although the structure of plastics makes them more resistant to degradation than many other natural and man-made materials, long-term exposure to environmental factors — such as ultraviolet light, oxidisers and other factors which we will touch upon below — may lead to embrittlement and fragmentation. Microplastics arising as a result of the breakdown of larger debris are categorised as *secondary*.

Fragmentation changes the mass-to-abundance ratio of plastics, increasing the number of plastic items while reducing their size. Not only may smaller plastics of this kind be more readily ingested by a wider range of species, but the formation of many smaller items of litter from a single source increases the number of organisms within an ecosystem that may be affected. At the microscale, these particles become available to taxa that would have remained predominantly unaffected by macroplastic debris, for example the planktonic organisms, and soil and freshwater macroinvertebrates (Egbeocha et al., 2018). Additionally, microplastics will experience different forces as a result of their smaller size.

The "parent" from which secondary microplastics are produced may include materials or equipment at use in the environment, for example fishing nets, mulch films or outdoor structures, or may form from existing discarded debris. Information regarding the scale of plastic sources in the environment varies in depth and reliability across geographic and spatial scales. For example, the area of land to which mulch films is applied has previously been estimated at over 4500 km^2 (Scarascia-Mugnozza et al., 2011). Mulch films are applied over soils each year in order to improve growing conditions (warming the soil, minimising water loss and reducing the application of pesticides and fertilisers). These films have been seen to result in contamination of terrestrial settings, however, without a comprehensive knowledge of the distribution and handling of mulch films, understanding the scale and distribution of microplastic inputs becomes challenging. Similarly, trawl fishing nets may be comprised of a combination of plastic polymers including, in benthic trawls, sacrificial ropes attached to the underside of the net to be abraded in place of the main net body. Regions with increased fishing activity are more likely to receive fisheries-sourced microplastics, however, the patterns of fisheries activity may be highly variable and are monitored more closely in some locations than others. Additionally, more intensive formation of microplastics may occur at sites of intentional fragmentation or breakdown, for example in the areas surrounding shipbreaking yards. Observations of coastal environments surrounding sites in India and Bangladesh have revealed an increased abundance of microplastics of a range of polymer types, including PU, nylon, polystyrene and polyester (Reddy et al., 2006).

1.2.1 Fragmentation versus degradation

Of course, the fragmentation of large plastic debris into smaller microplastics is different from the degradation of the polymer itself. As indicated above, plastics are composed of typically hydrophobic long-chain polymers of high molecular weight which are highly resistant to biodegradation. The formation of microplastics from larger plastic debris is the result of physical changes to the product such as embrittlement, cracking and flaking. Whereas the chemical breakdown of the plastic polymer to form smaller molecules, either by the action of microbes or other decomposers, is often reliant on prior oxidation or scission of the polymer chain or on the presence of a vulnerable functional group. As indicated above, this process may be initiated by a number of factors, such as UV light, temperature, water and oxygen, which are highly variable (Corcoran et al., 2009). For example, temperature and UV light may be affected by seasonal variation, latitude, burial, submersion and numerous other factors. This has previously been observed in observations of ropes exposed in subsea conditions (Welden & Cowie, 2017). Thus the recalcitrance of polymers enables both primary and secondary microplastics to persist in the environment for a significant period of time, resulting in their accumulation at identified sinks. Further discussion of the measures for accelerating the degradation of plastics, degradable plastics and bioplastics will be provided in Chapter 9.

1.2.2 Environmental contamination: understanding nomenclature

The size and chemical characteristics of microplastics have resulted in their widespread transport in the environment, the factors influencing which will be discussed in Chapter 2. Reporting the observed abundance of environmental microplastics at a location can be complex (Table 1.4). As indicated above, the most widely accepted definition of microplastics is plastic items measuring less than 5 mm along its longest axis, however, the lack of adherence to standard scales (those related to SI units) has resulted in some authors utilising a 1 mm cutoff in studies, leading to a subset of the literature being widely incomparable

Table 1.4 Approaches to the classification of environmental plastics.

Classification stage	Options
Size (class)	Macroplastics, mesoplastics (often unused), microplastics, nanoplastics
Morphology (class)	Bead, fibre, fragment
Morphology (subclass)	Sphere, grain, granule, pellet, bead, nib, nurdle, ball, filament, string, fibre, fibrous, bundle, film, foam,
Size (measurement)	No information beyond class (above), by subset (e.g. sieve size), 1-axis length, 2-axis length and width, 3-axis length, width and depth, volume
Colour	Secondary: Red, yellow, orange, green, blue, violet, black, white, clear

(Frias & Nash, 2019). While nanoplastics are typically regarded as smaller than either 1000 or 100 nm, few studies of microplastics include particles down to even the upper size limit (Gigault et al., 2018). Additionally, although larger fractions may be visible to the naked eye and smaller fraction visible under a microscope, there can be confusion regarding the identification of polymers. Thus the polymer present must be determined via chemical analysis methods. Indeed, many plastics may be indistinguishable from materials naturally occurring in the environment.

While plastics recovered from the environment share the previously described characteristics of durability and low density common in larger plastics, their reported mechanical properties and morphology are diverse. This diversity is driven both by the chemical composition of the material and by the lack of standardisation across the existing research. The combination of polymers and their additives results in plastics with a wide range of physical characteristics depending on the use to which they were put. Although a smaller proportion of microplastic debris, primary microplastics are themselves variable. For example, nurdles may be short cylinders, spheres or irregular ovoid shapes of approximately 0.5 cm in diameter, and granular abrasives vary in diameter in relation to their application. Secondary microplastics may be yet more challenging, with their characteristics dependent on the original structure of the parent plastic and the degree of degradation.

We may also look at the various forms that secondary microplastics exhibit, for example are they spherical, pelleted, granular or fibrous in nature? Additionally, there are a veritable riot of colours, sizes and textures which may inform us about the origin, fate and impacts of the material (Lusher et al., 2020). The colour and surface structure of both primary and secondary microplastics may also be affected by both weathering and biofouling, with yellowing and cracking reported in many environmental samples (Brandon et al., 2016). As with the definition of size, discussions regarding morphology have also resulted in disagreements between fields, for example environmental researchers frequently refer to microplastic fibres as 'microfibres' a term which has an existing definition within the textile industry, resulting in some confusion between those working on production, and those studying pollution.

Subsequently, there has been significant debate surrounding the way in which we report the characteristics of microplastics recovered from the environment. However, as a result of the above factors, the extent to which the properties of recovered microplastics are described varies. This variety poses challenges to the reliable and comparable classification of microplastics recovered from examples of both environmental media and biota. Many studies rely on simple categorisation based on sampling procedures, such as minimum and maximum retained sizes based on the limitations of the sampling process, or as fractions based on sample processing (e.g. sieving). Other publications classify either a subset or the whole sample using measurements of between 1 and 3 axes (Table 1.4). As we will see in later Chapter 3, this has significant implications for our understanding of microplastic distribution and impacts.

1.3 Observation to outcry

The increased abundance of microplastics has significant implications for the affected environment. Larger macro- and mesoplastics have been linked to ingestion, entanglement and injury in a variety of animals (Blettler & Mitchell, 2021). For example, uptake and injury as a result of misidentification of plastic as prey or the entanglement and subsequent mortality of chicks reared in nests made with litter recovered from the environment (Ryan, 2019). Many of these effects have been mirrored by microplastic, although their scale and ubiquitousness remain in question and will be discussed further in Chapters 4 and 5.

The issue of microplastic pollution, its pervasive nature and potential impacts, have received significant attention from both journalistic and documentary media, further amplified by nongovernmental organisations (NGOs), charities and special interest groups. Media content has had significant impacts on individual behaviour, industrial operating procedures and policy, raising public awareness by way of a combination of factual and emotive content (Borg, 2022). The work of NGOs and related groups has also contributed to the evidence base surrounding the plastics issue through a variety of citizen science studies or routing sampling events. For example, Fidra's The Great Nurdle Hunt, which invites participants to walk a stretch of local beach and record the approximate number of nurdles seen (based on predefined categories), using these cumulative observations from thousands of participants to lobby industry bodies to commit to best practices in the transport and handling of plastic feedstocks. As a result of this growing interest and of prior observation of large plastic debris in the environment, there is increasing movement towards local, national and international statutes for the prevention, monitoring and management of mishandled plastic wastes.

1.4 **Moving forward**

In order to understand microplastic pollution and its impacts, they and their parent plastics must be considered in a holistic manner. The issue of microplastic reduction should be set against a wider sustainability framework including the broader impacts of plastic use and disposal in relation to other key environmental challenges such as water availability, land use and global warming. In the following chapters, we will review what we know about the origins transport and impacts of plastics in the environment as well as the way in which we monitor that presence and abundance. In the second section, we will examine our reactions as individuals, innovators, businesses, NGOs, regulators and policymakers and attempt to create a forecast of microplastic management.

References

Blettler, M. C. M., & Mitchell, C. (2021). Dangerous traps: Macroplastic encounters affecting freshwater and terrestrial wildlife. *Science of the Total Environment, 798*. Available from https://doi.org/10.1016/j.scitotenv.2021.149317, http://www.elsevier.com/locate/scitotenv.

Borg, K. (2022). Media and social norms: Exploring the relationship between media and plastic avoidance social norms. *Environmental Communication, 16*(3), 371−387. Available from https://doi.org/10.1080/17524032.2021.2010783, http://www.tandf.co.uk/journals/titles/17524032.asp.

Brandon, J., Goldstein, M., & Ohman, M. D. (2016). Long-term aging and degradation of microplastic particles: Comparing in situ oceanic and experimental weathering patterns. *Marine Pollution Bulletin, 110*(1), 299−308. Available from https://doi.org/10.1016/j.marpolbul.2016.06.048, http://www.elsevier.com.

Corcoran, P. L., Biesinger, M. C., & Grifi, M. (2009). Plastics and beaches: A degrading relationship. *Marine Pollution Bulletin, 58*(1), 80−84. Available from https://doi.org/10.1016/j.marpolbul.2008.08.022.

Davies, G. (1987). Abysmal litter. *Marine Pollution Bulletin, 18*(2), 59−60. Available from https://doi.org/10.1016/0025-326X(87)90553-4, https://www.sciencedirect.com/science/article/pii/0025326X87905534.

Egbeocha, C. O., Malek, S., Emenike, C. U., & Milow, P. (2018). Feasting on microplastics: Ingestion by and effects on marine organisms. *Aquatic Biology, 27*, 93−106. Available from https://doi.org/10.3354/ab00701, https://www.int-res.com/articles/ab2018/27/b027p093.pdf.

Frias, J. P. G. L., & Nash, R. (2019). Microplastics: Finding a consensus on the definition. *Marine Pollution Bulletin, 138*, 145−147. Available from https://doi.org/10.1016/j.marpolbul.2018.11.022, http://www.elsevier.com/locate/marpolbul.

Gigault, J., Ter Halle, A., Baudrimont, M., Pascal, P. Y., Gauffre, F., Phi, T. L., El Hadri, H., Grassl, B., & Reynaud, S. (2018). Current opinion: What is a nanoplastic? *Environmental Pollution, 235*, 1030−1034.

Jewett, S. C. (1976). Pollutants of the northeast Gulf of Alaska. *Marine Pollution Bulletin, 7*(9), 169. Available from https://doi.org/10.1016/0025-326x(76)90213-7.

Lusher, A. L., Bråte, I. L. N., Munno, K., Hurley, R. R., & Welden, N. A. (2020). Is it or isn't it: The importance of visual classification in microplastic characterization. *Applied Spectroscopy*, *74*(9), 1139−1153. Available from https://doi.org/10.1177/0003702820930733, https://journals.sagepub.com/loi/asp.

Reddy, M. S., Basha, S., Adimurthy, S., & Ramachandraiah, G. (2006). Description of the small plastics fragments in marine sediments along the Alang-Sosiya ship-breaking yard, India. *Estuarine, Coastal and Shelf Science*, *68*(3−4), 656−660. Available from https://doi.org/10.1016/j.ecss.2006.03.018.

Ryan, P. G. (2019). Ingestion of plastics by marine organisms. *Handbook of Environmental Chemistry*, *78*. Available from https://doi.org/10.1007/698_2016_21, http://www.springer.com/series/698.

Scarascia-Mugnozza, G., Sica, C., & Russo, G. (2011). Plasticmaterials in European agriculture: Nactual use and perspectives. *Journal of Agricultural Engineering*, *42*, 15−28.

Seymour, R. B., & Kauffman, G. B. (1992). Polyuretanes: A class of modern versitile materials. *Journal of Chemical Education*, *69*.

Shahul Hamid, F., Bhatti, M. S., Anuar, N., Anuar, N., Mohan, P., & Periathamby, A. (2018). Worldwide distribution and abundance of microplastic: How dire is the situation? *Waste Management and Research*, *36*(10), 873−897. Available from https://doi.org/10.1177/0734242x18785730, https://journals.sagepub.com/home/WMR.

Welden, N. A., & Cowie, P. R. (2017). Degradation of common polymer ropes in a sublittoral marine environment. *Marine Pollution Bulletin*, *118*(1−2), 248−253. Available from https://doi.org/10.1016/j.marpolbul.2017.02.072, http://www.elsevier.com/locate/marpolbul.

Microplastics on the move

2

2.1 Introduction

In this chapter, we will move away from the scale and distribution of plastic sources to discuss one of the key concerns pertaining to microplastics and, increasingly, nanoplastics: these pollutants get everywhere! Initial observations made in even the most remote environments (such as at the poles, remote mountain ranges and in the deep sea) have revealed the presence of microplastics in some form. As a result, researchers are united in their outlook regarding the ubiquitousness of this type of pollution. It doesn't matter how far a location is from a source of plastics, they may be transported for long distances by multiple means, and aggregate in the receiving environment to high levels.

Only with the onset of the first nuclear tests and the introduction of chemicals now classed as persistent organic pollutants, has the potential for the global transport of pollutants become a significant source of concern. For example, in the globally observed doubling in the atmospheric relative carbon 14 concentrations observed following over a decade of intensive aboveground nuclear bomb tests, known as the bomb pulse (Zoppi et al., 2004). This effect was so globally ubiquitous and marked that it resulted in a new radiocarbon dating system, known as bomb pulse dating. Similarly, the global aquatic and atmospheric transport of chlorinated hydrocarbon pollutants, such as DDT, have been identified in remote locations such as the Antarctic Ocean for decades, a marker of intensive pesticide use (Tanabe et al., 1982; Zhang et al., 2022).

On a similar scale, microplastic transport and extensive aggregation have been reported worldwide, influenced by a range of potential transport pathways in air, water and the terrestrial environment (Fig. 2.1). These pathways result in clear distribution patterns (the mapping of which will be further addressed in Chapter 3), for example, increased microplastic concentrations are reported from locations such as Hawaii and the Great Garbage Patch. However, unlike the level of C-14 in the atmosphere — which reduces by approximately 4% per year — it is not known to what extent macro-, micro- and nanoplastics are removed or lost from the environment. As a result, it is essential that we better understand the movement of microplastics, in order to highlight at-risk environments and identify potential problem sources.

Microplastics. DOI: https://doi.org/10.1016/B978-0-443-13324-4.00002-9

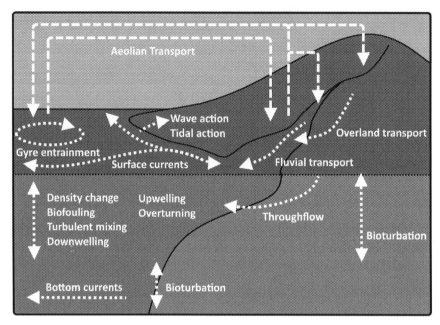

FIGURE 2.1

Summary of global microplastic transport. Diagram illustrates the movement of microplastics within and between environmental compartments

2.2 The effect of polymer structure on plastic transport

As mentioned in the previous chapter, the structure of the plastic polymer confers a range of effects on the final product. The transport and accumulation of both plastics and microplastics is typically associated with two key properties, low density and high molecular weight, the latter of which results in a polymer's resistance to degradation.

We have previously discussed the formation of polymers from repeating units (monomers), joined to create a long polymer chain. We may also describe these polymers as either thermoplastics or thermosets. Thermoplastics have very few links between polymer chains and, as a result, they may be melted and reformed. Conversely, in thermosets, strong bonds form between the polymer chains and, when heated, the plastic is degraded and loses its mechanical properties. In addition to the long polymer chain, side groups which stick out from the polymer backbone influence the structure and properties of the plastics. The number, types and positions of the various functional groups within a polymer affect where side groups sit and thus where bonds may form. These side groups may be arranged in one of three ways: either syndiotactic, with side groups on alternate sides; isotactic, with side groups all on one side; or atactic, with a random distribution of side groups (Fig. 2.2).

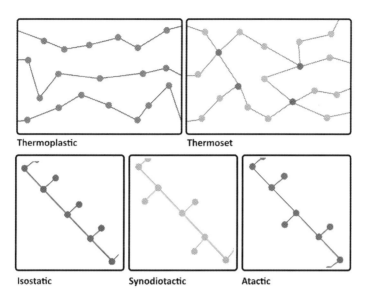

Thermoplastic   Thermoset

Isostatic   Synodiotactic   Atactic

FIGURE 2.2

Structural features of the plastic polymer. Figure illustrates the position of functional groups within the plastic polymer

The polymers formed from these chains may be described in two ways (although transitional states do occur); either *amorphous*, as a result of their atactic polymer chains, or *crystalline*, driven by the presence of syndiotactic and isotactic chains. Amorphous polymers have no organised structure and are typically low density, transparent (in pure form) and have weak resistance to solvents. Whereas crystalline polymers demonstrate a regular parallel or linear arrangement of molecules and have strong attraction between molecules, higher densities and are translucent and more rigid. We will come back to the differences between these crystalline and amorphic plastics in Chapter 3.

Whatever their form, the structure of a polymer typically results in a plastic of low density (in relation to many natural materials and substrates); subsequently, plastic litter may be light enough to be easily carried on air or water currents. This capacity for suspension, either in water or in air, is particularly true of plastics with a high surface area to volume ratio, such as films and fibres. Thus plastic may be redistributed from its initial point of disposal or loss to the environment.

In addition to being readily transported by wind and air, the low degradation rate of polymers increases the distance over which they may be moved. For example, natural materials may have a degradation period of a few months to a few years, limiting the distance over which they can be relocated by either air or water currents. Comparably, plastic polymers may take decades to fragment and many times that to fully degrade. As a result, the radius in which plastic materials may be moved is much greater.

2.3 **Airborne transport**

The atmospheric transport and deposition of plastics, particularly microplastics, has received only minimal attention to date, with no studies prior to 2015 (Wang et al., 2021). Within this period, published research has predominantly concerned the movement of microplastics from regions of high plastic input (with multiple sources) into otherwise 'pristine' environments by way of air currents. Such observations have been associated with the deposition of microplastics in snow and rain in locations including remote mountains (Allen et al., 2019; Bergmann et al., 2019; Parolini et al., 2021) and the Arctic (Bergmann et al., 2019) and Antarctic (Aves et al., 2022). Unsurprisingly, the air over even the remotest marine environments also holds microplastics and analysis of samples recovered from surveys in the North Atlantic revealed microplastics in just under 21%. These samples were collected from approximately 15 m above the water level and varied in volume between approximately 12 and 60 m^3 (Trainic et al., 2020).

Other observations of airborne plastic transport focus on the apparent concentrations of microplastics in urbanised, peri-urban and rural areas. Initial observations of atmospheric microplastic fallout revealed up to 355 items per m^2 per day, resulting in an estimate of between 3 and 10 tonnes of fibres deposited across urban Paris in a single year (Dris et al., 2016). More recently, air samples recovered within and at the fringes of five Chinese megacities revealed microplastic concentrations of between 104 and 650 items per m^3 with an average of 282 items per m^3. The size of these particles ranged from 5.9 to 1475.3 μm, although the majority of recovered particles (approximately 90%) were in the lower size categories of 5−30 and 30−100 μm. Perhaps most importantly, no significant differences were found between central and peripheral areas of the city (Zhu et al., 2021). Similar *numbers* of microplastics have been observed in central London, where the average abundance was 771 items per m^2 per day. This study also demonstrated the abundance of fibres, which were by far the dominant category of plastics, averaging 712 items per m^2 per day (Wright et al., 2020). However, as may be noted by the reported units in these studies, the comparability of these results may be affected by differing monitoring methods. In this case, the concentration of plastics in a known volume of air vs a deposited abundance of plastic over a known surface area and time period. When comparing airborne concentrations *and* depositional abundance in the same study, Kernchen et al. (2022) highlight the significance of proximity to plastic sources on microplastic concentrations. In their survey of six sites in the Weser River Catchment in Germany, the urban Bremerhaven, Bremen and Kassel sites demonstrated the highest deposition, while the rural and peri-urban Wasserkuppe, Solling and Weserbergland sites had the lowest. As with the observations outlined above, the majority of particles were below 100 μm, with larger fractions (> 1000 μm) dominated by fibres (Kernchen et al., 2022).

In addition to differences in microplastic abundance along the urban-rural continuum, microplastic concentrations have been seen to vary significantly within the built environment. Comparisons between indoor and outdoor settings in

Wenzhou, China have demonstrated reductions in the concentrations of microplastics in air samples of approximately an order of magnitude, from an average of 1583 items per m^3 indoors, to one of 189 items per m^3 outdoors. Again, this outdoor average varied between urban and rural environments, 224 and 101 items per m^3 respectively, and fibres continued to dominate the large microplastic fraction across all settings (Liao et al., 2021). Similarly, observations in São Paulo revealed increased deposition in indoor settings, from 123 items per m^2 per day to 309 items per m^2 per day (Amato-Lourenço et al., 2022).

2.3.1 Factors influencing airborne transport

The movement of air masses unsurprisingly has a significant effect on the movement of airborne pollutants. In the previously highlighted study of microplastic deposition in London, comparisons between the abundance of deposited particles and the wind direction were used to highlight the conditions under which microplastic inputs were increased. This revealed differing patterns in the deposition of fibrous and nonfibrous microplastics. Additionally, back trajectory analysis, in which the most likely origins of an air mass are predicted, suggests that variations in the dimensions of fibrous and nonfibrous plastics result in differences in their area of influence (Wright et al., 2020). Similarly, attempts were made to link the occurrence of microplastics in Antarctic snows highlighted above to the air movements in the proceeding 6.5 days. In this case, while the study indicates potential transport over distances of up to 6000 km, the authors also indicate the potential of the nearby scientific stations on Ross Island to have acted as a source of some of this material (Aves et al., 2022).

The deposition of microplastics is believed to be analogous to that of other airborne particulates. The dimensions (size, mass and surface area) of the material are recognised as influencing the settling rate of a particle (Wright et al., 2020). This settlement rate may also be influenced by changes in wind speed, with more energetic systems increasing transport distances and resuspension, as well as humidity precipitation, with rainfall increasing apparent deposition (Amato-Lourenço et al., 2022; Kernchen et al., 2022) as do dust storms (Abbasi et al., 2022).

2.4 Transport in water

Compared to the movement of small plastics in the air, much more has been written regarding the transport of microplastics in water. As with airborne plastics, proximity to sources has been recorded as having a significant impact on local microplastic abundance and settlement rates are linked to surface area, size and shape (Kowalski et al., 2016). From the first observations of microplastics in plankton samples, the potential for the movement of microplastics in the water column has been clear (Thompson et al., 2004). Since this period, researchers have sought to use our understanding of the dynamics of marine and riverine systems in order to map microplastic movement around the globe (Sherman & van Sebille, 2016). The ability of plastics of all sizes to ride in the water column is due to a key property of both plastic and water, density, but this is influenced by multiple additional factors which define these complex habitats.

2.4.1 Factors influencing vertical transport in water

In this section, we will discuss the factors which influence the level at which plastic sits within the water column, considering the very surface layers of water down to the epibenthic zone. The underpinning factor, the density of the polymer, is frequently described in terms of its specific gravity compared to that of pure water (which has a specific gravity of one). Those polymers with a specific gravity less than one will float in pure water. Conversely, those polymers with a higher specific gravity may sink in freshwater systems, in which they cannot be supported. However, adding salts to water increases the density of the solution. This means that plastics of high density, which may have been deposited in rivers, are able to be transported in estuarine and seawater. Seawater typically has a specific gravity of between 1.02 and 1.03. From Table 2.1, which provides the specific gravity of a number of common polymers, it can be seen that only a proportion of plastics will readily float in both fresh and saltwater environments. A larger proportion will float in saltwater settings, depending on the exact salinity of the environment. For example, estuarine zones are characterised by their salinity gradient, which may result in different plastic types in samples closer to the river mouth. Indeed, observations of plastics in the marine environment have reported higher concentrations of dense microplastics at depth (Zhao et al., 2022).

Unfortunately for those of us seeking simple measures by which we might determine the likelihood of polymer transport, the density of seawater is also

Table 2.1 The density of polymer types in relation to that of water in aquatic environments

Category	Polymer	Density
Floats in freshwater and seawater	LDPE	0.89–0.93
	PP	0.90–0.91
	HDPE	0.94–0.97
Floats in seawater	ABS	1.04
	PS	1.04–1.07
	PMMA (Acrylic)	1.09–1.20
Only buoyant in solutions denser than typical seawater	PA (Nylon)	1.13–1.15
	PVC	1.16–1.58
	PMA	1.17–1.20
	PE	1.38–1.39

influenced by temperature. Warmer water is typically less dense and cold water is denser. Fig. 2.3 shows the relationship between density and variation in salinity and temperature. The bars in the image allow us to identify regions of similar density, in the same way that isobars on a weather map indicate areas of the same pressure and contours on a map show the same height. For example, water with a salinity of 33.5 psu and a temperature of 5 degrees centigrade is approximately as dense as water with a salinity of 35 psu and a temperature of 12 degrees centigrade. As a result, water in temperate areas may carry a greater variety of polymers than in waters close to the equator. However, there are other factors which prevent the relationship between plastic density and water density from being consistent. Water turbidity, temperature and the influence of fouling organisms influence the buoyancy of plastics, and it has been hypothesised that long-term reductions in buoyancy result in deep-sea environments becoming substantial microplastic sinks.

2.4.1.1 Biofouling

Biofouling can significantly influence the buoyancy of plastics in aquatic environments, resulting in settlement. The term biofouling refers to a community of organisms that have developed on the surface of a material as well as their associated matrix of accessory chemicals and by-products. The process of fouling may begin as soon as an item is placed into water and is instigated by the settlement of the first microorganisms (Kooi et al., 2017; Welden & Cowie, 2017). Indeed, even early

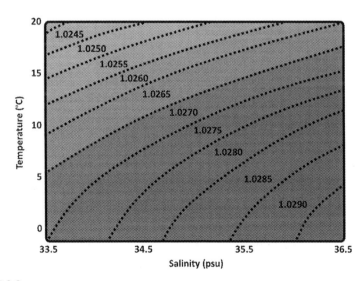

FIGURE 2.3

The combined influence of temperature (°C) and salinity (psu) on water density (g/cm^3). The figure identifies the way in which temperature and salinity affect the density of seawater

microbial colonisation may have a significant enough impact on plastic density should the item's surface area to volume ratio be sufficient (Amaral-Zettler et al., 2021). Some estimates of fouling and sinking rates put the period in which thin plastics remain in the surface waters at as little as 17 days (Fazey & Ryan, 2016), however, the period between introduction to the environment and the development of a biofouling community may be affected by various factors. For example, reduced biofouling and sinking rates have been seen in microplastic samples exposed to estuarine water as opposed to seawater (Kaiser et al., 2017). Additionally, the growth of biofouling communities may be dependent on light availability, temperature and the number of potential colonisers within the local environment (Barnes, 2002; Lobelle et al., 2021). The fouling and sinking rate may be especially relevant in freshwater and estuarine environments (particularly those with numerous or sizeable plastic inputs), where the relationship between sinking rate and residence time may influence whether microplastics are transported out of the waterbody or settle into local sediments (Semcesen & Wells, 2021).

The properties of the polymer may also influence the rate of colonisation (Liu et al., 2022), as well as the rate of sinking (Kooi et al., 2017). Many plastic polymers have smooth and comparatively hard surfaces which may prevent early colonisers from settling, however, other plastics are softer and have complex three-dimensional surfaces which provide easy sites for the colonisation by various organisms. Indeed, plastics and microplastics have been identified as sources of potentially harmful invasive species. Those organisms rafting on natural materials which quickly degrade may not successfully colonise new environments due to the degradation of the raft itself, the durability of plastics means that organisms stand a much higher chance of being carried to new locations. This has previously been seen in numerous observations of beach litter including in the study of debris released from the 2011 Japanese earthquake and tsunami, following which debris washed up as far away as Hawaii, Oregon and Washington (US) as many as 5 years later (Carlton et al., 2018; Hansen et al., 2018).

Despite the transoceanic movements of large debris, sufficiently heavy biofouling communities will sink floating plastic if they grow to a size up which the specific gravity of the whole becomes greater than that of the surrounding water. At this point, the whole item may descend below the water's surface and be deposited on the river, lake or seabed. It has been suggested that this process may result in a size-selective element to microplastic removal, with small particles of high surface area to volume ratios being more susceptible to the effects of fouling on position in the water column (Kaiser et al., 2017). This is due to the comparatively small mass of less dense polymer in relation to the mass of the encrusting biofilm (Kooi et al., 2017; Liu et al., 2022).

Organisms may also influence the vertical transport of microplastics by their inclusion in marine snow. Marine snow is a catch-all term for aggregations of organic particulates such as microbes, faecal matter and other particulate matter. Microplastics may be actively incorporated into marine snow following ingestion by organisms and excretion as faecal pellets, or by passive aggregation. As in our previous discussion of fouled plastics, marine snow has a combined specific gravity greater than the surrounding water and is thus drawn down into deeper zones. Observations of microplastic interactions with artificially produced marine snows showed significant levels of incorporation and increases in sinking rates across polymer types and morphologies (Porter et al., 2018). Modelling estimates of the significance of marine snow as a mechanism of exporting both microplastics to the sea floor indicates potential exports of between 7300 and 42000 metric tonnes per year (Kvale et al., 2020).

While both biofouling and the inclusion of microplastics into marine snow may have significant impacts of the transport of plastics, their influence may also be variable over time. For example, as plastic falls below the photic zone, photosynthesising organisms in the biofilm may die and be sloughed off, reducing the overall density of the polymer and enabling it to rise back to the surface (Lobelle et al., 2021). This process would result in cycles of sinking and ascension, adding further complexity to this picture.

2.4.1.2 Mixing

In addition to the influence of polymer density, water density and fouling, microplastics may be forced to different levels of the water column by either wind-driven or turbulent mixing, or by upwelling and overturning processes. In this case, rather than the plastic moving through the water column, it is the surrounding water is moved. The bulk movement of water transports any suspended plastics with it in the same way in which nutrients are moved to the surface at sites of upwelling. At the water's surface, the process of wind-driven mixing, by which frictional forces cause the cyclical displacement of particles and the formation of waves, may result in the movement of microplastic between layers in the water column (Kukulka et al., 2016). Simultaneous trawling at multiple depths has revealed that the typical decrease in the relative proportions of microplastics observed between the surface waters and 6 m in depth was reduced at higher wind speeds. This indicated that high winds reduce the differences in microplastic concentration at different levels in the surface water by wind-driven mixing (Reisser et al., 2015). The greater the frictional forces, the greater the depth which may be affected.

In coastal waters and in rivers, increased mixing may result in microplastics that have previously been deposited into subsurface layers or benthic sediments being resuspended. For example, the modelling of riverine deposition suggests that sufficiently high flows may be able to resuspend even the largest microplastics (Nizzetto et al., 2016). Changes in the surrounding conditions, such as the level of shear stress exerted by surrounding waterbodies, displacement of sediment or the occurrence of storms, act to resuspend plastics into the water column enabling them to be further transported, although for variable distances. For example, observations of microplastic abundance in surface, mid and bottom waters off the California coast were observed pre- and post storm. It was noted that, in offshore environments, the abundance of microplastic in benthic samples decreased following storm events whereas that in surface and midwaters increased, suggesting substantial resuspension of plastics due to the change in wave action (Lattin et al., 2004). This type of resuspension will be most relevant in shallow coastal zones (where the influence of increased wave action may be felt throughout the water column) as the deeper the water, the more significant the size of the waves must be to resuspend the sediment there.

While short-term changes in wind-driven mixing may occur frequently due to atmospheric conditions, the processes of upwelling, downwelling and overturning are comparatively more reliable and consistent over time. Upwelling and downwelling may be caused by currents running a parallel to the coastline in combination with the Coriolis effect, the effect of Coriolis force in relation to the position of the Intertropical Convergence Zone as well as by the strong winds of the Southern Ocean. In comparison, overturning is the result of the horizontal movement of warm water masses from the equator to the, and of deep cool water masses to, the additional effects of which will be discussed below. In Monterey Bay, observations of plastics across the meso- and epipelagic layers suggest that combined wind forcing and upwelling may be responsible for the inward movement of plastics (Choy et al., 2019). The action of mixing, upwelling and downwelling on both suspended and deposited sediments highlights that these zones are seldom permanent sinks for plastics. While vertical transport may aggregate plastics in benthic zones, the remaining material may be subject to redistribution.

2.5 Factors influencing horizontal transport in water

As with the vertical movement of microplastics, there are several factors which affect their horizontal movement in aquatic environments. However, many of these are specific to either still freshwater (lentic), flowing freshwater (lotic) or marine and brackish habitats. In these habitats, we may expect the influence of previously discussed factors, such as specific gravity, salinity and biofouling to remain relevant, however, the degree and drivers of water movements are more distinct.

2.5.1 Horizontal transport in lotic environments

The transport of plastics entering flowing waters is often thought of as unidirectional, with the movement of water primarily dictated by gravity. Nevertheless, riverine systems are highly diverse and complex, and the progress of microplastics will vary in relation to the morphology of the channel, the angle and rugosity of the riverbed, characteristics of the bedload and sinuosity of the channel, in addition to the strength of the current, and seasonal occurrence of both spate and low flow conditions and flooding (Kumar et al., 2021).

The retention of microplastics by sediments at the bottom of rivers has been highlighted as a plastic sink, however, increases in the velocity of the water increase the distance over which both sediment and microplastics may be transported (He et al., 2021). Additionally, modelling of microplastic entrainment in riverbed sediments, based on previous models of sediment transport, suggests that smaller microplastics are not retained, whereas larger, denser microplastics are (Nizzetto et al., 2016). Unfortunately, the assumptions of some field and modelling studies of the distribution of plastics in these systems fail to take into account the diversity of plastic materials, particularly those which are less dense than the surrounding water, and there is still much to learn about these environments (Cowger et al., 2021). Furthermore, in addition to the natural variation between rivers, man-made structures may also influence the movement of microplastics. Paired samples taken upstream and downstream of dams near Ithaca, New York, indicated significantly lower levels of microplastics below dams than in the reservoir samples above (Watkins et al., 2019).

2.5.2 Horizontal transport in lentic environments

The plastics entering lentic environments (particularly larger lakes) may be subject to surprisingly complex internal forces which may influence microplastic movement. In temperate zones, larger bodies of still water may develop complex seasonal stratification which can lead to layers of differing water density. These may be generalised as cooler layers at the bottom in summer and a thermally insulated warmer bottom layer in winter. Also, in lakes with a large surface area, such as the American Great Lakes, the long fetch (the distance over which the frictional forces of air currents can act on the water's surface) can also lead to the formation of wind-driven waves and circulating currents. This transport of water in lakes may also be influenced by lake bathymetry (the shape of the lakebed) and the presence of any inflows of water. As a result, the movement of suspended plastics and microplastics in large lakes can be challenging to predict. This intricacy has been modelled for both Lake Ontario and Lake Erie, comparing the resulting movement with the effect of either simplified estimates of biofouling or light-sensitive defouling. These models suggest that, at these sites, interactions between fouling methods and lake depth have the greatest effect on settlement (Daily et al., 2022).

2.5.3 Horizontal transport in marine environments

Suspended microplastics in the marine environment are affected by a variety of factors that define their horizontal movement. Some of these are analogous to transport on air currents and in lotic systems. Nevertheless, the transport of water around the globe may be considered at numerous scales, including the bulk water transport of thermohaline circulation, the basin-scale currents which form the great gyres, localised currents and tidal patterns.

Perhaps the most widely recognised influence on the horizontal transport of plastics and microplastics is that relating to the global and basin scale currents which form gyre systems. In our discussion of vertical transport, we highlighted the role of wind-driven mixing and overturning in the movement of plastics between layers of the water column. However, wind patterns and the Coriolis force also create five consistent, rotating currents which circle the northern and southern Pacific and Atlantic Oceans as well as the Indian Ocean. Many papers have studied the influence of these circulating currents in the creation of what is known as garbage patches, sites of increased concentrations of marine litter (Debroas et al., 2017; Lebreton, 2022). These are the sites to which floating debris is herded by circulating basin currents. Islands within these regions can also be susceptible to elevated levels of beached plastics (Rey et al., 2021). However, these patterns may not hold true below the surface. Modelling of deep water layers indicates that garbage patches become more 'leaky' at depth, as a result of differences in subsurface currents, and then disappear below 60 m (Wichmann et al., 2019). Indeed, observations of deep-sea floor plastics revealed that abundance was not based on patterns recorded in the surface waters, but rather on the thermohaline circulation, with apparent hotspots in canyons and trenches (Kane et al., 2020). At smaller scales, regions of upwelling and downwelling can also influence the horizontal distribution of plastics. Indeed, hotspots of microplastic abundance in the Ría de Vigo were observed to move in response to wind-driven shifts between upwelling and downwelling regimes (Díez-Minguito et al., 2020).

2.5.3.1 Windage

In addition to being influenced by the movement of the water around it, plastics floating partially above the water's surface may also be affected by a process known as windage. Windage describes the influence of the frictional forces exerted by the wind on an object, rather than on the water around it. Wind-driven currents typically travel at an angle to that of the prevailing wind due to the influence of the Coriolis effect. Plastics riding slightly above the water's surface are thus pushed to one side of the main current, resulting in their moving in a slightly different direction to that of the main flow. Those with a greater proportion of their area above the water's surface will be more affected by the frictional effects of the wind than those lower in the water (Fig. 2.4).

Fortunately for us, the proportion of plastic litter travelling in this manner will be comparatively low and the overall influence on the mass movement of plastics and microplastics may be minimal. Nonetheless, there are some areas in which the effect of the wind cannot be ignored, for example in the stranding of plastic material on beaches, which may be closely related to the presence of a prevailing onshore, offshore or alongshore breeze.

In marine settings, observations of manmade debris of all sizes can act as a basis for increasing our understanding of oceanic transport at variable scales. Thus the movement of anthropogenic debris has been used to determine the patterns of global water transport over many decades (Ebbesmeyer & Ingraham, 1992, 1994). Similarly, rafts of manmade debris may result in the distribution of

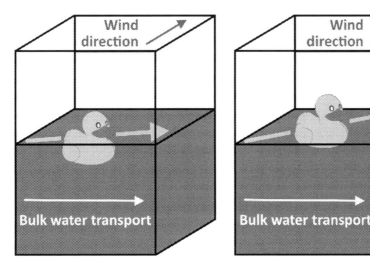

FIGURE 2.4

Differing effects of windage on floating plastics. The influence of the force of the wind on plastics sitting higher or lower in the water column

a range of alien species between countries and continents and are subsequently closely observed (Carlton et al., 2018; Hansen et al., 2018). In this regard, the study of marine microplastic pollution both benefits from and benefits wider ocean science, this is discussed in greater detail in Box 2.1.

Box 2.1 Mapping and modelling plastic movement

Objective

While it is relatively easy to identify areas of high plastic and microplastic input, the transport of these plastics and their resulting concentrations in the environment are not easy to predict. Below we will explore how observations of marine litter and computational methods have helped to shape our understanding of global circulation and the movement of microplastics.

Scope

An understanding of global circulation patterns is essential for a variety of marine industries, including shipping, recreational sailing and marine planning and construction. Thus the outputs of models which may reliably explain surface currents are of high value to private water users, commercial organisations, emergency services and military vessels, as well as to researchers in the field of ocean science. However important this information may be, until the mid-20th Century, the collection of data on which to base these models has been challenging. Initial mapping of currents was reliant on dropping identifiable material into the environment and hoping that it may be recovered at a later date. Subsequently, as we began to increase the testing of nuclear devices, the radioactive signals became an important tracer too. Today, the use of supercomputers and satellites has improved and refined our understanding of circulation at both local and global scales.

(Continued)

Box 2.1 Mapping and modelling plastic movement (Continued)

While a concern for many, influxes of plastic into the marine environment have presented researchers with an additional opportunity to understand both long-term global circulation and short-term fluctuating local currents.

Rationale

As indicated in the main text, the movement of ocean currents is dictated by a range of forces, including temperature and salinity, which influence the relative density of water masses. Interactions between prevailing wind and Coriolis force influences the formation of Eckman spirals and currents in the upper layers of water, and, as a result, the formation of geostrophic currents (Fossette et al., 2012). With the increasing availability of satellite altimetry of the topography of the ocean surface and information regarding the apparent strength of wind patterns (measured from scattering associated with surface roughness), we may remotely determine the height of sea-surface bulges and the apparent level of geostrophic flow. However, these are modelled measures, relying on both a preexisting understanding of circulation patterns and confirmation by in situ observations.

We may seek to predict the movement of water using two key approaches, either Eulerian or Lagrangian. I cannot think of a better example of the difference between Lagrangian and Eulerian methods than that set out in Ebbesmeyer and Scigliano's excellent book, *Flotsametrics and the Floating World* (Ebbesmeyer & Scigliano, 2009). In the text, the two approaches to ocean observation are compared to traffic enforcement: firstly, a Eularian speed trap, which stays at a fixed point, scanning the speed and direction of passing traffic at a single location. Secondly, a Lagrangian unmarked car, which moves with the traffic, experiencing changing conditions over differing spatial scales.

Historically, releases of marked (or tagged) drifters have acted as those mobile traffic enforcement units. However, such approaches frequently resulted in limited data due to costs, recovery rates and the impact of extreme spatial scales. Moreover, slight differences in position and timing can greatly affect their path, and individual drifters might or might not follow a 'typical' trajectory. It is also difficult to determine how representative the current or water mass that a buoy finds itself in is representative of that wider. Conversely, plastics and other debris are often present in large numbers in the environment. Given a known point of origin, the movement of this material may be back-projected to provide an in situ measurement of surface flows. Indeed, our appreciation of the vast distances travelled by plastics and the various factors that may influence this transport has led to attempts to determine the origins of observed plastic debris, as well as to predict the future trajectories of plastic material in the environment.

Learning from plastics

In addition to the daily diffuse inputs of plastics into the environment, extreme weather events, natural disasters and accidents at sea may result in high inputs of identifiable plastics with a known geographic origin. Such occurances include the loss of goods from container ships and the inputs of debris as a result of tsunami events. Researchers aware of inputs of floating marine debris may monitor its progress around the globe, examining both path and pace in order to learn about the bodies of water responsible for their transport. These methods have previously been applied in relation to bath toys, shoes lost from ships and materials lost from wrecks off the US coast.

For example, Nike shoes lost on the 27th of May 1990 in the mid-North Pacific Ocean travelled the ocean basins over the following 6 to 12 months, with thousands (approximately 2.5% of the original 80,000) making landfall on the northwest coast of the United States (Ebbesmeyer & Ingraham, 1992). Subsequently, observations of stranded shoes were compared with historical 'drift bottle' data. Drift bottles are simple sealed glass bottles containing information on how to contact the research team when it is found. Each bottle is dropped at a known location and, once

(Continued)

Box 2.1 Mapping and modelling plastic movement (Continued)

reported, the most likely track to its recovery point is determined. This combined data was compared to hindcast models and used as an opportunity to calibrate their outputs, the findings of which indicated variation from these predicted trajectories. A similar approach was also taken with the spill of almost 30 thousand '*Friendly Floatees*', also lost in the North Pacific in 1992. Within 2 years, these toys had travelled nearly 4,000 km, and they have continued to be recorded in the intervening years. Together with observations of other floating debris, the observations of toy strandings were included in projections using the Ocean Surface Current Simulator (OSCURS), which modelled geostrophic flow over the North Pacific and Bering Sea. Model outputs indicated orbital periods of up to 4 years, with proximity to coasts increasing the complexity of currents and the number of potential routes the *Floatees* might take. Using these models, we can identify both the sources of pollutants and their potential fates. For example, the OSCURS model and data of crab traps from vessels lost off the coast of Oregon have subsequently been used to identified the northwest coast of the US as a source of debris reovered from the Northwestern Hawaiian Islands (Ebbesmeyer & Ingraham, 1994).

Learning about plastics

As well as using plastics to validate models, we may use models to indicate the apparent risk posed by plastics. In their assessment of the optimal places in which to position microplastic capture devices, Sherman and van Sebille (2016) set out to utilise Lagrangian data on water transport gathered by the NOAA Global Drifter Program (Lumpkin, 2003). This data was used to develop probabilities that a plastic particle within a cell (a given area of ocean) would move into any of the adjacent cells over a 2-month period. The point of origin for each particle was determined based on prior estimates of input from riverine and coastal areas outlined in Jambeck et al. (2015) and the apparent level of input was linked to global plastic production, scaled to account for observed levels of microplastics in environmental samples for the year 2014. The time period simulated included 1965–2025, with the release of tracer particles every 2 months. The results of Sherman and van Sebille's study highlighted significant aggregations of particles at latitudes between 10 and 45N across the Pacific Ocean, and 20 and 40S in the Atlantic, suggesting these locations as potential sites of extraction.

Of course, this is not the only approach be which microplastic levels may be determined. In an earlier study, Van Sebille et al. (2015) compared modelled microplastic counts devised by employing three different methods by Maximenko et al. (2012), Lebreton et al. (2012) and the van Sebille et al. (2012) method applied above. Under each approach, MPs were deployed in ocean circulation models to map the effects of surface currents on redistribution.

Under the first of these alternative approaches, the Maximenko model (Maximenko et al., 2012) also used the probability of transport between individual cells (in this case half a degree distance) based on observations of drifting buoys. The microplastics deployed were advected through the ocean over a period of 10 years. They showed that, in 2–3 years, a high concentration of microplastics builds up in the five subtropical gyres with the potential to persist for hundreds of years before washing ashore.

The second alternative modelling approach laid out by Lebreton et al. (2012) uses ocean velocity fields from the 1/12-degree global Hybrid Coordinate Ocean Model. Here the input of plastics into the modelled environment occurred at river mouths, scaled according to their level of urbanisation (characterised as the impervious surface area) as well as on the level of shipping traffic (defined by the scale of local shipping routes). The number of microplastics released over the study period was based on global plastic production data, and their introduction and transport projected over 30 years. This revealed the importance of northern hemisphere accumulation zones as well as the impacts of densely populated and urbanised coasts on smaller seas on plastic concentrations.

(Continued)

> **Box 2.1 Mapping and modelling plastic movement (Continued)**
>
> **Learning and knowledge outcomes**
>
> Of course, the above studies are not the only ones to attempt to explore the relationship between plastic and water movement, but they do illustrate that plastics have the potential to tell us a lot about the physical processes occurring in marine settings, as well as providing a method by which to ground-truth novel modelling approaches. Additionally, modelling may assist in identifying candidate sites for plastic management interventions.
>
> Nevertheless, predicting the movement of plastics remains a significant research challenge. The factors affecting microplastic distribution occur at variable spatiotemporal scales: for example, local currents and global transport, season temperature-related density changes or the influence of the El Niño effect. Additionally, the highly heterogeneous nature of microplastic pollution can provide numerous confounding influences on the models and their apparent accuracy (Khatmullina & Chubarenko, 2019). Thus, while global models do help to capture the broad trends in plastic distribution, our ability to predict heterogeneity at smaller scales remains limited.

2.6 Transport in sediments and soils

As with our understanding of the airborne movements of microplastics, far less is known about the movement of microplastics through sediments and soils. Papers concerning soil microplastics, emerging around 2016, highlight highly variable levels of microplastic content both within and between sites (Helmberger et al., 2020). There are a number of routes by which these particles may enter the soil environment, such as the fragmentation of existing litter, airborne deposition, riverine or tidal inundation or the application of contaminated materials (Li et al., 2020). However, the movement of microplastics both horizontally and vertically within and out of the soil structure is comparatively understudied.

Nevertheless, several characteristics of the soil environment have been identified as significant in the transport and aggregation of microplastics, such as the apparent pore size (gaps in between sediments), their influence on plastic retention and horizontal transport, and the movement of water may result in the overground transport of microplastics or their movement into the soil. Additionally, microplastics at the interface between air and soil, or held within friable soils, may be transported via aeolian processes. In arid regions, observations from horizontal sediment traps placed have related the movement of microplastics to wind-driven shear stresses (Abbasi et al., 2022), highlighting not only surface movement but the potential for ongoing exchange with the atmosphere.

Observations have also been made of the impact of animals and plants on microplastic's redistribution via a process known as bioturbation. Here, the action of plants and animals serves to move microplastics into and through the sediment, altering microplastics abundance across the soil strata. This process has previously been demonstrated by invertebrates including earthworms, *Lumbricus terrestris* (Rillig et al., 2017), and springtails, *Folsomia candida* and *Proisotoma minuta* (Maaß et al., 2017). The mechanism by which bioturbation influences microplastic distribution is also variable, in addition to disrupting the soil layers in which microplastics are located, microplastics may adhere to or be ingested by organisms, resulting in their transport within the soil itself.

2.7 Conclusions

The number and variation in the factors responsible for the transport of microplastic in the environment represent a substantial challenge for researchers and environmental management. While our current level of understanding allows us to predict where elevated levels of microplastic aggregation may arise, we are currently ill-equipped to observe fine-scale changes in microplastic concentration or reliably determine the long-term fate of particles. Attempts have been made to establish back trajectories for plastics based on polymer type, abundance and environmental conditions but, again, uncertainty in the form of unclear deposition rates limits our ability to apportion microplastics to their sources or recent sinks. Additionally, there remains much more to learn about transport and partitioning at the interfaces between the realms of land, water and air. If combining these sources of uncertainty with the confounding effects of biota, it is clear that fine-scale mapping is highly site specific, and the relevant processes at any one site may vary significantly from those adjacent to them.

References

Abbasi, Sajjad, Rezaei, Mahrooz, Ahmadi, Farnaz, & Turner, Andrew (2022). Atmospheric transport of microplastics during a dust storm. *Chemosphere*, *292*, 133456. Available from https://doi.org/10.1016/j.chemosphere.2021.133456.

Allen, S., Allen, D., Phoenix, V. R., Le Roux, G., Durántez Jiménez, P., Simonneau, A., Binet, S., & Galop, D. (2019). Atmospheric transport and deposition of microplastics in a remote mountain catchment. *Nature Geoscience*, *12*(5), 339−344. Available from https://doi.org/10.1038/s41561-019-0335-5. Available from, http://www.nature.com/ngeo/index.html.

Amaral-Zettler, L. A., Zettler, E. R., Mincer, T. J., Klaassen, M. A., & Gallager, S. M. (2021). Biofouling impacts on polyethylene density and sinking in coastal waters: A macro/micro tipping point? *Water Research*, *201*. Available from https://doi.org/10.1016/j.watres.2021.117289. Available from, http://www.elsevier.com/locate/watres.

Amato-Lourenço, Luís Fernando, dos Santos Galvão, Luciana, Wiebeck, H. élio, Carvalho-Oliveira, Regiani, & Mauad, Thals (2022). Atmospheric microplastic fallout in outdoor and indoor environments in São Paulo megacity. *Science of The Total Environment*, *821*, 153450. Available from https://doi.org/10.1016/j.scitotenv.2022.153450.

Aves, A. R., Revell, L. E., Gaw, S., Ruffell, H., Schuddeboom, A., Wotherspoon, N. E., Larue, M., & Mcdonald, A. J. (2022). First evidence of microplastics in Antarctic snow. *Cryosphere*, *16*(6), 2127−2145. Available from https://doi.org/10.5194/tc-16-2127-2022. Available from, http://www.the-cryosphere.net/volumes_and_issues.html.

Barnes, D. K. A. (2002). Biodiversity: Invasions by marine life on plastic debris. *Nature*, *416*(6883), 808−809. Available from https://doi.org/10.1038/416808a.

Bergmann, M., Mützel, S., Primpke, S., Tekman, M. B., Trachsel, J., & Gerdts, G. (2019). White and wonderful? Microplastics prevail in snow from the Alps to the Arctic. *Science Advances*, *5*(8). Available from https://doi.org/10.1126/sciadv.aax1157. Available from, https://advances.sciencemag.org/content/5/8/eaax1157/tab-pdf.

Carlton, J. T., Chapman, J. W., Geller, J. B., Miller, J. A., Ruiz, G. M., Carlton, D. A., McCuller, M. I., Treneman, N. C., Steves, B. P., Breitenstein, R. A., Lewis, R., Bilderback, D., Bilderback, D., Haga, T., & Harris, L. H. (2018). Ecological and biological studies of ocean rafting: Japanese tsunami marine debris in North America and the Hawaiian Islands. *Aquatic Invasions*, *13*(1), 1−9. Available from https://doi.org/10.3391/ai.2018.13.1.01. Available from, http://www.aquaticinvasions.net/2018/AI_2018_JTMD_Carlton_etal.pdf.

Choy, C. A., Robison, B. H., Gagne, T. O., Erwin, B., Firl, E., Halden, R. U., Hamilton, J. A., Katija, K., Lisin, S. E., Rolsky, C., & Van Houtan, K. S. (2019). The vertical distribution and biological transport of marine microplastics across the epipelagic and mesopelagic water column. *Scientific Reports*, *9*(1). Available from https://doi.org/10.1038/s41598-019-44117-2. Available from, http://www.nature.com/srep/index.html.

Cowger, W., Gray, A. B., Guilinger, J. J., Fong, B., & Waldschläger, K. (2021). Concentration depth profiles of microplastic particles in river flow and implications for surface sampling. *Environmental Science and Technology*, *55*(9), 6032−6041. Available from https://doi.org/10.1021/acs.est.1c01768. Available from, http://pubs.acs.org/journal/esthag.

Daily, Juliette, Tyler, Anna Christina, & Hoffman, Matthew J. (2022). Modeling three-dimensional transport of microplastics and impacts of biofouling in Lake Erie and Lake Ontario. *Journal of Great Lakes Research*, *48*(5), 1180−1190. Available from https://doi.org/10.1016/j.jglr.2022.07.001.

Debroas, Didier, Mone, Anne, & Ter Halle, Alexandra (2017). Plastics in the North Atlantic garbage patch: A boat-microbe for hitchhikers and plastic degraders. *Science of The Total Environment*, *599−600*, 1222−1232. Available from https://doi.org/10.1016/j.scitotenv.2017.05.059.

Dris, R., Gasperi, J., Saad, M., Mirande, C., & Tassin, B. (2016). Synthetic fibers in atmospheric fallout: A source of microplastics in the environment? *Marine Pollution Bulletin*, *104*(1−2), 290−293. Available from https://doi.org/10.1016/j.marpolbul.2016.01.006. Available from, http://www.elsevier.com/locate/marpolbul.

Díez-Minguito, Manuel, Bermúdez, María, Gago, Jesús, Carretero, Olga, & Viñas, Lucía (2020). Observations and idealized modelling of microplastic transport in estuaries: The exemplary case of an upwelling system (Ría de Vigo, NW Spain). *Marine Chemistry*, *222*, 103780. Available from https://doi.org/10.1016/j.marchem.2020.103780.

Ebbesmeyer, C. C., & Ingraham, W. J. (1992). Shoe spill in the North Pacific. *Eos. Transactions American Geophysical Union*, *73*(34), 361−365. Available from https://doi.org/10.1029/91EO10273.

Ebbesmeyer, C. C., & Ingraham, W. J. (1994). Pacific toy spill fuels ocean current pathways research. *Eos, Transactions American Geophysical Union*, *75*(37), 425−430. Available from https://doi.org/10.1029/94EO01056.

Ebbesmeyer, C., & Scigliano, E. (2009). *Flotsametrics and the floating world: How one man's obsession with runaway sneakers and rubber ducks revolutionized ocean science*. Harper Collins.

Fazey, F. M. C., & Ryan, P. G. (2016). Biofouling on buoyant marine plastics: An experimental study into the effect of size on surface longevity. *Environmental Pollution*, *210*, 354−360. Available from https://doi.org/10.1016/j.envpol.2016.01.026. Available from, http://www.elsevier.com/inca/publications/store/4/0/5/8/5/6.

Fossette, S., Putman, N. F., Lohmann, K. J., Marsh, R., & Hays, G. C. (2012). A biologist's guide to assessing ocean currents: A review. *Marine Ecology Progress Series*, *457*, 285−301. Available from https://doi.org/10.3354/meps09581United. Available from, http://www.int-res.com/articles/theme/m457p285.pdf.

Hansen, G. I., Hanyuda, T., & Kawai, H. (2018). Invasion threat of benthic marine algae arriving on Japanese tsunami marine debris in Oregon and Washington, USA. *Phycologia*, *57*(6), 641−658. Available from https://doi.org/10.2216/18-58.1. Available from, http://www.phycologia.org/doi/pdf/10.2216/18-58.1.

Helmberger, M. S., Tiemann, L. K., & Grieshop, M. J. (2020). Towards an ecology of soil microplastics. *Functional Ecology*, *34*(3), 550−560. Available from https://doi.org/10.1111/1365-2435.13495. Available from, http://onlinelibrary.wiley.com/journal/10.1111/(ISSN)1365-2435.

He, Beibei, Smith, Mitchell, Egodawatta, Prasanna, Ayoko, Godwin A., Rintoul, Llew, & Goonetilleke, Ashantha (2021). Dispersal and transport of microplastics in river sediments. *Environmental Pollution*, *279*, 116884. Available from https://doi.org/10.1016/j.envpol.2021.116884.

Jambeck, J. R., Geyer, R., Wilcox, C., Siegler, T. R., Perryman, M., Andrady, A., Narayan, R., & Law, K. L. (2015). Plastic waste inputs from land into the ocean. *Science*, *347* (6223), 768−771. Available from https://doi.org/10.1126/science.1260352. Available from, http://www.sciencemag.org/content/347/6223/768.full.pdf.

Kaiser, David, Kowalski, Nicole, & Waniek, Joanna J. (2017). Effects of biofouling on the sinking behavior of microplastics. *Environmental Research Letters*, *12*(12), 124003. Available from https://doi.org/10.1088/1748-9326/aa8e8b. Available from, https://doi.org/10.1088/1748-9326/aa8e8b.

Kane, I. A., Clare, M. A., Miramontes, E., Wogelius, R., Rothwell, J. J., Garreau, P., & Pohl, F. (2020). Seafloor microplastic hotspots controlle by deep-sea circulation. *Science*, *368*(6495), 1140−1145. Available from https://doi.org/10.1126/science.aba5899. Available from, https://science.sciencemag.org/content/sci/368/6495/1140.full.pdf.

Kernchen, S., Löder, M. G. J., Fischer, F., Fischer, D., Moses, S. R., Georgi, C., Nölscher, A. C., Held, A., & Laforsch, C. (2022). Airborne microplastic concentrations and deposition across the Weser River catchment. *Science of the Total Environment*, *818*. Available from https://doi.org/10.1016/j.scitotenv.2021.151812. Available from, http://www.elsevier.com/locate/scitotenv.

Khatmullina, L., & Chubarenko, I. (2019). Transport of marine microplastic particles: Why is it so difficult to predict? *Anthropocene Coasts*, *2*(1), 293−305. Available from https://doi.org/10.1139/anc-2018-0024. Available from, http://www.nrcresearchpress.com/journal/anc.

Kooi, M., Van Nes, E. H., Scheffer, M., & Koelmans, A. A. (2017). Ups and downs in the ocean: Effects of biofouling on vertical transport of microplastics. *Environmental Science and Technology*, *51*(14), 7963−7971. Available from https://doi.org/10.1021/acs.est.6b04702. Available from, http://pubs.acs.org/journal/esthag.

Kowalski, N., Reichardt, A. M., & Waniek, J. J. (2016). Sinking rates of microplastics and potential implications of their alteration by physical, biological, and chemical factors. *Marine Pollution Bulletin*, *109*(1), 310−319. Available from https://doi.org/10.1016/j.marpolbul.2016.05.064. Available from, http://www.elsevier.com.

Kukulka, T., Law, K. L., & Proskurowski, G. (2016). Evidence for the influence of surface heat fluxes on turbulent mixing of microplastic marine debris. *Journal of Physical Oceanography*, *46*(3), 809−815. Available from https://doi.org/10.1175/JPO-D-15-0242.1. Available from, http://journals.ametsoc.org/doi/pdf/10.1175/JPO-D-15-0242.1.

Kumar, R., Sharma, P., Verma, A., Kumar Jha, P., Singh, P., Kumar Gupta, P., Chandra, R., & Vara Prasad, P. V. (2021). Effect of physical characteristics and hydrodynamic conditions on transport and deposition of microplastics in riverine ecosystem. *Water*, *13*(19), 2710. Available from https://doi.org/10.3390/w13192710.

Kvale, K. F., Friederike Prowe, A. E., & Oschlies, A. (2020). A critical examination of the role of marine snow and zooplankton fecal pellets in removing ocean surface microplastic. *Frontiers in Marine Science*, *6*. Available from https://doi.org/10.3389/fmars.2019.00808. Available from, https://www.frontiersin.org/journals/marine-science#.

Lattin, G. L., Moore, C. J., Zellers, A. F., Moore, S. L., & Weisberg, S. B. (2004). A comparison of neustonic plastic and zooplankton at different depths near the southern California shore. *Marine Pollution Bulletin*, *49*(4), 291−294. Available from https://doi.org/10.1016/j.marpolbul.2004.01.020.

Lebreton, L. (2022). The status and fate of oceanic garbage patches. *Nature Reviews Earth and Environment*, *3*(11), 730−732. Available from https://doi.org/10.1038/s43017-022-00363-z. Available from, https://www.nature.com/natrevearthenviron/.

Lebreton, L. C. M., Greer, S. D., & Borrero, J. C. (2012). Numerical modelling of floating debris in the world's oceans. *Marine Pollution Bulletin*, *64*(3), 653−661. Available from https://doi.org/10.1016/j.marpolbul.2011.10.027.

Liao, Z., Ji, X., Ma, Y., Lv, B., Huang, W., Zhu, X., Fang, M., Wang, Q., Wang, X., Dahlgren, R., & Shang, X. (2021). Airborne microplastics in indoor and outdoor environments of a coastal city in Eastern China. *Journal of Hazardous Materials*, *417*, 126007. Available from https://doi.org/10.1016/j.jhazmat.2021.126007.

Liu, S., Huang, Y., Luo, D., Wang, X., Wang, Z., Ji, X., Chen, Z., Dahlgren, R. A., Zhang, M., & Shang, X. (2022). Integrated effects of polymer type, size and shape on the sinking dynamics of biofouled microplastics. *Water Research*, *220*. Available from https://doi.org/10.1016/j.watres.2022.118656. Available from, http://www.elsevier.com/locate/watres.

Li, J., Song, Y., & Cai, Y. (2020). Focus topics on microplastics in soil: Analytical methods, occurrence, transport, and ecological risks. *Environmental Pollution*, *257*, 113570. Available from https://doi.org/10.1016/j.envpol.2019.113570.

Lobelle, D., Kooi, M., Koelmans, A. A., Laufkötter, C., Jongedijk, C. E., Kehl, C., & van Sebille, E. (2021). Global modeled sinking characteristics of biofouled microplastic. *Journal of Geophysical Research: Oceans*, *126*(4). Available from https://doi.org/10.1029/2020JC017098. Available from, http://agupubs.onlinelibrary.wiley.com/agu/jgr/journal/10.1002/(ISSN)2169-9291/.

Lumpkin, R. (2003). Decomposition of surface drifter observations in the Atlantic Ocean. *Geophysical Research Letters*, *30*(14). Available from https://doi.org/10.1029/2003GL017519. Available from, http://onlinelibrary.wiley.com/journal/10.1002/(ISSN)1944-8007/issues?year = 2012.

Maaß, S., Daphi, D., Lehmann, A., & Rillig, M. C. (2017). Transport of microplastics by two collembolan species. *Environmental Pollution, 225,* 456−459. Available from https://doi.org/10.1016/j.envpol.2017.03.009. Available from, http://www.elsevier.com/inca/publications/store/4/0/5/8/5/6.

Maximenko, N., Hafner, J., & Niiler, P. (2012). Pathways of marine debris derived from trajectories of Lagrangian drifters. *Marine Pollution Bulletin, 65*(1-3), 51−62. Available from https://doi.org/10.1016/j.marpolbul.2011.04.016.

Nizzetto, L., Bussi, G., Futter, M. N., Butterfield, D., & Whitehead, P. G. (2016). A theoretical assessment of microplastic transport in river catchments and their retention by soils and river sediments. *Environmental Science: Processes and Impacts, 18*(8), 1050−1059. Available from https://doi.org/10.1039/c6em00206d. Available from, http://www.rsc.org/publishing/journals/em/about.asp.

Parolini, M., Antonioli, D., Borgogno, F., Gibellino, M. C., Fresta, J., Albonico, C., De Felice, B., Canuto, S., Concedi, D., Romani, A., Rosio, E., Gianotti, V., Laus, M., Ambrosini, R., & Cavallo, R. (2021). Microplastic contamination in snow from western Italian alps. *International Journal of Environmental Research and Public Health, 18* (2), 1−10. Available from https://doi.org/10.3390/ijerph18020768. Available from, https://www.mdpi.com/1660-4601/18/2/768/pdf.

Porter, A., Lyons, B. P., Galloway, T. S., & Lewis, C. (2018). Role of marine snows in microplastic fate and bioavailability. *Environmental Science and Technology, 52*(12), 7111−7119. Available from https://doi.org/10.1021/acs.est.8b01000. Available from, http://pubs.acs.org/journal/esthag.

Reisser, J., Slat, B., Noble, K., Du Plessis, K., Epp, M., Proietti, M., De Sonneville, J., Becker, T., & Pattiaratchi, C. (2015). The vertical distribution of buoyant plastics at sea: An observational study in the North Atlantic Gyre. *Biogeosciences, 12*(4), 1249−1256. Available from https://doi.org/10.5194/bg-12-1249-2015. Available from, http://www.biogeosciences.net/volumes_and_issues.html.

Rey, S. F., Franklin, J., & Rey, S. J. (2021). Microplastic pollution on island beaches, Oahu, Hawai'i. *PLoS ONE, 16*(2). Available from https://doi.org/10.1371/journal.pone.0247224. Available from, https://journals.plos.org/plosone/article/file?id = 10.1371/journal.pone.0247224&type = printable.

Rillig, M. C., Ziersch, L., & Hempel, S. (2017). Microplastic transport in soil by earthworms. *Scientific Reports, 7*(1). Available from https://doi.org/10.1038/s41598-017-01594-7. Available from, http://www.nature.com/srep/index.html.

van Sebille, E., England, M. H., & Froyland, G. (2012). Origin, dynamics and evolution of ocean garbage patches from observed surface drifters. *Environmental Research Letters, 7*(4), 044040. Available from https://doi.org/10.1088/1748-9326/7/4/044040.

Van Sebille, E., Wilcox, C., Lebreton, L., Maximenko, N., Hardesty, B. D., Van Franeker, J. A., Eriksen, M., Siegel, D., Galgani, F., & Law, K. L. (2015). A global inventory of small floating plastic debris. *Environmental Research Letters, 10*(12). Available from https://doi.org/10.1088/1748-9326/10/12/124006. Available from, http://iopscience.iop.org/article/10.1088/1748-9326/10/12/124006/pdf.

Semcesen, P. O., & Wells, M. G. (2021). Biofilm growth on buoyant microplastics leads to changes in settling rates: Implications for microplastic retention in the Great Lakes. *Marine Pollution Bulletin, 170.* Available from https://doi.org/10.1016/j.marpolbul.2021.112573. Available from, http://www.elsevier.com/locate/marpolbul.

Sherman, P., & van Sebille, E. (2016). Modeling marine surface microplastic transport to assess optimal removal locations. *Environmental Research Letters, 11*(1), 014006. Available from https://doi.org/10.1088/1748-9326/11/1/014006.

Tanabe, S., Tatsukawa, R., Kawano, M., & Hidaka, H. (1982). Global distribution and atmospheric transport of chlorinated hydrocarbons: HCH (BHC) isomers and DDT compounds in the Western Pacific, Eastern Indian and Antarctic Oceans. *Journal of the Oceanographical Society of Japan, 38*(3), 137−148. Available from https://doi.org/10.1007/BF02110285.

Thompson, R. C., Olson, Y., Mitchell, R. P., Davis, A., Rowland, S. J., John, A. W. G., McGonigle, D., & Russell, A. E. (2004). Lost at sea: Where is all the plastic? *Science, 304*(5672), 838. Available from https://doi.org/10.1126/science.1094559.

Trainic, M., Flores, J. M., Pinkas, I., Pedrotti, M. L., Lombard, F., Bourdin, G., Gorsky, G., Boss, E., Rudich, Y., Vardi, A., & Koren, I. (2020). Airborne microplastic particles detected in the remote marine atmosphere. *Communications Earth and Environment, 1*(1). Available from https://doi.org/10.1038/s43247-020-00061-y. Available from, https://www.nature.com/commsenv/.

Wang, Y., Huang, J., Zhu, F., & Zhou, S. (2021). Airborne microplastics: A review on the occurrence, migration and risks to humans. *Bulletin of Environmental Contamination and Toxicology, 107*(4), 657−664. Available from https://doi.org/10.1007/s00128-021-03180-0.

Watkins, L., McGrattan, S., Sullivan, P. J., & Walter, M. T. (2019). The effect of dams on river transport of microplastic pollution. *Science of the Total Environment, 664*, 834−840. Available from https://doi.org/10.1016/j.scitotenv.2019.02.028. Available from, http://www.elsevier.com/locate/scitotenv.

Welden, N. A., & Cowie, P. R. (2017). Degradation of common polymer ropes in a sublittoral marine environment. *Marine Pollution Bulletin, 118*(1-2), 248−253. Available from https://doi.org/10.1016/j.marpolbul.2017.02.072. Available from, http://www.elsevier.com/locate/marpolbul.

Wichmann, D., Delandmeter, P., & van Sebille, E. (2019). Influence of near-surface currents on the global dispersal of marine microplastic. *Journal of Geophysical Research: Oceans, 124*(8), 6086−6096. Available from https://doi.org/10.1029/2019JC015328. Available from, http://agupubs.onlinelibrary.wiley.com/agu/jgr/journal/10.1002/(ISSN)2169-9291/.

Wright, S. L., Ulke, J., Font, A., Chan, K. L. A., & Kelly, F. J. (2020). Atmospheric microplastic deposition in an urban environment and an evaluation of transport. *Environment International, 136*, 105411. Available from https://doi.org/10.1016/j.envint.2019.105411.

Zhang, X., Zhang, X., Zhang, Z. F., Yang, P. F., Li, Y. F., Cai, M., & Kallenborn, R. (2022). Pesticides in the atmosphere and seawater in a transect study from the Western Pacific to the Southern Ocean: The importance of continental discharges and air-seawater exchange. *Water Research, 217*. Available from https://doi.org/10.1016/j.watres.2022.118439. Available from, http://www.elsevier.com/locate/watres.

Zhao, S., Zettler, E. R., Bos, R. P., Lin, P., Amaral-Zettler, L. A., & Mincer, T. J. (2022). Large quantities of small microplastics permeate the surface ocean to abyssal depths in the South Atlantic Gyre. *Global Change Biology, 28*(9), 2991−3006. Available from https://doi.org/10.1111/gcb.16089. Available from, http://onlinelibrary.wiley.com/journal/10.1111/(ISSN)1365-2486.

Zhu, X., Huang, W., Fang, M., Liao, Z., Wang, Y., Xu, L., Mu, Q., Shi, C., Lu, C., Deng, H., Dahlgren, R., & Shang, X. (2021). Airborne microplastic concentrations in five megacities of northern and southeast China. *Environmental Science and Technology*, 55(19), 12871−12881. Available from https://doi.org/10.1021/acs.est.1c03618. Available from, http://pubs.acs.org/journal/esthag.

Zoppi, U., Skopec, Z., Skopec, J., Jones, G., Fink, D., Hua, Q., Jacobsen, G., Tuniz, C., & Williams, A. (2004). Forensic applications of 14C bomb-pulse dating. *Nuclear Instruments and Methods in Physics Research, Section B: Beam Interactions with Materials and Atoms, 223-224*, 770−775. Available from https://doi.org/10.1016/j.nimb.2004.04.143

Monitoring environmental accumulation

3.1 Introduction

In addition to the many observations demonstrating the ubiquitous nature of microplastic pollution, the regular reports of apparently alarming concentrations of microplastics have become a hallmark of the research (Fig. 3.1). Following the first records of microplastics in the environment which outlined their apparently increasing abundance (Thompson et al., 2004), studies have repeatedly revealed that microplastic abundance is not evenly spread, demonstrating high levels of heterogeneity (difference) at both temporal and spatial scales. Such variation has been consistent across air (Kernchen et al., 2022), soils (Zhang et al., 2022), rivers (Zhang et al., 2021) and marine and coastal settings (Ovide et al., 2022). Reported microplastic concentrations may vary by several orders of magnitude, the apparent scale of which is dependent on a variety of environmental and anthropogenic factors, *as well as the monitoring methods employed*. However, our understanding of this variability may only scratch the surface of what is

FIGURE 3.1 Stranded microplastics.

Beached microplastics on a sandy shore.

Image Credit: Matthew Cole.

Microplastics. DOI: https://doi.org/10.1016/B978-0-443-13324-4.00003-0

found in the environment. We are limited by the coverage of our sampling, as well as its coherence and rigour. For example, little is known about the distribution of the smallest categories of microplastics due to the relative challenges in their sampling and identification. This is borne out in a review which compared the minimum size ranges reported in environmental sampling programmes with an analysis of currently unregulated sources of primary microplastics which reported significant volumes of particles below this cut-off (Conkle et al., 2018), and supported by experimental comparisons of microplastic captured in nets of differing mesh size (Lindeque et al., 2020). To better understand the distribution of the risks posed by microplastic pollution we must have robust survey schemes. In the following sections, we will highlight the implications of these observations to environmental managers, researchers and other interested groups, discuss the diversity of methods available to those working on this topic and consider how best to account for this diversity.

3.2 The call for microplastic monitoring

The leakage of plastics and microplastics into the environment highlighted in Chapters 1 and 2 (in addition to their uptake and impacts on organisms and their habitats discussed in Chapters 4 and 5) has led to significant global pressure to monitor and minimise inputs of microplastics. The identification of at-risk habitats, pollution events and even culpable parties is reliant on the implementation of a routine, reliable and comparable approach to sampling, analysis and reporting. As such, a number of parties at both national and international scales have sought to implement monitoring programmes, for which a number of frameworks have been proposed. For example, under the OSPAR Convention (the Convention for the Protection of the Marine Environment of the North-East Atlantic) the level of plastic litter in a bioindicator (stranded Fulmar, *Fulmarus glacialis*) serves as a bellwether for the region's marine debris. More recently, the UN's Group of Experts on the Scientific Aspects of Marine Environmental Protection's (GESAMP) WG 40: Plastics and Microplastics in the Ocean, has sought to not only suggest an approach to the design and presentation of environmental sampling regimes but also to define the terminology applied (GESAMP, 2019).

3.3 Modes of microplastic monitoring

3.3.1 The requirements of researchers versus the requirements of monitoring agencies

There are numerous methods for the isolation, quantification and reporting of microplastics, and there remains a significant ongoing diversification of approaches. The drivers of this are twofold. First, in the infancy of the field of microplastic research, the determination and reporting of microplastic pollution was often carried out in an ad hoc manner, alongside existing work, and the materials used were frequently dependent on the available resources of the institute or individual. Many of these methods have subsequently been normalised by the increasing number of researchers turning to existing publications when designing their own sampling regimes. Second, as the field has specialised (e.g. into smaller plastic fractions or highly challenging environments) the abundance of highly novel new methods has significantly increased (Rist et al., 2021). Lack of adherence to a single set of methods has resulted in tension both between the requirements of individual researchers as well as between researchers and monitoring agencies. The level of detailed information required to answer specific research questions, for example the abundance of small microplastic and nanoplastic fractions, may vary significantly from that needed by, for example, an individual conducting routine monitoring of wastewater (Rist et al., 2021). Thus although the demand for a comprehensive approach to routine microplastic monitoring is widespread and a number of methodologies have been proposed, the form (methods and frequency) and requirements (targets, reliability and data standards) of this monitoring remain a source of debate (Hartmann et al., 2019; Lu et al., 2021; Sridharan et al., 2021).

Unfortunately, this diversity of methods has also resulted in inconsistencies in reporting, the decisions underpinning which are highlighted in Box 3.1. As seen in previous chapters, some sampling methods result in the reporting of *concentrations*, for example items per litre or per m^3, others necessitate the presentation of data in *abundances*, often in relation to a known 2-dimensional area, but sometimes on a "per sample" basis (Rist et al., 2021). However, paucity of available information (especially as new areas of interest have developed) has frequently driven researchers and other interested groups to make inappropriate comparisons between studies in an attempt to make wider assessments of the impacts of microplastics, frequently to increase the spatial or temporal range considered.

Similarly, the distribution of sampling locations and frequency of sampling have a significant impact on the usefulness of survey results. Sites may be selected by randomised, stratified, transect or other systematic approaches, further details of which are available in Box 3.2, depending on environmental characteristics, research aims and prior knowledge of the habitat. The selection of sampling sites is routinely defined by the research question under consideration, for example studies to determine the influence of a point source of plastics on their abundance in the environment may use a series of transects to establish the relationship between microplastic concentration and distance from source, whereas a study comparing the abundance of microplastics in fields which have received treatment by sewage sludge with those that have not might use stratified sampling methods. While each is valid according to its aims, they may result in inconsistent coverage of the true microplastic distribution within the environment studied.

Of course, an entirely controlled approach to sampling is not always achievable due to the realities of environmental monitoring and research. The confounding effects of time, weather, site access and finances continue to frustrate researcher efforts to design an idealised sampling regime.

Box 3.1 A mix of methods — designing sampling regimes for microplastic monitoring

Objective

In this section, which may be useful to students, new researchers and environmental managers, we will design a hypothetical sampling regime, following a pre-determined workflow in order to establish the most appropriate sampling and processing techniques.

Rationale

Rapid expansion in the field of microplastic monitoring and impact assessment have resulted in the development of numerous potential measures for both sampling and analysis, not all of which are suitable for broad purposes or easy comparison. As a result, the accessibility and comparability of the current evidence base is in question. The step-by-step guide to choosing a sampling plan laid out below is designed to demystify this process and assist the reader in planning or interpreting monitoring regimes. These measures may not be suitable to all settings and care must be taken by the reader to identify where differences at their own sites may invalidate the proposed protocols. A summary of the below methods is given in Fig. 3.2.

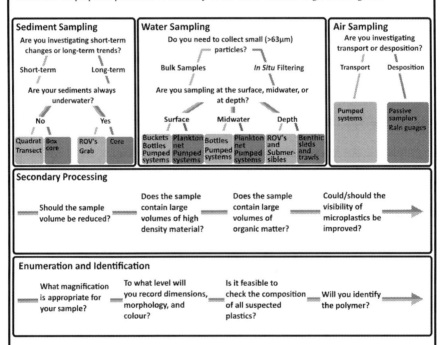

FIGURE 3.2 Flow chart for the collection and analysis of environmental samples.

A decision tree for the selection of sampling and processing steps.

Workflow

As with any sampling regime, the method or combination of methods applied will be defined by the research question at hand and bounded by the resources available to the researcher (not

(Continued)

Box 3.1 A mix of methods — designing sampling regimes for microplastic monitoring (Continued)

everyone has ready access to an ROV!). The methods outlined in Section 3.3 and flagged in Fig. 3.2 summarise many of these core approaches, however, there are numerous controllable aspects of both sampling and analysis that may be altered to suit the requirement of the work undertaken, for example the use of bulk samples, or the mesh and pore sizes of filtration steps.

Example: the area downstream of a wastewater treatment facility.

In this case, the researcher is interested in undertaking a short-term, *preliminary* assessment of the impact of wastewater outflows on microplastic concentration of a riverine site. The river is in its lower course, with a silty bed. It is also assumed that the impact of the outfall of microplastic concentration will diminish with distance downstream, however, this must be confirmed through the proposed observations. By asking a few questions, we can develop a suitable sampling regime.

1. *What compartments of the river system am I interested in?*

   In this case, the researcher may be interested in both the sediment and water column to establish whether any observed increase in abundance in microplastic in the water column arising from the water treatment facility translates to that in the sediments.

2. *What sampling strategy?*

   As the study will explore the increase in microplastic concentration as a result of the wastewater outfall, sites have been selected above and below the outfall itself, enabling a comparison with 'normal' contamination levels. Additionally, as the researcher wishes to determine how far the impact of the outfall extends, sites below the outfall will be arranged systematically along a transect, in order to establish longitudinal changes.

3. *Do you need to collect samples below 63 µm?*

   In this case, our hypothetical laboratory *is* equipped to deal with microplastics at the lower end of the microplastic scale, thus bulk sampling is an option. As an exploratory study, the researcher may wish to take a smaller number of samples, either net-collected or filtered in situ, to enable any temporal variability in larger microplastic output to be captured. In this case, we will opt to include lower-size classes and take the bulk sampling approach.

4. *Benthic sediment sampling*
   a. Am I interested in short-term differences or long-term trends?

   In this case, which is a preliminary study, sampling may only occur over a short period of time. Nevertheless, examination of deeper sediments may highlight the presence of longer-term trends.

   b. Are the sediments underwater?

   Yes, and, as the sediments are silty, collection options include both grabs and cores. The removed depth of sediment and the number of horizons would be dependent on the exact interest of the researcher. Alternatively, settlement of microplastic from the water column may be determined using traps.

   c. Does the sample contain large volumes of high-density material?

   In this case, it is likely that the researcher must pretest any density separation measures to determine the best solution and number of replicates. Silty sediments contain many small particles which are easily resuspended and may take an extended period of time to settle. The use of spiked samples to confirm extraction efficiency would be of high importance.

(Continued)

Box 3.1 A mix of methods — designing sampling regimes for microplastic monitoring (Continued)

---

d. Does the sample contain large volumes of organic matter?

Rivers in their lower course often contain high volumes of organic matter. Digestion may be essential in order to enable successful extraction of microplastics and their observation under the microscope.

e. Could the visibility of plastics be improved?

In the event that silts and organic matter remain, staining may assist in the later enumeration of samples

5. *Surface water sampling*
   a. Are you sampling at the surface, midwater or at depth?

In order to determine the abundance of suspended plastics released from the outfall, samples will be taken at the water's surface. Options for bulk sampling include bucket and bottle sampling, as well as the use of pumped samples and plankton nets if not analysing the smallest fraction.

   b. Does the sample contain large volumes of high-density material?

While water samples frequently contain minimal sediment or other high-density matter, this may be altered by high rainfall periods or the presence of the sediments in the effluent from the outfall. A preliminary assessment of the sample content should be made to determine whether microplastics may be easily distinguished in filtered samples.

   c. Does the sample contain large volumes of organic matter?

Similarly, the presence of large numbers of diatoms or algae or accessive levels of particulate organic matter may necessitate a digestion stage.

   d. Could the visibility of plastics be improved?

In the event that sediment or organic matter remains, staining may assist in the later enumeration of samples.

6. *Sample processing and analysis*
   a. What magnification is appropriate for your sample?

If bulk sampling has been used, researchers may prefer to stretch their analysis to their absolute lower limit of detection. Luckily, our hypothetical laboratory is well equipped to identify small microplastic fractions, however, the time required for this type of processing may be prohibitive, and only applied to a subsample. In whatever way the enumeration of plastics is undertaken, sample handling and the limits of detection should be clearly outlined.

7. *To what level will you record dimensions, morphology and colour?*

The size classes used should represent the lower limits of detection and be matched to as wide a range of studies as feasible. In general, the finer the detail, the more easily results can be interpreted without risk of error. At a minimum, plastics should be described according to main classes (nibs, fragments, films and fibres), ascribed to secondary colour (red, orange, yellow, green, blue, purple, black, white and clear) groups.

8. *Is it feasible to check the identity of all suspected plastics?*

In some studies, the abundance of recovered microplastics may preclude the confirmation of all suspected microplastics. In such cases, a subsample of plastic may be sent for analysis and the

(Continued)

Box 3.1 A mix of methods — designing sampling regimes for microplastic monitoring (Continued)

resulting misidentification rate applied to the full data set. Additionally, prior staining using Nile Red or similar dye may have reduced the risk of misidentification, reducing the requirement for a full secondary analysis. As this is a pilot study, a full analysis may not be required, however, it should be remembered that results intended for publication or inclusion in reports should suitably validated.

9. *Will you identify the polymer?*

   As above, the nature of this study may not require the identification of plastics to polymer level. It may be sufficient to confirm that the recovered particles are plastics rather than other man made or natural materials. Additionally, the scale and intended audience of the study may dictate the level of analysis undertaken.

Challenges and solutions

As with all sampling regimes, the distribution, number and size of samples, the number of repeat sampling events and the suitability and efficacy of the collection and analytical processes will be depending on the availability of funding and resources. The use of quality control and quality assurance processes (e.g. the application of procedural blanks) is vital in providing the reader with the necessary information to accurately interpret and apply findings.

3.3.2 Extracting plastics from the environment

3.3.2.1 Sediments and soils

In this section, we will explore the collection and processing of sediments (including benthic and shoreline) from marine and freshwater environments and soils. As indicated above, the available methods for the collection of microplastics from sediments or soils are highly variable, and may be selected in accordance with the aims of the practitioner (Rist et al., 2021). As a result, samples may be recovered solely from surface deposits (the top few centimetres of the soil or sediment) or may seek to establish the abundance of microplastics at depth. The majority of these are *bulk* samples, in which the entire sediment volume is removed from the environment to a second location for processing, although some researchers try to isolate larger microplastics fractions on site (Hidalgo-Ruz et al., 2012), and may be kept as individual site samples or combined into composite samples which represent a single location (Möller et al., 2020).

3.3.2.2 Terrestrial and exposed intertidal sediments

While both terrestrial and exposed intertidal sediments may be less challenging to access than the submerged sediments discussed below, the methods applied may be equally complex. Sampling design remains of utmost importance in order to account for the numerous drivers of variability. For example intertidal

---

Box 3.2 Sampling regimes

Randomised sampling

Simple random sampling methods uses number generators (or similar) to identify sampling points. For example, a numbered grid may be overlaid above a map of a site and squares selected by drawing lots. These measures may be applied in situations in which the researcher has minimal information about the site, or the site or population is homogenous in nature. Randomised sampling may also be applied after the application of specific conditions (see below).

Stratified sampling

Existing information about the site, such as areas of specific land-use or habitat should be used to create subgroups. Random locations within these subgroups are selected for. These methods ensure that highly heterogenous sites or species of limited distribution are accounted for. An example of this process includes separating a beach into low, medium and high intertidal zones prior to sampling.

Systematic or grid sampling

Here an initial point (e.g. time, date or location) is selected at random, and the remaining sampling points are assigned in relation to a specific pattern or regular intervals. This selection method is suitable for studies in which uniform coverage of the site is of high importance, limited information is available regarding the distribution of the study subject. For example, if observing the distribution of invasive species along a river, points of interest may be selected based on random sampling, with 9 subsites at 50-m intervals downstream of the first location.

Transect sampling

Transect sampling is a form of linear sampling, in which regular observations are made over a set distance. Transects are of particular use in measuring environmental gradients. For example, the effect of trampling on heathland vegetation may be explored by using a series of strip transects at a tangential angle to the path direction.

---

environments will be subject to distinct patterns of exposure and submersion, typically aggregating microplastics within the strandline, which should be recognised within the sampling strategy (Hidalgo-Ruz et al., 2012).

In sampling these environments, some studies report only surface deposition of microplastics in terrestrial or exposed intertidal environments, representing short-term microplastic deposition, whereas others attempt to assess long-term changes or transport through the sediment by examining deeper sediment horizons.

When surface sampling, the perimeter of the area of interest is commonly defined using a box core, quadrat or similar boundaries (Piehl et al., 2018). Subsequently, visible material may be hand-gathered, using spoons or tweezers, or as a sample of surface sediments collected using a trowel or similar prior to in situ or lab-based plastic separation (Besley et al., 2017; Möller et al., 2020). Studies of microplastic abundance buried below the surface sediments often rely on bulk sampling methods. Again, a box core, soil auger (Corradini et al., 2019) or similar may be used, from which sediments may be removed as a single sample, or as subsamples from predefined depth ranges (Yang et al., 2021).

On-site separation of plastics may involve either passing samples through a sieve to exclude particles of a size *above* that of interest to the researcher or on-site floatation of microplastics by mixing sand with seawater and decanting the floated material through a sieve to retain plastics above the minimum size of interest to the researcher. Parallel laboratory extraction measures and subsequent secondary processing will be covered in more detail in Section 3.4.

3.3.2.3 Submerged intertidal and benthic sediments

The sampling of submerged sediments, particularly those in offshore zones and the deep-sea, is some of the most taxing and expensive to undertake. This is reflected in the comparatively low number of replicates reported in associated publications (Van Cauwenberghe et al., 2013; Woodall et al., 2014). Nevertheless, these observations have proved essential in broadening our understanding of plastic transport and sinks, highlighting impacts in inaccessible environments.

The collection of sediment samples is most commonly achieved by way of grabs and cores which may be retained in part or in entirety (Pagter et al., 2018; Van Cauwenberghe et al., 2013). Less frequently, smaller samples may also be taken using sediment traps, sleds or remotely operated vehicles (Barrett et al., 2020) (Fig. 3.3). Depending on the nature of the study, the difference between these

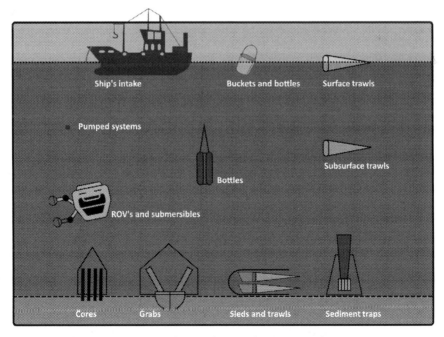

FIGURE 3.3 Methods for the collection of microplastics from aquatic samples.

Sampling methods and their operating depths.

sampling approaches may be substantial, however, comparisons between grabs and cores deployed in the same environment have previously demonstrated no apparent change in the recovery rates of plastic (Pagter et al., 2018). As above, samples taken in this manner may be subject to a range of secondary processing steps in order to separate plastic material from the matrix of sediments and organic matter and to assist in the enumeration and quantification of microplastics.

3.3.2.4 Characterisation of sediments and soils

In order to achieve maximum microplastic extraction and give context to the results generated, the researcher or observer must characterise the sediment samples. Complementary analytical approaches applied to sediments include the proportion of organic matter and water content, soil type and/or granulometry (Pagter et al., 2018). Loss on Ignition may be applied in order to determine factors including wet and dry weight/moisture content of terrestrial or intertidal sediments and organic matter. Whereas granulometry may be used to identify the grain size of sediments (particularly the proportion of silts and clay). As previously indicated, these factors may influence the efficiency of microplastic extraction, as well as settlement, burial and retention of plastics in the environment.

3.3.2.5 Water column

Similarly to sediments and soils, the extraction of microplastics from water samples may be via either bulk sampling (removing a specific volume of water from a water body and establishing the abundance of microplastics within said sample) or utilise in situ separation (via net, filter or sieve) in order to separate microplastics from the water on site (Hidalgo-Ruz et al., 2012). These methods are summarised in Fig. 3.3. Unlike in soils and sediments, bulk sampling of the water column is by far the less commonly reported approach, driven by the relative ease by which water may be excluded from a sample. Nevertheless, there are drawbacks to this approach, as the smallest size fractions are not easily retained for analysis (Covernton et al., 2019).

Net sampling methods are by far the most widely employed and abundant measures of isolating microplastics from water samples (Rist et al., 2021). Nets differ by type, dimensions and mesh size and may be deployed horizontally within a water layer, or vertically through multiple water levels to average microplastic abundance between depths (Table 3.1). Alternatively, bulk sampling may take place at the surface by way of buckets or bottles (Covernton et al., 2019), or at depth using Remotely Operated Vehicles, Niskin bottles or similar apparatus (Di Mauro et al., 2017). Alternatively, water sampling by way of pumped systems can be used to take bulk samples or may be combined with filters or sieves in order to combine the benefits of multiple methods. Pumped systems give greater control over the depth of sampling than net deployments, providing researchers with an accurate measure of the volume processed while maintaining the reduction in sample volume (Zobkov et al., 2019). Where specialist pumps are not

Table 3.1 Commonly used water sampling apparatus, methods and their variants.

Sample collection	Bulk		Filtration	
	Methods	**Lower Limit of Detection (post-processing)**	**Methods**	**Lower Limit of Detection**
Equipment broad class	Niskin bottles Pumped systems Bottles Bucket	Filter pore size (0.45–0.2 μm) Sieve mesh (1, 0.5, 0.25, 0.125 mm)	Plankton net Hand net Pond net Bongo net Water intake/ Pumped systems	Net mesh (333, 250, 162, 150, 63 μm) Sieve mesh (1, 0.5, 0.25, 0.125 mm)

commonly available existing systems may be adapted, for example the use of water intake systems on board research vessels (Lusher et al., 2015).

The acquisition, transport and storage of bulk samples may limit the feasible total volume of water that may be sampled using these systems. Additionally, sampling in this manner may not take into account the heterogeneity in distribution that may otherwise be encompassed in a larger number of net samples samples. However, these methods afford easy determination of microplastic concentrations by known volume, whereas net sampling may limit the reporting of recovered plastics. In some studies, the volume of water sampled using nets is determined using a flow meter or other estimation methods (e.g. multiplying net opening by distance travelled and adjusting for current), enabling concentrations to be determined. Conversely, if sampling at the water's surface, partial emergence of the net may prevent accurate estimation, resulting in reporting by two-dimensional area alone.

Both bulk samples and filtered samples are commonly subjected to secondary processing in the laboratory. This may include separation into fractions, the removal of organic matter and/or the preparation of plastics for enumeration or identification, discussed in more detail in Section 3.4.

3.3.2.6 Airborne plastics

As with water and sediment sampling, the methods used to assess airbourne plastics may be separated into two distinct approaches: firstly, direct (active) sampling of a known volume of air in order to determine microplastic concentrations or, secondly, passive sampling of deposited particles to identify settlement over a known area and time period. The distinction between these methods was previously made in Chapter 2, where studies could be separated by their reporting of plastic either as items per m^3 (e.g. Zhu et al., 2021) or items per m^2 per day (e.g.

Wright et al., 2020), or, in a few cases, both (Kernchen et al., 2022), although recent reviews suggest that depositional studies are currently the most abundant (Sridharan et al., 2021).

The active sampling of known volumes of air is usually achieved by way of either homemade or specialist pumped sampling equipment, which draws air through a filter at a predetermined rate. For example, Zhu et al. (2021) employed an intelligent middle flow total suspended particulate sampler (LB-120F, Lubo Co., Qingdao) at a predetermined intake flow of 100 ± 0.1 L/min. The lower detection limit of microplastics is influenced by the filter type, in this case, a glass fibre filter with a pore size of 0.7 μm. In this study, conditions were also standardised to ensure that replicates were taken at 1.6 meters, three times during the day, following rain events. Alternative sampling methods include the observation of airborne plastics in a marine setting undertaken by Trainic et al. (2020) who utilised a vacuum pump set to a flow rate of approximately 20 L/min. Air was drawn in through a funnel mounted on the backstay of their research vessel, through tubing, to be deposited on a polycarbonate filter with a pore size of 0.8 μm. A similar setup (albeit without boat) was used by Gaston et al. (2020) to quantify the difference in indoor and outdoor microplastic concentrations in coastal California. In this case, airflow through the 1.6 μm glass fibre filter was not controlled at the pump but was measured using a totaliser.

As an alternative route to enable the quantification of microplastic exposure, deposition has been explored in both indoor and outdoor environments, with outdoor deposition determined under both wet and dry conditions. Wet deposition refers to the mass of plastic deposited during precipitation events, whereas dry deposition considers that in fair-weather conditions. As with the airborne fraction, the abundance of deposited microplastics has been determined by both specialised traps and ad hoc methods. In their study of microplastic deposition in London, Wright and colleagues deployed rain gauges over a period of approximately 4 days, followed by transfer onto an alumina-based membrane filter with a pore size of 0.2 μm (Wright et al., 2020). Whereas levels of microplastic deposition at landfill sites were established using a 250 mm glass funnel inserted into a glass bottle. This was placed at 3 m above the ground and allowed to gather both dry fallout and rainwater over a period of either 3 or 4 days. The retained dust and liquid were then passed through a 1 mm sieve before secondary treatment, the final filter was a 1.6 μm glass fibre filter (Thinh et al., 2020).

3.3.2.7 Snow and ice

Plastics recovered from the cryosphere are commonly associated with the wet deposition of airborne plastics. However, unlike wet and dry fallout as described above, these microplastics may be retained in deposited ice and snow, which may be held in place for as long as the environment remains below the melt-point. The number of studies exploring microplastics in the cryosphere is minimal and methods are highly variable. In just one wide ranging study, a combination of both researcher-led and citizen science methods were employed, as a result sampling equipment included a combination of glass jars, mugs, spoons and ladles. Analysis was performed on 1 mL samples of melted snow, diluted with 9 mL of Milli-Q, subsequently reduced onto aluminium oxide filters with a pore size of 0.2 μm (Bergmann et al., 2019). Similarly, in the Antarctic, snow samples were collected using a stainless-steel spoon, funnel and flask. These were passed through a nitrocellulose membrane with a pore size of 0.45 μm pore size, prior to secondary processing (Aves et al., 2022). Similar collection methods have also been employed in the Alps and Northern Iran and the Tibetan Plateau, during which stainless steel utensils were used to fill glass jars prior to secondary processing and passage through either a cellulose filter with a pore size of 1 μm (Abbasi et al., 2022; Parolini et al., 2021) or an Anodisc filter with a pore size of 0.2 μm (Y. Zhang et al., 2021).

One issue with the handling of snow samples is extrapolation from the concentration of plastics in melted snow. Depending on the manner in which snow is packed, the initial surface area and volume of snow may substantially exceed that of the melted water volume. By condensing the initial volume of the substrate, recording microplastics in this manner may artificially increase microplastic concentrations and limit comparability between samples in a single study.

Microplastics in ice are typically recovered using corers. For example, the occurrence of microplastic in the East Antarctic coastal face was explored using an archived core collected under trace metal conditions using an electro-polished stainless-steel corer. As with a number of the snow samples collected above, samples were filtered using a 0.2-μm Whatman Anodisc filter (Kelly et al., 2020). Similarly, analysis of microplastics in sea ice in the Fram Strait in the Arctic utilised a Kovacs 9-cm diameter corer, also subsequently filtering using a 0.2-μm Anodisc (Peeken et al., 2018). Analysis of microplastic at different horizons (depths) of the cores may then be used to establish both abundance and apparent fluctuations in environmental concentration.

Finally, following the collection of snow and ice samples, some authors utilised secondary processing (as set out in Section 3.7) in order to isolate plastic from other material in the sample. However, this was not always deemed a necessary step, with filter contents often being immediately enumerated, stained, subjected to spectral identification methods or combination of approaches (outlined in Section 3.4.4).

3.3.2.8 Citizen science

Beyond the realms of traditional research, citizen science methods have also been used to establish the distribution of surface microplastics. As previously mentioned, Fidra's Great Nurdle Hunt, invites members of the public to count the number of nurdles observed while walking a 1 km stretch of beach (Owen & Parker, 2018). The observations are recorded at the interval level (0, 1−30, 30 100, 100−1000 or more than 1000) to reduce the impact of individual error by recorders. However, observations of plastics smaller than nurdles by members of the public are challenging, and the viability of citizen science in microplastic monitoring may be limited. However, the distribution of macro and meso parent materials as a proxy for microplastic contamination may be feasible. At macro-scale, photographic surveys of litter surveys at city scales carried out by citizen scientists over five dates in 2021/22 resulted in over 3500 records, over 80% of which were plastics (Winton et al., 2023).

3.4 Secondary processing of microplastic samples

The sampling processes outlined above commonly require secondary extraction and processing steps to enable the enumeration, classification and characterisation of the recovered microplastics. The steps applied may include separation by weight to exclude sediments and other materials of high density, the elimination of low-density organic material by way of digestion or oxidation and the sorting of plastic fractions by size. The following sections will explore the various processes by which this may be achieved, highlighting less common additional processing steps as well as alternative approaches which may be relevant to specific samples or experimental aims, and the decision making steps behind appropriate method selection are set out in Fig. 3.2 in Box 3.1.

3.4.1 Exclusion of high-density material

As we have previously indicated, many polymers have a low specific gravity, a property commonly exploited in order to separate plastics from samples. Unwanted material within a sample (such as sediments, shells, carapaces and other anthropogenic litter) may have a higher density than the plastic contained therein. As a result, a number of methods, including gravity-based separation using high-density solutions, fluidisation and elutriation, may be applied to extract the lower-density material (plastics) from a wet sample. For example, the Munich Plastic Sediment Separator uses a series of stirring and settlement phases, followed by passage through a dividing chamber to remove plastics (Imhof et al., 2012). Alternatively, researchers have attempted to separate plastics from dry environmental media using mechanical separators and the electrostatic properties of polymers (Felsing et al., 2018). These steps may be applied to sediment and soil samples, as well as to particulates previously recovered during sampling of air and water.

When utilising gravity-separation methods, the volume of material treated may represent a whole sample or a subsample selected to enable efficient microplastic extraction (overfill your container with soil or sediment, and microplastics at the bottom may not be recovered). Samples are then mixed with a solution of sufficient density to float out plastic material while leaving unwanted objects at the bottom of the vessel. After a settlement period, the floating material may be skimmed off and/or filtered in order to separate out the microplastics and other low-density objects. Most commonly used is sodium chloride (Rist et al., 2021), the density of which is 1.2 g/cm^3 in a saturated solution (Table 3.2); however, the

Table 3.2 Commonly applied density separation solutions, ranked by highest reported operating density and their effective range.

Reagent	Density g cm^{-3}	Effective polymer range							
		LDPE	PP	HDPE	ABS	PS	PA	PVC	PET
Sodium polytungstate	1.5–3.1	X	X	X	X	X	X	X	X
Sodium Iodide	1.6–1.8	X	X	X	X	X	X	X	X
Zinc chloride	1.5–1.7	X	X	X	X	X	X	X*	X
Sodium bromide	1.5	X	X	X	X	X	X	X*	X
Calcium chloride	1.39	X	X	X	X	X	X	X*	X*
Sodium hexametaphosphate	1.3	X	X	X	X	X	X	X*	
Sodium chloride	1.2	X	X	X	X	X	X	X*	

*Efficacy is dependent on formulation of the polymer

selection of a solution must be guided by the properties of the sample and the plastics. For example, fine sediments and organic matter may take a long time to settle when ultra-high-density solutions are used, however, these may be necessary in samples where a high proportion of dense polymers (such as acrylonitrile butadiene styrene (ABS) and PVC) are predicted. Indeed, observations of recovery rates in sand, soil and compost revealed consistently higher recoveries from sand and lower from compost (Schütze et al., 2022).

In addition to simple mixing and agitation of a solution, aeration or elutriation processes may be employed. During these processes a current of air or fluid is forced up through the sample to drive lighter particles to the surface causing them to overflow into a second vessel. Alternatively, high-density solutions may also be used in conjunction with mechanical plastic separators (Fig. 3.4), such as the Munich Plastic Sediment Separator (Imhof et al., 2012), the Sediment Microplastics Isolation Unit (Coppock et al., 2017) and the JAMSTEC glass separator (Nakajima et al., 2019).

Cofounding elements in the separation of plastic include the properties of the sample (e.g. the size and depth of sediments), the polymers present in the sample, the number of repeats of the extraction process and the manner in which microplastics are recovered from the resulting solutions. For example when comparing methods for the recovery and isolation of microplastics using different methods, saturated salt solutions were found to be more effective than elutriation at extracting microplastics from sediments (Pagter et al., 2018). As a result, it is essential that researchers conduct preliminary testing or utilise positive controls in order to establish the effectiveness of their extraction measures.

3.4.2 Removal of organic matter

Many environmental samples contain a high proportion of organic material. This may include dead organic matter and particulates, plant or algal material and

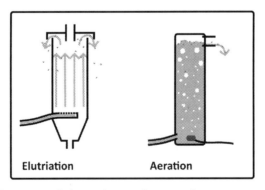

FIGURE 3.4 Aeration and elutriation during density separation.

Passage of either air or water through a sample to enhance density separation.

small biota (such as plankton and soil invertebrates). The density of this material may be similar to, or less than that of plastic, impeding its removal using the methods outlined above. As a result, the efficient enumeration and secondary analysis of plastics in a sample may be prevented. Subsequently, samples believed to be high in organic material may be subject to a digestion step in order to eliminate this material.

Digestion may be carried out using acids (including nitric acid, hydrochloric acid and sulphuric acid), strong bases (including sodium hydroxide and potassium), hydroxide and enzymes (including protease, lipase, cellulase, amylase and chitinase), however, the most widely applied method utilises hydrogen peroxide, often in combination with Fenton's reagent. Samples are typically mixed with the digestion agent, and either left to process either at ambient temperatures or heated in a water bath or similar. While the temperature at which enzymes are applied is defined by the apparent optimum range, the digestion efficiency of many chemical reagents may be improved by heating, resulting in a quicker digestion time. After digestion, the supernatant is filtered, and the filter may be treated to neutralise or remove residues of the digestion agent. Table 3.3 indicates the reagents and associated conditions for digestion which may be used alone or in combination. Nevertheless, some of these more aggressive processes can result in damage to the polymer structure. Conversely, some of the 'gentler' methods, such as enzyme digestions may be associated with increased costs, particularly when processing large numbers of samples. Thus researchers should consider the balance of time, cost and impact on recovery when selecting more aggressive measures (Lusher, Welden et al., 2020).

3.4.3 Staining

In addition to removing nonplastic material from the sample, researchers have also employed methods to make plastics stand out. This has previous been attained by way of selective staining or labelling, for example using dyes which fluoresce under UV light. As previously discussed, many plastics have a hydrophobic structure, as a result they can be marked using lithophilic stains such as Nile red (Shim et al., 2016), Rose-Bengal (Ziajahromi et al., 2017), Safarine T and flourescin isophosphate (Lv et al., 2019). This process has previously been used to assist in the enumeration of plastics, as well as to confirm their identification. Results have been seen to be mixed, however, accuracy may be improved by subsequent counterstaining (Maxwell et al., 2020), as well as the adjustment of ambient conditions and camera settings (Prata et al., 2020). Nonetheless, while this method benefits from being relatively cheap and quick compared to spectroscopy and similar high-tech approaches to the identification of polymer types, staining may be influenced by the presence of other hydrophobic substances in the sample and may underestimate less-hydrophobic polymers such

Table 3.3 Examples of commonly reported digestion agents and their associated protocols.

Class	Reagent	Formula	Concentration	Incubation	Temp °C
Acids	Acetic acid	CH_3COOH	10%	48 h	60
	Hydrochloric acid	HCl	3.7% (1 M)	24 h	Ambient – 80
			7.3% (2 M)		
	Nitric acid	HNO_3	4%–37%	2–24 h	Ambient – 100
	Sulphuric acid	H_2SO_4	50%	48 h	
Oxidisers	Hydrogen peroxide	H_2O_2	30%	<3 weeks	Ambient – 70
	Fenton's reagent	$FeH_4O_6S^{+2}$	30%		Ambient – 50
Bases	Potassium hydroxide	KOH	10%	<3 weeks	Ambient – 80
	Sodium hydroxide	NaOH	5%–18%	>48 h	Ambient – 60
Enzymes	Corolase 7089		1%	24 h	60
	Proteinase-K		0.05%	2 h	50
	Tripsin		0.31%	0.5 h	38–42

as polyacrylamide, polyurethanes and polyvinyl chloride (Erni-Cassola et al., 2017). These polymers do not bind as well to the stain and are thus less likely to be detected by observers or automated systems.

3.4.4 Enumeration and identification

Microplastics in samples arising from the methods outlined above are typically counted under microscope, with the maximum and minimum magnification determined by the apparent limit of detection of the sampling and analytical steps, or microscope-based automated systems. Nevertheless, this may still be a challenging process. Due to the diversity of polymers, morphologies, fillers and uses, microplastics in samples can look like a great many things that *aren't* plastic, such as lenses, bone and minerals, resulting in underestimation. Similarly, many things that aren't plastics can *appear* like them under the microscope, resulting in overestimation. Thus despite multiple sets of recommendations regarding the visual assessment of plastic (Lusher, Bråte, et al., 2020), the use of secondary analysis to confirm polymer identity is of high importance.

Methods to establish the presence of plastics may be grouped as those that tell the researcher that a particle is indeed plastic, and those that can tell the researcher the extact polymer and its associated chemical structure. Staining methods may be included in the first category, along with melt point analysis, birefringence under polarised light microscopy and charge-based sorting. During melt point analysis suspected plastics are steadily heated, and the point at which the material melts is recorded and compared to known values for different polymers. Birefringence refers to the double refraction, bending, when travelling through a material, splitting it into an 'ordinary' ray and an 'extraordinary ray' (Fig. 3.5). In plastics, birefringence (also known as 'anisotropy') is caused by manufacturing processes, during which, isotropic plastics are subjected to stressing and forcing which changes the orientation of the polymer chains, changing them from an amorphous state, resulting in the formation of crystalline structures within the polymer (Fig. 3.6). This effect is frequently used in the identification of fibres but may be seen in many other plastic products.

The identity of the polymer and the associated chemical composition of suspected microplastics may be determined using spectroscopic approaches, such as (μ)Fourier Transformed Infrared (IR), Raman, Short Wavelength IR, Near IR and mass spectrometry and colorimetry. Each of these methods seeks to determine the polymer structure by directing a beam onto a sample and analysing the light that is subsequently absorbed or emitted. Changes between the original light and which is emitted, reflected or scattered are used to determine the composition of the sample. For example, IR methods identify plastics by assessing how much energy is absorbed by the bonds in a polymer as a reduction energy at key wavelengths (Fig. 3.7). Each bond within a molecule absorbs light at a specific wavelength, and polymers the presence of key functional groups help to create a spectrum that is unique to the polymer. The spectrum formed by a sample is then

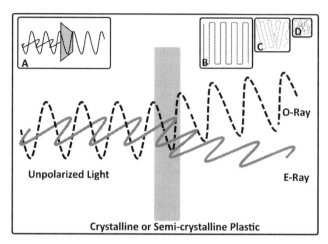

FIGURE 3.5 Birefringence of crystalline plastics.

Splitting of light beams as they pass through a plastic structure, resulting in identifiable patterns.

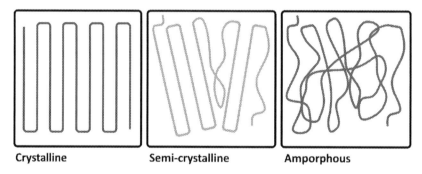

FIGURE 3.6 Structural factors influencing the optical properties of plastics.

The arrangement of polymer chains within plastics of varied crystallinity.

compared to a library of existing spectra in order to identify the material. Raman methods are similar but use lasers of a single wavelength. In this case, the returned light is scattered by the molecules in the sample, altering its frequency. As with IR methods, the frequencies are determined by the bonds within the polymer, each of which has its own vibrational frequency.

Mass spectrometry approaches analyse the mass distribution of ionised molecules within a sample when passed through a vacuum within a mass analyser to strike a detector. The heavier the molecule, the slower it moves and thus the molecular weight may be determined. As with the above methods, a fingerprint is created which enables the identification of the polymer. Nevertheless, misidentification may still occur when using the above methods. Exposure to environmental

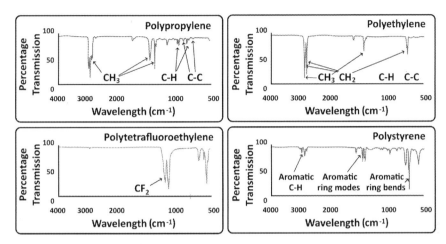

FIGURE 3.7 FT-IR spectra of polymers.

Spectra of polypropylene, polyethylene, polytetrafluoroethylene and polystyrene.

conditions can result in ageing of the polymer structure, stretching and distorting bonds and altering the fingerprint that they create. This is particularly true in plastics like PVC, the polymer structure of which is quickly altered in the environment (Fernández-González et al., 2022). We can improve our identification by developing more details libraries of weathered polymers; however, the availability of these resources is currently limited.

3.4.5 Recovery efficiency, quality assurance and quality control

Despite the measures outlined above, it is still challenging to ensure that we are truly representing the abundance of microplastics in our samples. In addition to the impact of plastics heterogeneity on the number of samples required for a representative analysis, the collection of large or complex samples may result in low microplastic extraction efficiency, and the elimination of unwanted material may be impossible. Indeed, interlaboratory comparisons have revealed underestimation of microplastics below 1 mm in size of up to 20% (Isobe et al., 2019). Additionally, contamination by unwanted environmental plastics during sample retrieval, transport, processing and analysis may result in the artificial inflation of some plastic categories (Prata et al., 2021). These issues are typically controlled by the use of negative and positive controls (samples spiked with a known number of microplastics), as well as assessment of airborne contamination in the work area (Box 3.3). When employing positive controls, a mix of polymer types and morphologies is preferred in order to accurately assess the efficacy of the method for all microplastic types. Nevertheless, these measures are inconsistently applied and recent comparisons of quality assurance and quality control methods applied to correct microplastics data sets have been less

Box 3.3 Coping with contamination — lessons from trace element analysis.

Objective

Contamination by ambient microplastics is an ever-present concern during the collection and analysis of environmental samples, and the results of many projects have been called into question due to their lack of quality assurance procedures. However, researchers in the field of microplastics are certainly not the first to come across this issue. Here, we will explore the management of contamination in trace element analysis, highlighting where parallels may be drawn between the two fields.

Scope

The following sections consider the quality assurance measures not associated with the accuracy and precision of individual sampling and analytical methods but, rather, those concerned with the broader management of secondary contamination. These assurance measures frequently consider preparation for sampling, sampling acquisition and preprocessing, in addition to the analytical step. As with observations of microplastic pollution, trace metal analysis may target biota (in biomonitoring programmes) as well as environmental media. Areas of concern may include the species (biomonitoring, uptake and toxicological studies) or environmental matrix (sediment, water and air), storage, tools and ambient environment(s).

Rationale

Trace element analysis is used to determine the presence and concentration of elements found at low levels in the environment. These may include metals such as mercury, cadmium, zinc, lead, arsenic, selenium, chromium, nickel, silver and tin. At such low concentrations, the potential for accidental contamination resulting in overestimation is high. Thus contamination control is key in ensuring the reliability and comparability of results.

Workflow

The process of sample collection, preprocessing and analysis presents numerous potential sources of metal contamination. In the flow chart below, we identify key stages in the analytical workflow and potential metal inputs (Fig. 3.8).

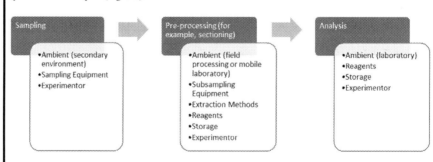

FIGURE 3.8

Sources of contamination during sampling and analysis.

Challenges and solutions

Tools and storage

The selection of tools should made in such a way that their impacts on their analysis are minimised. In the case of trace metals, there is a preference form plastics, quartz and borosilicate glass,

(Continued)

Box 3.3 Coping with contamination — lessons from trace element analysis. (Continued)

---

however, metals may be acceptable where the element to be analysed is not present. Impacts of the wrong metal choice have previously been observed in spikes in nickel and chrome associated with the preparation of biological samples using stainless steel (Kramer, 2007).

Unsurprisingly, prior decontamination of equipment is key in preventing contamination, first by rinsing, followed by a nitric or hydrochloric acid bath where appropriate and subsequent rinse in ultrapure water. For metal analysis, tools may then be stored in clean polythene bags until required.

Sampling

Sampling is often the step at which researchers have the least control over potential contamination sources (Wagner, 1995). Sampling equipment should be selected that does not interfere with the apparent levels of the target element. Where the construction of equipment may influence the concentration of key elements, subsamples should be removed taken from within the sample, away from points of contact.

The laboratory

As with microplastics, trace metal analysis may be influenced by ambient conditions in the laboratory, as a result, it is recommended that preparation and analytical steps are undertaken under 'clean-bench' conditions or, if possible, in clean-room conditions. In this latter case, it is recommended that laboratory air is passed through a class 100 filter in order to lower the concentration of airborne particulates. Researchers also don low-shedding outerwear and talc free gloves, which should be changed whenever a researcher comes into contact with a potential source of contamination.

Applicability in microplastic research

Sampling regimes for microplastic must be carefully planned and documented to minimise and monitor sources of error. The issues outlined in relation to trace element analysis are very similar to those influencing the reliability of microplastic analysis (Wagner, 1995). Clean air and clean bench procedures are both highly relevant to the recovery and enumeration of microplastics, particularly in relation to highly mobile fibres and smaller plastic fractions. However, where the above measures call for the use of plastics in order to minimise metal input into a sample (such as the selection of tools and equipment, storage of samples and maintenance of decontaminated equipment until use), when analysing plastic concentrations, metals, quartz and borosilicate glass become the most favourable options. Where this is not possible, spectral analysis of the composition of the equipment or container may assist in eliminating accidental contamination. If complete exclusion of potential plastic sources is not feasible, the efficacy of their contamination control procedures may be confirmed by way of negative controls.

Of course, these methods are insufficient to ensure accuracy in isolation, researchers must consider the importance of sufficient replication (number of samples, spatial resolution and temporal resolution) in order to capture apparent heterogeneity, and that the methods chosen represent suitable limits of detection (by way of positive controls).

than positive, with just 7 of the 51 methods tested proving to provide accurate results. Preferred methods are those based on LOD/LOQ principles, which take into account the limits of detection and/or limits of quantification of the applied approach (Dawson et al., 2023).

3.5 **Reporting**

The final substantive barrier to our successfully determining the distribution of microplastics in the environment is ensuring accurate and comparable reporting. The impact of the sampling regime on the reporting of either concentration or abundance has been highlighted above. A recent review of studies indicates that reports of the outcomes of both water and sediment sampling frequently provide microplastic observations using variable units. For example, a small proportion (less than 2.5%) of studies opt to report by microplastic mass rather than number of microplastic items. Similarly, reports of sampled sediments describe plastic abundance by either dry weight, wet weight or sample volume, and there remains the previously mentioned persistent rift in reporting by two or three-dimensional area (Rist et al., 2021). However, there are a great number of inconsistencies beyond these. For example, concerning microplastic dimensions, the size of particles may only be reported in terms of their longest dimension, giving minimal information regarding the potential mass of the microplastics recovered. Similarly, the terminology and morphology used in the description of microplastics is inconsistently reported in the literature (Lusher et al., 2020).

3.6 Conclusions

Microplastics are always on the move and, while many sites accumulate elevated levels of plastic, few of these are permanent sinks. Nevertheless, the observed aggregation at key sites may represent a significant threat to ecosystem function, and it is of high importance that reliable estimates of microplastic type, morphology and abundance may be generated. Unfortunately, we still lack comparable datasets over sufficient geographic scales to enable substantial interpretation or impactful management. The number and diversity of methods for the enumeration of plastics across environments (driven by the combined limiting factors of time, cost and available resource) have resulted in inconsistent reporting. Similarly, while quality assurance methods are widely acknowledged as essential, the way in which they are applied may not be helping researchers to achieve the reliability in their results that they might wish. On the whole, those utilising existing data should handle with care, ensuring that they understand the impacts of dictation rates and size, extraction efficiencies, and identification in order to avoid running afoul of this inconsistency.

References

Abbasi, S., Alirezazadeh, M., Razeghi, N., Rezaei, M., Pourmahmood, H., Dehbandi, R., Mehr, M. R., Ashayeri, S. Y., Oleszczuk, P., & Turner, A. (2022). Microplastics captured by snowfall: A study in Northern Iran. *Science of the Total Environment, 822*. Available from https://doi.org/10.1016/j.scitotenv.2022.153451, http://www.elsevier.com/locate/scitotenv.

Aves, A. R., Revell, L. E., Gaw, S., Ruffell, H., Schuddeboom, A., Wotherspoon, N. E., Larue, M., & Mcdonald, A. J. (2022). First evidence of microplastics in Antarctic snow. *Cryosphere, 16*(6), 2127−2145. Available from https://doi.org/10.5194/tc-16-2127-2022, http://www.the-cryosphere.net/volumes_and_issues.html.

Barrett, J., Chase, Z., Zhang, J., Holl, M. M. B., Willis, K., Williams, A., Hardesty, B. D., & Wilcox, C. (2020). Microplastic pollution in deep-sea sediments from the great Australian bight. *Frontiers in Marine Science, 7*. Available from https://doi.org/10.3389/fmars.2020.576170, https://www.frontiersin.org/journals/marine-science#.

Bergmann, M., Mützel, S., Primpke, S., Tekman, M. B., Trachsel, J., & Gerdts, G. (2019). White and wonderful? Microplastics prevail in snow from the Alps to the Arctic. *Science Advances, 5*(8). Available from https://doi.org/10.1126/sciadv.aax1157, https://advances.sciencemag.org/content/5/8/eaax1157/tab-pdf.

Besley, A., Vijver, M. G., Behrens, P., & Bosker, T. (2017). A standardized method for sampling and extraction methods for quantifying microplastics in beach sand. *Marine Pollution Bulletin, 114*(1), 77−83. Available from https://doi.org/10.1016/j.marpolbul.2016.08.055, http://www.elsevier.com/locate/marpolbul.

Conkle, J. L., Báez Del Valle, C. D., & Turner, J. W. (2018). Are we underestimating microplastic contamination in aquatic environments? *Environmental Management, 61*(1), 1−8. Available from https://doi.org/10.1007/s00267-017-0947-8, https://www.springer.com/journal/267.

Coppock, R. L., Cole, M., Lindeque, P. K., Queirós, A. M., & Galloway, T. S. (2017). A small-scale, portable method for extracting microplastics from marine sediments. *Environmental Pollution, 230*, 829−837. Available from https://doi.org/10.1016/j.envpol.2017.07.017, http://www.elsevier.com/inca/publications/store/4/0/5/8/5/6.

Corradini, F., Meza, P., Eguiluz, R., Casado, F., Huerta-Lwanga, E., & Geissen, V. (2019). Evidence of microplastic accumulation in agricultural soils from sewage sludge disposal. *Science of the Total Environment, 671*, 411−420. Available from https://doi.org/10.1016/j.scitotenv.2019.03.368, http://www.elsevier.com/locate/scitotenv.

Covernton, G. A., Pearce, C. M., Gurney-Smith, H. J., Chastain, S. G., Ross, P. S., Dower, J. F., & Dudas, S. E. (2019). Size and shape matter: A preliminary analysis of microplastic sampling technique in seawater studies with implications for ecological risk assessment. *Science of the Total Environment, 667*, 124−132. Available from https://doi.org/10.1016/j.scitotenv.2019.02.346, http://www.elsevier.com/locate/scitotenv.

Dawson, A. L., Santana, M. F. M., Nelis, J. L. D., & Motti, C. A. (2023). Taking control of microplastics data: A comparison of control and blank data correction methods. *Journal of Hazardous Materials, 443*. Available from https://doi.org/10.1016/j.jhazmat.2022.130218, http://www.elsevier.com/locate/jhazmat.

Di Mauro, R., Kupchik, M. J., & Benfield, M. C. (2017). Abundant plankton-sized microplastic particles in shelf waters of the northern Gulf of Mexico. *Environmental Pollution, 230*, 798−809. Available from https://doi.org/10.1016/j.envpol.2017.07.030, http://www.elsevier.com/inca/publications/store/4/0/5/8/5/6,

Erni-Cassola, G., Gibson, M. I., Thompson, R. C., & Christie-Oleza, J. A. (2017). Lost, but found with Nile red: A novel method for detecting and quantifying small microplastics (1 mm to 20 μm) in environmental samples. *Environmental Science and Technology*,

51(23), 13641−13648. Available from https://doi.org/10.1021/acs.est.7b04512, http://pubs.acs.org/journal/esthag.

Felsing, S., Kochleus, C., Buchinger, S., Brennholt, N., Stock, F., & Reifferscheid, G. (2018). A new approach in separating microplastics from environmental samples based on their electrostatic behavior. *Environmental Pollution*, *234*, 20−28. Available from https://doi.org/10.1016/j.envpol.2017.11.013, http://www.elsevier.com/inca/publications/store/4/0/5/8/5/6.

Fernández-González, V., Andrade-Garda, J. M., López-Mahía, P., & Muniategui-Lorenzo, S. (2022). Misidentification of PVC microplastics in marine environmental samples. *TrAC Trends in Analytical Chemistry*, *153*, 116649. Available from https://doi.org/10.1016/j.trac.2022.116649.

Gaston, E., Woo, M., Steele, C., Sukumaran, S., & Anderson, S. (2020). Microplastics differ between indoor and outdoor air masses: Insights from multiple microscopy methodologies. *Applied Spectroscopy*, *74*(9), 1079−1098. Available from https://doi.org/10.1177/0003702820920652, https://journals.sagepub.com/loi/asp.

GESAMP. (2019). Guidelines or the monitoring and assessment of plastic litter and microplastics in the ocean. In P. J. Kershaw, A. Turra, & F. Galgani (Eds.), *Rep. Stud. GESAMP No. 99* (p. 130). IMO/FAO/UNESCO-IOC/UNIDO/WMO/IAEA/UN/UNEP/UNDP/ISA Joint Group of Experts on the Scientific Aspects of Marine Environmental Protection.

Hartmann, N. B., Hüffer, T., Thompson, R. C., Hassellöv, M., Verschoor, A., Daugaard, A. E., Rist, S., Karlsson, T., Brennholt, N., Cole, M., Herrling, M. P., Hess, M. C., Ivleva, N. P., Lusher, A. L., & Wagner, M. (2019). Are we speaking the same language? Recommendations for a definition and categorization framework for plastic debris. *Environmental Science and Technology*, *53*(3), 1039−1047. Available from https://doi.org/10.1021/acs.est.8b05297, http://pubs.acs.org/journal/esthag.

Hidalgo-Ruz, V., Gutow, L., Thompson, R. C., & Thiel, M. (2012). Microplastics in the marine environment: A review of the methods used for identification and quantification. *Environmental Science and Technology*, *46*(6), 3060−3075. Available from https://doi.org/10.1021/es2031505.

Imhof, H. K., Schmid, J., Niessner, R., Ivleva, N. P., & Laforsch, C. (2012). A novel, highly efficient method for the separation and quantification of plastic particles in sediments of aquatic environments. *Limnology and Oceanography: Methods.*, *10*(JULY), 524−537. Available from https://doi.org/10.4319/lom.2012.10.524Germany, http://aslo.org/lomethods/locked/2012/0524.pdf.

Isobe, A., Buenaventura, N. T., Chastain, S., Chavanich, S., Cózar, A., DeLorenzo, M., Hagmann, P., Hinata, H., Kozlovskii, N., Lusher, A. L., Martí, E., Michida, Y., Mu, J., Ohno, M., Potter, G., Ross, P. S., Sagawa, N., Shim, W. J., Song, Y. K., ... Zhang, W. (2019). An interlaboratory comparison exercise for the determination of microplastics in standard sample bottles. *Marine Pollution Bulletin*, *146*, 831−837. Available from https://doi.org/10.1016/j.marpolbul.2019.07.033, http://www.elsevier.com/locate/marpolbul.

Kelly, A., Lannuzel, D., Rodemann, T., Meiners, K. M., & Auman, H. J. (2020). Microplastic contamination in east Antarctic sea ice. *Marine Pollution Bulletin*, *154*, 111130. Available from https://doi.org/10.1016/j.marpolbul.2020.111130.

Kernchen, S., Löder, M. G. J., Fischer, F., Fischer, D., Moses, S. R., Georgi, C., Nölscher, A. C., Held, A., & Laforsch, C. (2022). Airborne microplastic concentrations and deposition across the Weser River catchment. *Science of the Total Environment*, *818*.

Available from https://doi.org/10.1016/j.scitotenv.2021.151812, http://www.elsevier.com/locate/scitotenv.

Kramer, K. J. M. (2007). *Quality assurance of sampling and sample handling for trace metal analysis in aquatic biota. Quality assurance in environmental monitoring: Sampling and sample pretreatment* (pp. 179−214). Netherlands: Wiley Blackwell. Available from http://onlinelibrary.wiley.com/book/10.1002/9783527615216, https://doi.org/10.1002/9783527615216.ch08.

Lindeque, P. K., Cole, M., Coppock, R. L., Lewis, C. N., Miller, R. Z., Watts, A. J. R., Wilson-McNeal, A., Wright, S. L., & Galloway, T. S. (2020). Are we underestimating microplastic abundance in the marine environment? A comparison of microplastic capture with nets of different mesh-size. *Environmental Pollution*, *265*. Available from https://doi.org/10.1016/j.envpol.2020.114721, https://www.journals.elsevier.com/environmental-pollution.

Lu, H. C., Ziajahromi, S., Neale, P. A., & Leusch, F. D. L. (2021). A systematic review of freshwater microplastics in water and sediments: Recommendations for harmonisation to enhance future study comparisons. *Science of the Total Environment*, *781*. Available from https://doi.org/10.1016/j.scitotenv.2021.146693, http://www.elsevier.com/locate/scitotenv.

Lusher, A. L., Tirelli, V., O'Connor, I., & Officer, R. (2015). Microplastics in Arctic polar waters: The first reported values of particles in surface and sub-surface samples. *Scientific Reports*, *5*. Available from https://doi.org/10.1038/srep14947, http://www.nature.com/srep/index.html.

Lusher, A. L., Bråte, I. L. N., Munno, K., Hurley, R. R., & Welden, N. A. (2020). Is it or isn't it: The importance of visual classification in microplastic characterization. *Applied Spectroscopy*, *74*(9), 1139−1153. Available from https://doi.org/10.1177/0003702820930733, https://journals.sagepub.com/loi/asp.

Lusher, A. L., Welden, N. A., Sobral, P., & Cole, M. (2020). Sampling, isolating and identifying microplastics ingested by fish and invertebrates. *Informa UK Limited*, 119−148. Available from https://doi.org/10.1201/9780429469596-8.

Lv, L., Qu, J., Yu, Z., Chen, D., Zhou, C., Hong, P., Sun, S., & Li, C. (2019). A simple method for detecting and quantifying microplastics utilizing fluorescent dyes − Safranine T, fluorescein isophosphate, Nile red based on thermal expansion and contraction property. *Environmental Pollution*, *255*, 113283. Available from https://doi.org/10.1016/j.envpol.2019.113283.

Maxwell, H., Melinda, F., & Matthew, G. (2020). Counterstaining to separate Nile red-stained microplastic particles from terrestrial invertebrate biomass. *Environmental Science & Technology*, *54*(9), 5580−5588. Available from https://doi.org/10.1021/acs.est.0c00711.

Möller, J. N., Löder, M. G. J., & Laforsch, C. (2020). Finding microplastics in soils: A review of analytical methods. *Environmental Science and Technology*, *54*(4), 2078−2090. Available from https://doi.org/10.1021/acs.est.9b04618, http://pubs.acs.org/journal/esthag.

Nakajima, R., Tsuchiya, M., Lindsay, D. J., Kitahashi, T., Fujikura, K., & Fukushima, T. (2019). A new small device made of glass for separating microplastics from marine and freshwater sediments. *PeerJ*, *7*(10), e7915. Available from https://doi.org/10.7717/peerj.7915.

Ovide, B. G., Cirino, E., Basran, C. J., Geertz, T., & Syberg, K. (2022). Assessment of prevalence and heterogeneity of meso- and microplastic pollution in icelandic waters. *Environments*, *9*(12), 150. Available from https://doi.org/10.3390/environments9120150.

Owen, R. P., & Parker, A. J. (2018). Citizen science in environmental protection agencies. *JSTOR*, 284−300. Available from https://doi.org/10.2307/j.ctv550cf2.27.

Pagter, E., Frias, J., & Nash, R. (2018). Microplastics in Galway Bay: A comparison of sampling and separation methods. *Marine Pollution Bulletin*, *135*, 932−940. Available from https://doi.org/10.1016/j.marpolbul.2018.08.013, http://www.elsevier.com/locate/marpolbul.

Parolini, M., Antonioli, D., Borgogno, F., Gibellino, M. C., Fresta, J., Albonico, C., De Felice, B., Canuto, S., Concedi, D., Romani, A., Rosio, E., Gianotti, V., Laus, M., Ambrosini, R., & Cavallo, R. (2021). Microplastic contamination in snow from western Italian alps. *International Journal of Environmental Research and Public Health*, *18*(2), 1−10. Available from https://doi.org/10.3390/ijerph18020768, https://www.mdpi.com/1660-4601/18/2/768/pdf.

Peeken, I., Primpke, S., Beyer, B., Gütermann, J., Katlein, C., Krumpen, T., Bergmann, M., Hehemann, L., & Gerdts, G. (2018). Arctic sea ice is an important temporal sink and means of transport for microplastic. *Nature Communications*, *9*(1). Available from https://doi.org/10.1038/s41467-018-03825-5, http://www.nature.com/ncomms/index.html.

Piehl, S., Leibner, A., Löder, M. G., Dris, R., Bogner, C., & Laforsch, C. (2018). Identification and quantification of macro-and microplastics on an agricultural farmland. *Scientific reports*, *8*(1), 1−9.

Prata, J. C., Alves, J. R., da Costa, J. P., Duarte, A. C., & Rocha-Santos, T. (2020). Major factors influencing the quantification of Nile red stained microplastics and improved automatic quantification (MP-VAT 2.0). *Science of the Total Environment*, *719*. Available from https://doi.org/10.1016/j.scitotenv.2020.137498, http://www.elsevier.com/locate/scitotenv.

Prata, J. C., Reis, V., da Costa, J. P., Mouneyrac, C., Duarte, A. C., & Rocha-Santos, T. (2021). Contamination issues as a challenge in quality control and quality assurance in microplastics analytics. *Journal of Hazardous Materials*, *403*. Available from https://doi.org/10.1016/j.jhazmat.2020.123660, http://www.elsevier.com/locate/jhazmat.

Rist, S., Hartmann, N. B., & Welden, N. A. C. (2021). How fast, how far: Diversification and adoption of novel methods in aquatic microplastic monitoring. *Environmental Pollution*, *291*, 118174. Available from https://doi.org/10.1016/j.envpol.2021.118174.

Schütze, B., Thomas, D., Kraft, M., Brunotte, J., & Kreuzig, R. (2022). Comparison of different salt solutions for density separation of conventional and biodegradable microplastic from solid sample matrices. *Environmental Science and Pollution Research*, *29*(54), 81452−81467. Available from https://doi.org/10.1007/s11356-022-21474-6, https://www.springer.com/journal/11356.

Shim, W. J., Song, Y. K., Hong, S. H., & Jang, M. (2016). Identification and quantification of microplastics using Nile red staining. *Marine Pollution Bulletin*, *113*(1−2), 469−476. Available from https://doi.org/10.1016/j.marpolbul.2016.10.049, http://www.elsevier.com/locate/marpolbul.

Sridharan, S., Kumar, M., Singh, L., Bolan, N. S., & Saha, M. (2021). Microplastics as an emerging source of particulate air pollution: A critical review. *Journal of Hazardous Materials*, *418*, 126245. Available from https://doi.org/10.1016/j.jhazmat.2021.126245.

Thinh, T. Q., Sang, T. T. N., Viet, T. Q., Tam, L. T. M., Dan, N. P., Strady, E., & Chung, K. L. T. (2020). Preliminary assessment on the microplastic contamination in the atmospheric fallout in the Phuoc Hiep landfill, Cu Chi, Ho Chi Minh city. *Vietnam Journal of Science, Technology and Engineering, 62*(3), 83−89. Available from https://doi.org/10.31276/vjste.62(3).83-89.

Thompson, R. C., Olson, Y., Mitchell, R. P., Davis, A., Rowland, S. J., John, A. W. G., McGonigle, D., & Russell, A. E. (2004). Lost at sea: Where is all the plastic? *Science (New York, N.Y.), 304*(5672), 838. Available from https://doi.org/10.1126/science.1094559.

Trainic, M., Flores, J. M., Pinkas, I., Pedrotti, M. L., Lombard, F., Bourdin, G., Gorsky, G., Boss, E., Rudich, Y., Vardi, A., & Koren, I. (2020). Airborne microplastic particles detected in the remote marine atmosphere. *Communications Earth and Environment, 1*(1). Available from https://doi.org/10.1038/s43247-020-00061-y, https://www.nature.com/commsenv/.

Van Cauwenberghe, L., Vanreusel, A., Mees, J., & Janssen, C. R. (2013). Microplastic pollution in deep-sea sediments. *Environmental Pollution, 182*, 495−499. Available from https://doi.org/10.1016/j.envpol.2013.08.013, http://www.elsevier.com/inca/publications/store/4/0/5/8/5/6.

Wagner, G. (1995). Basic approaches and methods for quality assurance and quality control in sample collection and storage for environmental monitoring. *Science of the Total Environment, 176*(1−3), 63−71. Available from https://doi.org/10.1016/0048-9697(95)04830-8.

Winton, S., Roberts, K. P., Bowyer, C., & Fletcher, S. (2023). Harnessing citizen science to tackle urban-sourced ocean plastic pollution: Experiences and lessons learned from implementing city-wide surveys of plastic litter. *Marine Pollution Bulletin, 192*. Available from https://doi.org/10.1016/j.marpolbul.2023.115116, http://www.elsevier.com/locate/marpolbul.

Woodall, L. C., Sanchez-Vidal, A., Canals, M., Paterson, G. L. J., Coppock, R., Sleight, V., Calafat, A., Rogers, A. D., Narayanaswamy, B. E., & Thompson, R. C. (2014). The deep sea is a major sink for microplastic debris. *Royal Society Open Science, 1*(4). Available from https://doi.org/10.1098/rsos.140317, http://rsos.royalsocietypublishing.org/content/royopensci/1/4/140317.full.pdf.

Wright, S. L., Ulke, J., Font, A., Chan, K. L. A., & Kelly, F. J. (2020). Atmospheric microplastic deposition in an urban environment and an evaluation of transport. *Environment International, 136*, 105411. Available from https://doi.org/10.1016/j.envint.2019.105411.

Yang, L., Zhang, Y., Kang, S., Wang, Z., & Wu, C. (2021). Microplastics in soil: A review on methods, occurrence, sources, and potential risk. *Science of the Total Environment, 780*, 146546. Available from https://doi.org/10.1016/j.scitotenv.2021.146546.

Zhang, M., Tan, M., Ji, R., Ma, R., & Li, C. (2022). Current situation and ecological effects of microplastic pollution in soil. *Reviews of Environmental Contamination and Toxicology, 260*(1). Available from https://doi.org/10.1007/s44169-022-00012-y, https://www.springer.com/journal/44169.

Zhang, Y., Gao, T., Kang, S., Allen, S., Luo, X., & Allen, D. (2021). Microplastics in glaciers of the Tibetan Plateau: Evidence for the long-range transport of microplastics. *Science of The Total Environment, 758*, 143634. Available from https://doi.org/10.1016/j.scitotenv.2020.143634.

Zhang, Z., Deng, C., Dong, L., Liu, L., Li, H., Wu, J., & Ye, C. (2021). Microplastic pollution in the Yangtze River Basin: Heterogeneity of abundances and characteristics in different environments. *Environmental Pollution, 287*, 117580. Available from https://doi.org/10.1016/j.envpol.2021.117580.

Zhu, X., Huang, W., Fang, M., Liao, Z., Wang, Y., Xu, L., Mu, Q., Shi, C., Lu, C., Deng, H., Dahlgren, R., & Shang, X. (2021). Airborne microplastic concentrations in five megacities of northern and southeast China. *Environmental Science and Technology, 55*(19), 12871−12881. Available from https://doi.org/10.1021/acs.est.1c03618, http://pubs.acs.org/journal/esthag.

Ziajahromi, S., Neale, P. A., Rintoul, L., & Leusch, F. D. L. (2017). Wastewater treatment plants as a pathway for microplastics: Development of a new approach to sample wastewater-based microplastics. *Water Research, 112*, 93−99. Available from https://doi.org/10.1016/j.watres.2017.01.042, http://www.elsevier.com/locate/watres.

Zobkov, M. B., Esiukova, E. E., Zyubin, A. Y., & Samusev, I. G. (2019). Microplastic content variation in water column: The observations employing a novel sampling tool in stratified Baltic Sea. *Marine Pollution Bulletin, 138*, 193−205. Available from https://doi.org/10.1016/j.marpolbul.2018.11.047, http://www.elsevier.com/locate/marpolbul.

Exploring interaction, uptake and impacts

4

4.1 Introduction

Interactions between plastic debris and organisms have been reported for many decades. This is particularly true of animals in the marine environment where entanglement has been seen to influence both the health and survivorship of a range of species including fish, mammals and seabirds. As with microplastics, macroplastic debris has a great many sources; for example, abandoned, lost and discarded fishing gear (ALDFG), which is typically made of synthetic polymers and may continue ensnaring biota long after its initial loss to the environment. Ingestion of plastics from terrestrial sources, including bags, straws and other single-use items, has also been seen to have a negative effect on species, either influencing nutritional state or resulting in direct damage to marine mammals, turtles, birds and other fauna.

As indicated in Chapter 1, a much wider range of species are able to take up microplastics than are able to ingest larger plastic debris. The reason for this may be explained by way of a simple comparison with humans, it is easy for us to swallow a marble, but we would struggle to swallow a beach ball! So it is with smaller species in the wider environment, the reduced size of meso-, micro- and nanoplastics makes them available for ingestion by an increased number of species in addition to making them more likely to adhere to body structures. These interactions are also influenced by the diversity in shape and form of microplastic pollution, believed to result in apparent misidentification of plastics as food in some species. As a result, the uptake of micro- and nanoplastics has been observed in a range of taxa, including key invertebrates groups such as planktonic organisms, terrestrial detritivores and species for human consumption. Such observations have resulted in widespread concern regarding individual animal health, ecosystem structure and functioning, as well as secondary effects on both human health and the reliability of ecosystem services.

4.2 Modes of exposure

The prevalence of microplastics in the environment may increase the regularity with which organisms interact with plastic (one microplastic becomes many

Microplastics. DOI: https://doi.org/10.1016/B978-0-443-13324-4.00004-2

microplastics, increasing local abundance). Subsequently, the uptake of microplastics has been extensively recorded in both wild-caught organisms and those exposed to microplastics during laboratory trials. However, interactions between organisms and microplastics may also be via entanglement, adherence, or other extended contact.

4.2.1 Direct ingestion

4.2.1.1 Observations of ingested microplastics in wild-caught animals

Early "unofficial" records of microplastic uptake — observations of small plastic particles in reports and studies dating from before a distinction was made between microplastics and larger plastic debris — date from as early as the 1970s. However, since 2010, the number of records of microplastics in wild-caught organisms has increased substantially (Lusher et al., 2017) and includes vertebrate and invertebrate taxa of substantial variability in both morphology and feeding mode.

Key vertebrate groups in which microplastic ingestion has been frequently recorded include surface-feeding seabirds and pelagic fish. Observations in surface-foraging seabirds may be the result of a number of factors, including their selective feeding behaviours (De Pascalis et al., 2022), in addition to links to land during the breeding season, where microplastic levels may be higher than those in pelagic zones (Young et al., 2009). In and around breeding colonies, microplastics effects on fitness and survivorship may be more easily observed and measured, facilitating the inclusion of these species in routine monitoring schemes (Van Franeker et al., 2011). Thus microplastic uptake has been observed in albatross (Diomedea) (Young et al., 2009), petrels (De Pascalis et al., 2022), fulmars and shearwaters (Terepocki et al., 2017). It has subsequently been suggested that the uptake of plastics by this group is the result of changes to the polymer when exposed to environmental conditions. These changes result in the production of dimethyl sulphide, the scent of which may serve to attract the above species (Savoca et al., 2016). In addition to reported ingestion by surface foraging seabirds in coastal waters, microplastic uptake has been recorded in wading birds, such as rails (Rallidae) (Weitzel et al., 2021), gulls (Laridae) and gannets (Sulidae) (Basto et al., 2019), and ducks and geese (Anatidae) (Reynolds & Ryan, 2018).

Although less studied, microplastics have also been reported in species feeding in marginal and wholly terrestrial habitats. For example, an examination of 36 seaside sparrows, *Ammospiza maritima*, foraging in three estuarine marshland regions of Mississippi, USA, revealed the presence of plastics in 25 individuals (69%), most of which were fibres (Weitzel et al., 2021). Fibres were also the most comment form of microplastics identified in adults and nestlings in a number of migratory birds recovered from locations in Wisconsin and Milwaukee, USA (Hoang & Mitten, 2022). Apparent accumulation of microplastics has also been reported in birds of prey. In Florida, USA, 63 individuals of eight different species recovered by the Audubon Centre for Birds of Prey in 2018 were found to contain microplastics of varied morphologies and polymers. Notably, where the number of individuals was suitably large to enable a statistical comparison, fish-feeding species were seen to contain less plastic than those individual feeding on terrestrial prey (Carlin et al., 2020).

In another well-studied vertebrate group, the bony fish, early observations of microplastic ingestion initially included commercially important species and those commonly caught as by-catch (Lusher et al., 2013; Neves et al., 2015). Subsequently, microplastic uptake has been reported in an increasing number of species of both bony and cartilaginous fish, including both sharks (Valente et al., 2019) and rays (Pinho et al., 2022), spread across multiple aquatic environments.

Observations of uptake by other vertebrates are rarer but increasing. For example, there is now extensive literature on the uptake of microplastics by marine mammals including baleen (Besseling et al., 2015) and toothed whales (Lusher et al., 2015), dolphins (Hernandez-Gonzalez et al., 2018) and pinnipeds (Perez-Venegas et al., 2020), as well as turtles (Duncan et al., 2019). Emerging evidence also indicates the presence of microplastics in a range of terrestrial mammals (Thrift et al., 2022) and reptiles (Gül et al., 2022), however, the impacts of exposure routes and environmental contamination are poorly understood.

When compared to the vertebrates discussed above, records of microplastic uptake by invertebrates are considerably more numerous and include representatives of many of the major taxa. Examples from aquatic environments include molluscs, such as mussels, oysters and sea-slugs, crustaceans, including lobsters, crabs and shrimp (Lusher et al., 2017) and cnidarians (Devereux et al., 2021) and numerous holo- and meroplanktonic organisms (Cole et al., 2013).

As can be seen from the examples above, while microplastics may be ingested by a variety of organisms the distribution of research effort across these studies is not even, with some taxa being over- or underrepresented. Biases toward certain taxonomic groups may occur for a variety of reasons, for example the accessibility of samples as a result of commercial harvesting, geographic distribution, legal protections, or other barriers to sampling (such as damage incurred during the sampling process).

Another drawback of the existing literature is that, while the studied species are seen to have taken up microplastics from the environment, they frequently give little information regarding the route of uptake, the factors influencing that uptake and the period over which plastics are retained. On occasion, where sample sizes are large enough and individuals processed sufficiently early postmortem, the abundance of microplastics may be linked to factors such as individual morphology, feeding patterns or transfer through specific prey species. However, in most cases, these questions have been addressed instead through laboratory exposure experiments and observations of individuals in captivity.

4.2.1.2 Observations of ingested microplastics following controlled exposure

In addition to the observed uptake by organisms in the environment, the ingestion of microplastics has been explored in controlled trails. Organisms have been exposed to multiple polymers individually or as mixtures, and with or without secondary contaminants. The majority of these studies are laboratory-based, under highly controlled in vitro conditions or, more rarely, complex mesocosm systems; however, semicontrolled exposure of captive individuals at rehabilitation studies have also been used.

The delivery of microplastics is often achieved by mixing polymers with environmental media (water or sediment), particularly in establishing uptake by filter feeding organisms (e.g. Hall et al., 2015) and detritivores (Porter et al., 2023). Alternatively, microplastics maybe introduced into processed feeds (e.g. gelatine blocks) (Farrell & Nelson, 2013) or natural prey (Nelms et al., 2018). The type and concentration of plastics used may vary significantly between studies, for example the size range may be selected to correspond to that of the typical diet and the choice of polymer and concentrations may be chosen based on those reported from the environment. Additionally, the morphology of plastics used can also be highly variable, although (as will be discussed below) critical voices have highlighted the overuse of readily available shapes, such as microspheres, and concentrations far above those routinely observed in the environment.

4.2.1.3 Factors influencing ingestion and retention

Observations of retained microplastic in both wild-caught organisms and animals exposed under controlled conditions indicate that long-term contamination by microplastics may be affected by a range of common factors. Environmental concentration has a significant influence on microplastic uptake, and the size and morphology of plastics must fall within a viable range for ingestion. It has also been noted above that biofouling and degradation of plastics may increase the likelihood of plastic uptake and that visual similarities between plastics and preferred prey species may result in the increased uptake of microplastics exhibiting characteristic colours or shapes.

In addition to the characteristics and abundance of plastic, there are a variety of factors affecting uptake that may vary between taxa. For example, in species of marine worms, feeding mode significantly influences the uptake of material, with filter feeders ingesting more fibres than deposit feeders (Porter et al., 2023). A similar effect has also been observed in freshwater species, with intake varying in relation to size and feeding mode. Here larger animals and those engaged in pelagic filtering were seen to have the highest plastic burden (Scherer et al., 2017). This is not true of all species, as between guild differences in plastic uptake were *not* observed in coastal fish off Fortaleza, Brazil (Dantas et al., 2020).

The retention period of plastic will significantly affect the capacity of microplastics to bioaccumulate. Bioaccumulation is the increase in the body load of a substance, harmful or not, over time and is typified by an organism taking up a substance at a faster rate than it is eliminated. Although much less studied than microplastic uptake, there have been some analyses of depuration either in wild-caught organisms, as a precursor to experimental exposures, or following exposure trials. Furthermore, it may be that some species reach an equilibrium with their environment, managing to eliminate a proportion of internalised microplastic over time (Porter et al., 2023). Bioaccumulation potential will be further discussed in Chapter 5.

Life history stage, sex and morphology may also influence microplastic ingestion and retention. Observations of wild-caught langoustine, *Nephrops norvegicus,* have revealed that females, who moult less frequently once mature, and smaller individuals, contained larger aggregations of microplastics than larger/male counterparts (Welden & Cowie, 2016a) (Fig. 4.1). As a result of this and the above factors, microplastic concentrations may be difficult to predict.

4.2.2 **Other observed interactions**

Although ingestion is the most routinely reported interaction between wild-caught organisms and microplastics, frequent contact between microplastics and biota may result in other associations. As with ingestion, the nature and duration of these associations is dependent on the characteristics of both organism and plastic.

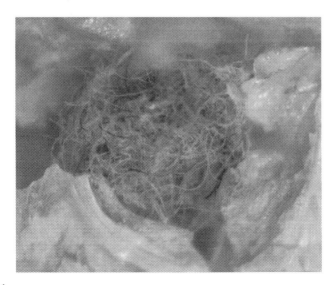

FIGURE 4.1

Aggregation of plastic fibres recovered from the stomach of *N. norvegicus*. A bolus of fibres retained due to the presence of the *Nephrops* gastric mill.

Image Credit: Natalie Welden.

The adherence of microplastics to exterior structures such as gills and carapaces, and observations of microplastic intake and translocation have been recorded in a number of species, including crabs (Watts et al., 2016) and fish (e.g., Koongolla et al., 2020). Similarly, in molluscs, microplastics have been observed to adhere to the foot and mantel (Kolandhasamy et al., 2018). However, adherence and subsequent uptake has been seen to be limited in some species (Batel et al., 2018).

4.2.3 Plastic, plants and algae

The presence of microplastics in both aquatic and terrestrial ecosystems has also resulted in interactions with plants. As with other biofouling species, photosynthesisers (particularly microalgae) may also colonise the surface of plastics in the environment (Nava & Leoni, 2021), resulting in their long-distance transport. In addition, the integration of microplastics into sediments has led to their coming into contact with roots and holdfasts, and in the water they may adhere to leaves and thaluses.

The average number of microfibres recovered from samples taken from *Cladophora* mats in Lakes Michigan and Erie were found to be approximately 34,000 and 32,000 for each kilogram of dried algae. Subsequent observations of the interactions between Cladophora and fibres in a laboratory setting revealed both entanglement and apparent internalisation (Peller et al., 2021). Similarly, it has also been shown that microplastics may be retained on the surface of fucoid algae (Gutow et al., 2016; Sundbæk et al., 2018), facilitating their uptake by grazing species (Gutow et al., 2016) and by the red algae, *Gelidium*, which is used in both agar and for human consumption (Bilbao-Kareaga et al., 2023). In a number of studied species, the abundance of microplastics entangled with or adhering to seaweeds may be much greater than that in the surrounding environment (Feng et al., 2020).

While terrestrial plants are less well studied, both micro- and nanoplastics have been seen to enter tissues by way of uptake through the roots. While many species consider this movement in relation to species of agricultural importance (Liu et al., 2022; Qi et al., 2018), recent observations have also indicated uptake by birch tree roots exposed to fluorescently marked microbeads (5–50 μm) over a period of 5 months (Austen et al., 2022).

4.3 Quantifying exposure

As with the extraction of microplastics from various environmental media discussed in the previous chapter, the methods by which researchers establish uptake of plastics may be highly specific to the studied taxa. Table 4.1 below outlines options for microplastic extraction and enumeration: from this we can observe multiple potential method combinations. Below we will compare these methods and discuss their benefits and drawbacks in relation to their focal species. However, we must first spare a moment to consider the way in which samples are collected and the role that this may have in steering our understanding of microplastic uptake and impacts in numerous species. The research discussed above has varied from opportunistic sampling of organisms collected for purposes other than microplastic research (such as in the course of other research projects or species caught or farmed for human consumption), to specific wild organisms and those exposed during laboratory trials. The results arising from these varied approaches are often compared relatively uncritically. In the past this may have been driven by the limited number of studies regarding the abundance of microplastics in wild-caught organisms (indeed, in many taxa, particularly those in terrestrial and remote settings, there remains precious little data); nevertheless, disregarding the method of both collection and storage may result in the unintentional elevation or reduction in apparent uptake. For example, rough handling of some species may result in the egestion of recently consumed material (Lusher et al., 2017). Conversely, contact with plastics during capture, handling, or storage (or extended periods of exposure to air) may result in artificial increases in adhered material.

4.3.1 Dissection or digestion?

In this section, we will begin with the assumption that we are observing unlabelled plastics (those not stained or dyed before interaction with the organism in question) which are thus difficult to detect without a prior extraction step. As is evident from Table 4.1, there are numerous ways by which microplastics may be separated from animal, plant or fungal tissues prior to enumeration or secondary analysis. In some animal species, we may attempt to depurate a live individual or otherwise observe egested microplastics by sampling either faecal or regurgitated material. For example, the uptake of microplastics in both ducks and geese (Anatidae) has been determined by the analysis of faecal samples, with microplastics present in up to 17% of samples (Reynolds & Ryan, 2018). In some cases, the presence of ingested microplastics may be confirmed by induced regurgitation, for example in storm petrels (De Pascalis et al., 2022). However, more frequently, the options are to digest the obscuring tissues, to remove specifically targeted structures by dissection, or a combination of the two. The choice of method by which plastics are separated from organic material may be influenced by a number of factors, for example the focus of the research undertaken, the number of samples or the limit of detection required. Is the researcher interested in the partitioning of plastics between specific organs, or simply in whole microplastic body load? If the former then dissection, or dissection followed by digestion may be in order. Similarly, do they need to know the microplastic load of each individual, or is an average sufficient? In the latter case, they may opt for the digestion of grouped or pooled samples (Lusher et al., 2017).

In addition, the approach selected is often dependent on the characteristics of the organism in question: its size, morphology and composition. What is the mass

Table 4.1 Methods for the determination of adhered, ingested or otherwise internalised microplastics, organised by appropriateness to individual taxa or functional groups.

Taxa	Common pretreatment steps	Dissection methods	Digestion methods	Justification/ suitability
Diatoms and unicellular algae	Filtering of bulk samples/ reduction in sample volume grouped subsamples		X	Not readily suited to the manual manipulation required for individual dissection too small to warrant individual digestion
Macroalgae and plants	Rinsing of individuals to remove external plastics	X	X	Adhered plastics may be removed for observation via an initial rinsing step, the retained liquid from which is to be filtered and observed A subsample of tissues may be excised and digested to identify internal plastic aggregation
Soft-bodied organisms below c. 3 cm	Filtering of bulk samples/ reduction in sample volume Aggregated samples		X	Not readily suited to the manual manipulation required for individual dissection Too small to warrant individual digestion
Organisms with hard structures below c. 3 cm	Filtering of bulk samples/ reduction in sample volume Aggregated samples		X	Not readily suited to the manual manipulation required for individual dissection Too small to warrant individual digestion
Worms and other soft-bodied invertebrates above c. 3 cm	Rinsing of individuals to remove external plastics Aggregated samples	X	X	Organisms with limited differentiation between tissues or which present handling challenges may be digested rather than dissected Aggregated samples may be employed dependent on the research question

(Continued)

Table 4.1 Methods for the determination of adhered, ingested or otherwise internalised microplastics, organised by appropriateness to individual taxa or functional groups. *Continued*

Taxa	Common pretreatment steps	Dissection methods	Digestion methods	Justification/ suitability
Carapaced invertebrates and vertebrates above c. 3 cm	Rinsing of individuals to remove external plastics Aggregated samples	X	X	Removal of hard structures and/or excess tissues is recommended to prevent plastics being obscured during direct observation or from increasing the volume of reagent and associated time required for digestion

of the individuals? Those over 10 cm in length may simply require too great a volume of reagent to enable the digestion of sufficient samples without either excessive waste or cost. In such cases, it may be simpler and cheaper to undertake a dissection to excise target organs or tissues (such as the gills or stomach) which may then be fixed and prepared for visual analysis or flushed to remove their content which may then be retained for analysis. In many studies of microplastics ingestion in stranded seabirds and mammals, levels of contamination are determined via the dissection of stranded individuals (e.g., Basto et al., 2019); similarly, snake samples obtained from museum collections (Gül et al., 2022). Comparable approaches have also been used in the examination of inhaled plastics, with enzymatic digestion of excised lung tissue used to identify airborne contaminants (Tokunaga et al., 2023).

Also of concern to the researcher is the body composition of the individual. Protein-rich tissue, fatty tissue, shells, carapaces, fish scales, reptile scales, fur, hair and plant tissues must be treated appropriately, being more susceptible or resistant to digestion in either acids, bases, oxidisers or enzymes. Indeed, the mix of pretreatment steps, selection of the reagent, application of heat and duration of exposure are of utmost importance when designing a digestion methodology. The use of inappropriate reagents, insufficient exposure periods and colder temperatures may result in incomplete digestion and underestimation of microplastic loads due to the obscuring effect of undigested organic material. However, overly aggressive digestion steps may result in the damage to, or loss of, plastics, also driving underestimation of apparent microplastic load.

The precise method applied may also vary in relation to the research question considered, if the research question concerns whole body plastic loads averaged across a whole population, and the organism itself is small and soft bodied, then the researcher may choose to employ grouped samples of whole organisms (as in Wright et al., 2013) to increase the efficiency of their sampling and analysis. While Table 4.1 give example justifications for the methods employed by the researchers, it must be noted that each processing step is a possible point of contamination by airborne plastics and that clean technique and both positive and negative controls should be used at all stages of the process (as outlined in Box 3.3).

4.3.2 Observation, enumeration and identification

If a digestion step is applied, the sample is typically sieved or filtered in a similar manner to that used on the supernatant retained during the density separation of microplastics from sediments. A full description of this process and necessary controls may be found in Chapter 3. Filter papers of retained undigested material and the flushed inorganic contents of digestive tract may then be analysed in much a similar way as well, either by prior observation under light-microscope, followed by a secondary characterisation step such as micro-FTIR, or by a combined enumeration and characterisation stages (Rist et al., 2021).

4.3.2.1 In situ observations

The use of fluorescently labelled plastics, typically employed in laboratory uptake trials, may enable the identification and enumeration of plastics without the need for prior extraction – either in smaller whole organisms (such as planktonic species) or in prepared histological samples (e.g. of gut or liver tissues). Examples of such observations may be seen in Fig. 4.2, here plastics were fluorescently labelled and observed under fluorescence microscopy. Such methods enable the distrubution of microplastics within an animal, organ or tissue to be observed. In observing plastics in wild caught species, staining has also been used to aid in microplastic detection. The use of stains and dyes, such as Nile red, fluorescein and others, has the benefit of making plastics more visible due to their preferential binding to hydrophobic materials, however, these stains may also adhere to other lithophilic particles and result in artificial inflation of the plastic count. A full account of staining, fluorescent techniques and their uses can be found in Table 4.2.

4.4 Determining impacts

Prior to the current interest in the uptake of microplastics, observations of the impacts of *macro*plastics on biota from the previous *c.*40 years had already demonstrated the potential negative effects that interactions could cause. For example, the entanglement of marine fauna in pots, traps and nets result in incidents of direct injury or mortality or decreasing motive or feeding capacity resulting in

FIGURE 4.2

Copepod containing labelled microplastics. Fluorescing microplastics may be easily identified under the microscope.

Image Credit: Matthew Cole.

Table 4.2 Dyes and stains previously used in the identification of microplastics.

Stain	Uses	Excitation wavelength (nm)
Nile red	Previously used as a stain for intracellular lipids.	460 and 543
Rose Bengal	Used for the staining of dead or damaged cells. As well as the staining of foraminifera.	525
Safranine T (basic Red 2)	Often used as a counterstain for nuclei, cartilage, mucin, and mast cells.	520
Rhodamine B	Often used as a fluorescent tracer.	50−490 and 515−565
Rit blue	Textile dye.	470
Rit pink	Textile dye.	545
iDye blue	Textile dye.	470
iDye pink	Textile dye.	545

strandings and death. Similarly, the consumption of macroplastic litter was widely known to result in damage or blockage in the gastrointestinal tracks of numerous species, influencing growth, mobility and survivorship. Some of the most common or widely reported of these interactions include the consumption of plastic bags by species of sea turtle, the entanglement and stranding of large marine mammals as a result of abandoned lost and discarded fishing gear, and the high chick mortality rate of the Laysan albatross, *Phoebastria immutabilis*, fed alarming levels of plastic litter by surface foraging adults. Outside of the marine environment, birds, mammals and snakes are often reported as becoming ensnared in terrestrial plastic litter, to much the same end. Thus the rapid increase in observations of uptake and adherence after 2010 unsurprisingly led to extensive concern regarding the subsequent impacts of microplastics on affected species. Since this time, the number of studies seeking to determine the ability of micro- and nanoplastics to translocate into various tissues, as well as their ability to affect a variety of endpoints have grown significantly (Yin et al., 2021), the methods and findings of which we will consider below.

4.4.1 Physical injury and structural changes

Of the impacts identified in prior studies of damage as a result of *macro* plastic litter, physical injury is perhaps the most apparent. As with macroplastics, it has also been suggested that the uptake of microplastics may result in irritation, inflammation and damage to the gastrointestinal lining of many species, as well as histological changes to a variety of tissues. In the terrestrial snail, *Achatina fulica*, exposure to microplastic fibres at a concentration of 0.71 g/kg of soil, resulted in damage to villi in the gastrointenstinal tract of 40% of the exposed individuals (Song et al., 2019). Similarly, analysis of the intestines of brine shrimp exposed to plastic revealed the presence of lipid droplets among the epithelia and, later, deformation of intestinal epithelia (Y. Wang et al., 2019).

Barramundi, *Lates calcarifer*, exposed to PE fibres were able to eliminate fibres following ingestion, however, slight histological damage was observed as were a range of secondary issues including oxidative stress (Xie et al., 2021).[1] In addition to direct impacts to the gastrointestinal structures, gills and other gas exchange structures may also come into direct contact with plastics. This contact may result in inflammation, mucous cell proliferation and cell death, in addition to apparent oxidative stress, noted in the gills of a number of fish species, including carp sp. and goldfish, *Carassius aurtatus* (Cao et al., 2023; Khosrovyan et al., 2023).

Concerningly, as microplastics are subject to further fragmentation, studies of *nano*plastic PS uptake by *Mytilus edulis* and *Crassostrea virginica* have indicated that nanoparticles remain in the body longer than microplastics and are readily translocated into the digestive gland (Ward & Kach, 2009). *M. edulis* exposed to PS nanoplastics (30 nm) in the presence of food indicated minor decreases in filtering (Wegner et al., 2012). A reduction in feeding following PS nanoplastic exposure has also been observed in the brine shrimp, *Artemia fransiscana*, which also demonstrated "massive" retention in the gut lumen, reduced motility and increased molting (Bergami et al., 2016). In freshwater systems,

---

[1] Oxidative stress is the imbalance between oxidants and antioxidants, during which the generation of free radicals exceeds the antioxidant system's ability to deactivate them.

nanoplastics have also been seen to influence mortality, and mobility in *Daphnia manga,* although these affects were absent in with some polymers (Booth et al., 2016).

4.4.2 **Feeding and nutrient exchange**

Under the category of "energetics," we will consider the potential impact that microplastics may have on food consumption, nutrient uptake, metabolism and related effects. One of the most dramatic ways in which microplastics may influence feeding and nutrient uptake is by their presence within the gastrointestinal tract. This effect has long been recognised in macroplastics and their effect on marine fauna such as whales and turtles, as well as in seabirds. Where large aggregations of microplastics are observed, it is predicted that the feeding impulse is reduced. This is as a result of a process known as false-satiation (to feel full despite the stomach content having no nutritional value). If plastics are not expelled, the reduced feeding impulse may persist for an extended period of time, resulting in a reduction in energy uptake. This effect has previously been observed in the Langoustine, *Nephrops norvegicus*, in which exposure to microplastic fibres over an eight-month period was seen to reduce food consumption (Welden & Cowie, 2016b), we will discuss this study in greater detail in Box 4.1. A second similar impact of plastic ingestion is that of nutrient dilution. In species which continue to feed despite the presence of microplastics, the ratio of valuable nutritive material vs nonnutritive material in the gut is altered, resulting in an energetic cost to the organism. Finally, plastic-caused damage to the function of the gut wall, already indicated in many species, has been identified as disrupting the uptake of nutrients across the gut.

The secondary effects of reduced in nutrient uptake as a result of microplastic exposure have been identified in both marine and terrestrial invertebrate species. The long-term retention of fibres and subsequent reduction in feeding resulted in lower energy stores and limited growth in langoustine (Welden & Cowie, 2016b), and exposure to microplastic fibres over a period of up to 8 weeks reduced growth in female house crickets, *Gryllodes sigillatus*. However, in this latter species, negative effects were only observed in fibre ($<$5 mm) fed individuals, not in those fed microbeads (75–105 μm) (Fudlosid et al., 2022). Subsequently, false satiation and nutrient dilution may be increased in situations in which plastics are misidentified and preferentially consumed. For example, in riverine shrimp for, Chongming Island in the mouth of the Yangtze River, it has been seen that microplastics that resembled typical prey were most regularly selected (T. Wang et al., 2023).

Of course, these impacts are not limited to invertebrates, a study of the impacts of small ($<$125 μm) and large (3 mm) microplastic ingestion on body mass, oxidative stress, cytokine levels, blood-biochemical parameters and reproductive hormones in Japanese quail, *Coturnix Japonica*, revealed that exposure to a cumulative weight of 600 mg of microplastics, fed over 5 weeks, had a range of effects on the individual. Microplastics were observed to be retained in the

---

Box 4.1 Exploring the impact of long-term microplastic exposure in langoustine, *Nephrops norvegicus*

Objective

Previous observation of extensive uptake of microplastic fibres by wild-caught N. norvegicus indicated potentially substantial risks to this species. In the study outlined below, researchers attempted to determine the impacts of long-term dietary exposure to microplastic fibres by way of comparison to starved and fed controls. The results of this study were intended not only for the research community, but also for NGOs, policymakers and the fishing industry, who may be directly impacted by any negative effects on this important fishery.

Scope

The exploration of MP impacts on *N. norvegicus* followed the following steps:

- Acquisition of an initial sample of *N. norvegicus*
- A month-long depuration stage
- Determination of remaining plastic levels in a subsample of individuals
- Separation of lobsters into individual flow-through tanks and assignment to either the experimental group or the fed or unfed controls
- Initial length and weight measurement and assessment of haemolymph protein
- Initiation of the exposure period
- Routine feeding of the experimental group and fed control, and recovery of remaining material to determine daily food intake
- End of the exposure period
- Final length and weight measurement and assessment of haemolymph protein
- Determination of hepatic index and water content in the hepatopancreas

Rationale

High levels of microplastic fibre retention have previously been observed in *N. norvegicus* from the Clyde Sea, Scotland. This retention was believed to be the result of the species structural biology as well as the concentration of local plastic sources, and the resulting boluses of fibres represented a high proportion of the available stomach volume. This species therefore presented all the signs of a species potentially at risk of the negative effects of MP ingestion, however, *N. norvegicus* is also capable of surviving for extended period without food. Thus an eight-month experiment was devised that sought to compare mortality and various indices of body condition (growth, body mass, haemolymph protein, hepatopancreas water content and hepatic index) between a fed control, a plastic-fed experimental group and a starved control. It was intended that the study enable researchers to predict the impact of microplastic uptake on mortality and a variety of sublethal endpoints in *N. norvegicus* and closely related species, in addition to secondary impacts on both the environment and the fishery.

Workflow

This experiment explored the impact of long-term (8-month) MP exposure on the following endpoints:

- Mortality
- Growth/body mass
- Protein content of the haemolymph, which indicates the early effects of reduced nutrient uptake
- Hepatosomatic index, which compares the size of the hepatopancreas (liver and site of fat storage) with the mass of the whole organism
- Water content of the hepatopancreas, which increases as fat stores are used

(Continued)

Box 4.1 Exploring the impact of long-term microplastic exposure in langoustine, *Nephrops norvegicus* (Continued)

All animals were kept individually in recirculating flow-through tanks. Both fed groups (with and without plastic) were given squid mantle at a weight in excess of maximum dietary intake reported in prior studies. This enabled the before and after weight of food to be taken, enabling the average food consumption of individual in each group to be determined.

Those individuals in the experimental (plastic-fed) condition were given squid mantel that was preseeded with plastic fibres of similar dimensions to those reported in observations of wild-caught *N. norvegicus*. The starved control was provided no food throughout the experimental period. In addition to length and weight at the start of the experiment, the initial level of protein in the haemolymph was determined using a Bradford protein assay.

Following the 8-month experimental period, the Bradford assay was repeated, and the individuals were again measured and weighed before being humanely euthanised and dissected to remove the the stomach and hepatopancreas removed. The stomach contents of all individuals were then observed to determine the level of intentionally added and exogenous plastics. A subsample of hepatopancreas was dried to constant weight to determine its water content, and subsequently examined for its fat content, after which the hepatosomatic index calculated.

Results

Comparisons between the experimental (plastic-fed) group and the fed and unfed controls revealed changes akin to, but not at the same scale as, the unfed control. Food consumption was below that of the fed control, growth was reduced and mortality increased. Additionally, blood protein and the hepatosomatic index were lower, and the level of water in the hepatopancreas was raised, suggesting that individuals in the plastic-fed group were, to an extent, relying on these energy stores.

Learning and knowledge outcomes

In addition to their contribution to our wider understanding of microplastic impacts in crustaceans, the results of this study show the importance of accounting for long-term exposure when seeking to determine the impact of microplastic uptake. Additionally, not often utilised in publications, the use of starved controls gives context to the apparent scale of any negative effects arising as a result of microplastic exposure.

---

stomachs of quail, however, while small microplastics increased the activity of antioxidant hormones, those quail exposed to large microplastics demonstrated changes in estradiol levels, and exposure to microplastics also reduced the growth rates (Monclús et al., 2022).

In addition to the observed effects of microplastics reported above, the impact of uptake may be complicated by environmental factors. The growth of biofilm communities may mitigate impacts on filter-feeding organisms, with reduced mortality seen in *D. magna*-offered microplastics with developed biofilm communities (Amariei et al., 2022).

4.4.3 Respiration

Microplastic may also affect the ability of organisms to respire efficiently. In its simplest form, this may be caused by physical disruption of oxygen uptake, whereby plastics adhering to the surface of gas exchange structure block the movement of oxygen and induce respiratory stress. Alternatively, oxygen uptake may be affected by damage to these membranes, causing irritation and inflammation.

Early observations of the effects of PS microspheres on the gill structures of shore crabs, *Carcinus maenas*, compared the level of oxygen in experimental tanks containing crabs to that of 6 seawater-only control tanks, revealing a significant reduction in oxygen consumption (Watts et al., 2016). Other studies have used before-after measures of oxygen in sealed systems. Exposure of copepods, *Calanus helgolandicus*, to PS beads of varied diameter revealed both preferential uptake of the smallest size category, as well as behaviour similar to that of starved individuals and a reduction in the basal metabolic rates. Metabolic rate was determined by measuring the respiration rate of active and anaesthetised individuals placed in sealed 2.0 mL syringes for either one or two hours, after which a dissolved oxygen sensor was used to determine the volume of oxygen taken up (Isinibilir et al., 2020). Similarly, the respiration rates of the sponge, *Petrosia ficiformis*, exposed to PE powder at 500 ng/mL, or approximately 26.1 PE-MPs/L, for 24 or 72 hours were determined by enclosing sponges in a sealed container over a 2-hour period prior to analysis of the water oxygen content. Results of the study indicated a significant decrease in both respiratory and filtration in sponges exposed for 72 hours (De Marchi et al., 2022).

It has also been suggested that microplastics may affect oxygen uptake and transport by way of inhibition of respiratory pigments or other secondary effects. Although, haematological analysis of the blood of juvenile rainbow trout, *Oncorhychus mykiss*, exposed to PS demonstrated no significant difference in either red blood cell count, haemoglobin concentration, haematocrit value, mean erythrocyte volume, mean erythrocyte haemoglobin, or mean corpuscular haemoglobin concentration; although a decrease in lymphocyte counts and increase in neutrophils was observed (Hollerova et al., 2023). Similarly, assessment of the changing haematological characteristics of mice exposed to PS particles indicated revealed decreased white blood cell counts but no significant change in red blood cell counts or haemoglobin levels (Sun et al., 2021).

In those species inhabiting sediments, there may be more complex influences on species respiration, such as this impact of microplastic contamination of sediments on oxygen diffusion and availability.

4.4.4 Behavioural change

The impact of microplastics on behaviour has already been lightly touched upon above, in which microplastics which resemble typical prey or other foods may be preferentially selected. However, the presence of microplastics and their interaction with organisms may change the behaviour of biota in further ways. For example, accidental ingestion may result in false satiation or the fouling or obstruction of feeding structures, particularly in filterers, inhibiting feeding behaviours. It has also been indicated that microplastic uptake may influence predator-prey interactions, such as predator avoidance behaviours. For example, edible periwinkles, *Littorina littorea*, responses to chemical cues usually associated with a crab predator were examined with or without PP leachates. It was seen that snail's ability to right themselves when dislodged was reduced (took more time), the time to explore when placed in a new habitat was less (making it comparable to conditions with no apparent predators) and the rate of withdrawal into the shell was lower (suggesting decreased avoidance). Snails exposed to virgin plastic leachate were also seen to spend more/less time in refugia (perceived safe spaces) (Seuront, 2018). Indeed, both micro- and nanoplastics have been seen to influence conditions in the brain, inducing oxidative stress, AChE inhibition (reduction in the enzyme that breaks down acetylcholine) and the regulation of neurodevelopmental genes (Yin et al., 2021).

Nevertheless, not all species demonstrate measurable changes in behaviour. The spiny chromis, *Acanthochromis polyacanthus*, is a plantivorous coral reef fish. Juvenile *A. polyacanthus* exposed to either *c.* 2 mm or <300 μm PS particles at different concentrations over 6 weeks demonstrated size-dependent differences in uptake, with smaller plastics being more readily ingested and retained. Changes in behaviour were measured using two sets of video observations at the 6-week point. Characteristic monitored included swimming, aggression and foraging behaviours. Variable responses to plastic ingestion were observed between groups and individuals; although there appeared to be more aggression at moderate microplastic exposures, this was not found to be significant (Critchell & Hoogenboom, 2018). Similarly, the woodlouse, *Porcellio scaber*, fed pellets seeded with microplastic demonstrated no difference in energy reserves or feeding behaviour (Jemec Kokalj et al., 2018).

Again, subsequent breakdown of microplastics may lead to additional impacts. In reviewing the effects of nanoplastics on the behaviour of freshwater organisms, Ferreira et al. (2023) noted reports of changes in swimming behaviour, feeding, increases in interactions with conspecifics, reduction in aggressiveness and variable impacts on predator avoidance. In many cases, these behavioural changes emerged at lower concentrations than other endpoints.

4.4.5 Reproduction

The uptake of plastics and associated changes in fitness have also been seen to affect reproductive capacity, these impacts include damage to reproductive organs, oocytes and sperm, as well as developmental abnormalities and a shift in the sex ratios of offspring (Yin et al., 2021). Additionally, in many species, fecundity is linked to size or body mass. Those species in experiencing a reduction in growth as a result of micro- or nanoplastic uptake may see a resulting decrease in the reproductive output.

Oocyte[2] damage has been observed in a number of vertebrate and invertebrate species, for example 2-month exposure to 2- and 6-μm PS microplastics at a

---

[2] Immature egg cell.

concentration of 0.023 mg/L resulted in reductions in the size and number of oyster oocytes as well as the velocity of sperm (Sussarellu et al., 2016). Cnidaria, *Hydra attenuata,* exposed to PE flakes for a period of up to 96 hours showed decreases in the number of hydranths, feeding polyps, within a colony (Murphy & Quinn, 2018). In mice, 35-day exposure to *c.* 0.8 μm diameter PS microplastics at 30 mg/kg of body mass have been seen to affect the extrusion rate and the survival rate of oocytes (Liu et al., 2022), and in 90-day exposure to PS microplastics at concentrations of either PE-MPs 0.125, 0.5 or 2 mg per mouse per day resulted in reduced growth in males and increases neutrophils in both sexes (Park et al., 2020). However, such impacts may be concentration linked, as studies exploring the effect of 1 μm PS microplastics on *Danio rerio* over 21 days revealed no impacts at the lowest levels (10 mg/L) (Qiang & Cheng, 2021). Gamete damage may be the result of inflammation or damage to the gonads, metabolic changes, altered hormone levels and oxidative stress, as well as direct adherence to the chorionic membrane of embryos changing both metabolism and egg hatchability (Chen et al., 2020).

As indicated above, microplastic ingestion or uptake by adults may result in reduced feeding capacity, nutrient dilution and limited growth, this may also affect the production of eggs or offspring as well as their health and survivorship. Exposure of the sea urchin, *Paracentrotus lividus*, to PS nanoplastics has previously indicated embryo toxicity and abnormal gene expression (Della Torre et al., 2014). Observations of both female mice exposed to PE over 90 days and of their resulting offering revealed that the number of live births, the sex ratio of pups and body weight of pups was notably altered in plastic-exposed individuals and that there was a change in the subpopulation of lymphocytes within the spleen of the offspring, similar to that seen in females (Park et al., 2020). Micro- and nanoplastics may also be transferred from parent to offspring, for example via the translocation of nanoplastics into the yolk sac, gastrointestinal tract and other organs (Pitt et al., 2018).

4.4.6 Survivorship

Whether directly, indirectly, or via a combination of means, micro- and nanoplastic uptake and the changes described above may increase the probability of mortality in an individual organism. Direct mortality has been seen to arise as a result of acute damage to key structures, whereas chronic exposure may lead to a variety of secondary effects which may increase mortality over extended timescales.

In addition to directly affecting the health and survivorship of an organism, micro- and nanoplastic exposure may increase susceptibility to other environmental impacts, including those of macroplastic debris (Rivers-Auty et al., 2023). For example, mussels, *Mytilus galloprovincialis*, exposed to PS nanoplastics measuring 50 nm, 100 nm and 1 μm in diameter were used to determine the effects of similar particles on different cells within the immune system. It was observed that the microscale plastics provoked an immunological response, with large granular cells exhibiting the largest changes (Sendra et al., 2020).

Additionally, rainbow trout, *Oncorhynchus mykiss*, exposed to a combination of plastic (PA or PS) and infectious hematopoietic necrosis virus (IHNV) were seen to experience higher mortality than those exposed to either microplastics or IHNV alone (Seeley et al., 2023). Similarly, exposure to microplastics has been seen to affect the toxicity of associated pollutants (Yin et al., 2021).

4.4.7 Impacts in plants, algae and associated taxa

As the fundamental driving force of productivity in almost all environments, the potential negative effects of plastics on primary producers have garnered substantial concern. Laboratory assessments of the impacts of microplastics on microalgae have sought to establish changes in key endpoints including growth (Lagarde et al., 2016; Sjollema et al., 2016), cell counts and biomass, in addition to cell morphology (Yokota et al., 2017), photosynthesis (Sjollema et al., 2016) and enzyme activity (Fu et al., 2019; Seoane et al., 2019). However, exposure of the sea grass, *Zostera marina*, to microplastic was seen to result in limited effect on photosynthesis and respiration (Molin et al., 2023). The concentration of microplastic used in the above studies ranges from less than 1 particle per mL to several thousand microplastics per mL, and the diameter of the plastics used was between 0.5 μm and 1 mm (Nava & Leoni, 2021). Additionally, only a handful of studies consider impacts on the same model organism, species widely used in studies of toxicology, an issue that will be further considered in Box 4.2, the use of model organisms and standardise methodologies enbale. Typically, changes in growth hae only been observed in cases where the concentration of particles was at the higher end of the range outlined above, or the dimensions of plastics were of just a few microns. Conversely, changes to the capacity of algae to photosynthesis have been reported. Algae, *Chlorella* sp. and *Scenedesmus* sp., exposed to positively and negatively charged PS nanoplastics revealed both preferential binding with positively charged PS beads and disruption to photosynthesis (Bhattacharya et al., 2010), an effect previously described in *Scenedesmus by* (Besseling et al., 2014). Of course, as with *Z. marina*, some species do not show significant responses to exposure to microplastics, for example of the 28 studies on MP effects on phytoplankton in a recent review by Amaneesh et al. (2023), nine showed no measurable response in the studied parameters.

As with animals, the impacts of microplastic on terrestrial primary producers are typically less studied than those in aquatic systems, studies of vascular plants have attempted to assess translocation of plastics and the effect of interactions on factors including germination (Boots et al., 2019), growth (Gao et al., 2019; Urbina et al., 2020), morphology (Maity et al., 2020), photosynthesis (Dong et al., 2020; Shen et al., 2023), transpiration (Urbina et al., 2020) and oxidative stress (López et al., 2022; Shen et al., 2023). As mentioned above, many of these species preferentially consider the effects on species for human consumption such as rice, wheat, barley, cabbage, lettuce and maize (Li et al., 2022). Also as above,

Box 4.2 The role of the model organism in microplastic research

Objective

Model organisms are used in toxicological studies to provide researchers with comparable results across studies and target substances, however, microplastics are unlike many other pollutants in their non-soluable nature and that they are often not readily internalised following exposure/ingestion. Larger and mid-sized MP may not readily translocate from the digestive system and gill structures into tissues, as a result, the apparent risks posed by uptake and retention may be related to structural and morphological characteristics not currently represented by existing models. This case study will consider the diversity of existing models used in toxicological testing, and their potential to interact with various fractions of MPs. The selection and use of model organisms in toxicological research and routine monitoring is relevant to academics, regulators, policymakers and associated industries.

Rationale

Model organisms are widely used in the study of genetics, biological processes, disease, behaviour and a host of other fields. Our depth of understanding these organisms and the availability of genetically similar strains of these models enable the development of a consistent body of literature which is applicable to similar taxa in the wider environment (Ankeny & Leonelli, 2020). Current model organisms for aquatic systems include animals such as the water flea, *Daphnia*, cnidarian, *Hydra* spp., great pond snail, *Lymnaea stagnalis*, madaka, *Oryzias latipes,* and zebrafish, *Danio rerio,* and plants and alga such as the duck weed, *Lemna gibba,* and single-celled alga, *Chlamydomonas reinhardtii*. As can be seen from text elsewhere in this volume, many of these organisms have previously been used in studies of microplastic toxicity in relation to a variety of endpoints, however, despite the benefits of utilising more widely available models such as *Daphnia*, *D. rerio* and *O. latipes*, these species differ in biology, morphology and habit than those seen to be at greatest risk of micoplastic uptake and retention in typical studies of both wild-caught and lab exposed animals. By identifying the apparent susceptibility of existing models to microplastic uptake and retention and identifying alternative models for susceptible taxa, we may better understand the impacts of mid-large end (nontranslocating) microplastics in the environment. Below we will consider some of the methods and models widely used in toxicity testing, in this case using examples drawn from the OECD Guidelines for the Testing of Chemicals.

Invertebrates

Invertebrate model species and associated protocols which may assist in furthering our understanding of the problems associated with microplastic uptake include reproduction in the snail *L. stagnalis*, toxicty in Chironomid larvae, and bioaccumulation in benthic oligochates. Test No. 243, the *L. stagnalis* Reproduction Test, seeks to determine the impact of prolonged exposure on the reproduction and survival of *L. stagnalis*. Over 28 days, the impacts of chemical concentration on both mortality and and reproduction (measured as number of egg clutches) is determined, along with eggs per clutch and shell length. The results are used to determine the effective concentration (ECx) based on the number of clutches per individual-day, establishing the relationship between concentration and reduction in reproductive output, as well as the highest concentration which has no effect.

In Test No. 219: Sediment-Water Chironomid Toxicity Using Spiked Water, chironomid larvae are exposed to at least five concentrations of the test chemical in sediment-water systems. Comparisons are then made between the emergence and development rate between the various concentrations, as well as larval survivorship after 10 days. The relationship between concentration and measured endpoints is then determined in order to identify relevant effective concentrations.

(Continued)

Box 4.2 The role of the model organism in microplastic research (Continued)

Similarly, over longer periods, Test No. 315: Bioaccumulation in Sediment-dwelling Benthic Oligochaetes seeks to determine the bioaccumulation of sediment-associated chemicals in endobenthic oligochaetes worms. These tests are specific to stable, neutral organic chemicals having log K_{ow} values between 3.0 and 6.0, superlipophilic substances that show a log K_{ow} of more than 6.0, or stable metallo-organic compounds which tend to associate with sediments. These tests are composed of an initial 28-day uptake phase, during which animals are exposed to spiked sediments, followed by a maximum 28-day depuration phase to determine the potential for elimination. Using this data, the uptake rate constant (ks), the elimination rate constant (ke) and the kinetic bioaccumulation factor (BAFK = ks/ke) are calculated.

Fish

Selection of fish species may be undertaken with regulatory requirements or exposure scenarios (for example, fresh or salt water, or temperature or tropical climates). These models include *D. rerio, Pimepjales promelas, Cyrinus carpio, O. latipes, Peocilia reticulata* and *Lepomis macromis*. Unlike invertebrate testing, many of these species are already widely represented in microplastic research, although the aims of the related experiments vary. As above, Test No. 203, the Fish, Acute Toxicity Test exposes the target to the test chemical for a period of up to 96 hours. Groups are assigned to one of a minimum of 5 test concentrations, with the number of dead individuals recorded at 24, 48, 72 and 96 hours. The concentrations resulting in 50% mortality are then determined. Animals may also be exposed to a "limit test" at one dose of 100 mg/L. Again, mortality is recorded at 24, 48, 72 and 96 hours and the relationship between duration of test and mortality plotted.

As with invertebrates, short-term acute exposure may be compared to extended tests of bioaccumulation and elimination. In test No. 305: Bioaccumulation in Fish: Aqueous and Dietary Exposure, uptake as a result of via either the water column or feed over 28 days is determined, following which fish are transferred to a medium free of the test substance, or fed un-spiked feed. A depuration phase is always necessary unless uptake of the substance during the uptake phase has been insignificant. Following the test period, a bioconcentration factor or biomagnification factor are determined, based on the total concentration of the contaminant in fish, (concentration by wet weight of the fish), normalised to a fish with a 5% lipid content. However, tests do not need to include whole organisms, Test No. 249: Fish Cell Line Acute Toxicity – The RTgill-W1 cell line assay, utilised the permanent cell line from rainbow trout (*Oncorhynchus mykiss*) gill, RTgill-W1, in w 24 well-plate. The effect of the chemical is tested over 24 hours of acute exposure, following which fluorescent dyes are added which indicate cell viability, following which the effective concentrations which result in 50% loss in cell viability are determined.

Choice of endpoint

In addition to considering the apparent susceptibility of the taxonomic group, the choice of endpoint will also influence the model selected. Some species are more suited to certain research questions due to housing requirements or reproductive cycles, and their use may be supported or negated by existing technological methods such as the use of automated video analysis and tracking.

The duration of toxicological tests

Alongside the issue of selecting the right organisms, we must also consider that fact that the ingestion and uptake of chemically inert plastics may result in very different impacts than typical pollutants, thus short-term, acute toxicity testing may fail to capture the long-term effects of chronic exposure and retention.

(Continued)

Box 4.2 The role of the model organism in microplastic research (Continued)

Learning and knowledge outcomes

Despite their benefits of consistency and comparability, the requirements of the tests above and the organisms widely used as models may not be suited to determining the effects of plastics in the environment. Taxa most at risk from plastic uptake have previously been represented to be smaller species and those with complex internal morphologies. Such characteristics are only partially covered by the common models highlighted here. In order to develop our understanding of the impacts of microplastics of all sizes, we must utilise models that are representative of the taxa at greatest risk of long-term elevated exposure. Thus other suggested models, such as the copepod *Tigriopus*, may be more useful in identifying threats to the wider environment. Additionally, it may be more effective if a wider variety of invertebrate taxa from a greater range of feeding guilds is consistently utilised in order to understand the impacts of plastic polymers, size and morphology on individuals and populations.

---

these factors were not seen to be significantly affected across all trials, with some soil amendments occasionally inducing positive changes in growth, and the conditions and polymers to which plants were exposed varying significantly.

4.4.8 Human-level effects

Early observations of the presence of microplastics in drinking water, food stuffs and the atmosphere, as well as the effects observed in impacted species, have rapidly led to concern over potential human exposure and impacts. In comparison with many wild species, humans come into routine contact with plastic materials that represent a vast store of potential micro- and nanoplastics, including our clothes, linens, carpets, food packaging (Hussain et al., 2023) and preparation surfaces (Yadav et al., 2023), as well as in the workplace (Shahsavaripour et al., 2023). Additionally, there are numerous dietary routes, both plant and animal, that would enable the trophic transfer of micro- and nanoplastics to humans (W. Wang et al., 2022). Thus as we have already mentioned, early observations of microplastic uptake by humans have been noted in samples of stool (Braun et al., 2021; Schwabl et al., 2019; Zhang et al., 2021) as well as tissues of the lung, which revealed 31 microplastics from 20 individuals (Amato-Lourenço et al., 2021), colon, 3641 microplastics from 11 individuals (Ibrahim et al., 2021), testes, in four of six tested (Zhao et al., 2023), placenta, 12 microplastics from 4 individuals (Ragusa et al., 2021), and blood, 17 out of 22 individuals (Leslie et al., 2022). In Pacet, East Java, Indonesia, screening of both consumables and stool samples has revealed plastics in over 60% of the tested individuals, with PP levels in faeces as high as 10.19 micrograms per gram (μg/g), followed by HDPE, PS and PET. Microplastics were also recorded in food, seasoning and hygiene products (toothpaste), with notable levels in table salts (2.6 μg/g) and toothpaste (15.42 μg/g) as well as in tempeh (Wibowo et al., 2021). Those studies concerning lung tissue showed a higher proportion of particles, whereas others, such as those in the colon demonstrated a predominance of fibres. Those microplastics observed within the tissues may have passed across the gut wall, a route available to particles below 150 μm (Campanale et al., 2020), or through the lining of the lungs. At the lowest size ranges, microplastics below 10 μm may translocate into the cells themselves (Wu et al., 2019). The mix of polymers observed is as diverse as is observed in both environmental media and wild-caught organisms; however, PP, PS, PET, PA, and PE were the most routinely reported (Table 4.3).

The methods used to isolate microplastics from human tissues are predominantly the same as those discussed above and in the previous chapter (e.g., Amato-Lourenço et al., 2021). As indicated in the table, these studies commonly focus either on one target organ, tissue or similar sample or a group of closely related targets (Kutralam-Muniasamy et al., 2023). However, as with the studies on nonhuman animals and those on environmental media, the methods employed in such studies have significant implications regarding the reliability of the findings and confidence in predictions of toxicity drawn as a result. Lack of apparent or incomplete application of quality assurance or control measures (or failure to report these measures) has already been highlighted in relation to some published studies, reducing the reliability, or perceived reliability, of reported microplastic concentrations. In particular, lack of standardisation when dealing with the collection and handling of samples from human subjects has been identified as a potential source of error (Malafaia, 2023). Similarly, criticisms have also been made regarding low levels of replication and apparent contamination control (Kuhlman, 2022).

As a result of these concerns, and the current dearth of information, the level of uptake and retention of microplastics by humans remain poorly understood, and there remains little concrete evidence for any acute effects of microplastic exposure in humans. Nevertheless, numerous studies have attempted to identify and model the potential damage that may arise from these interactions. For example, a combined experimental and modelling approach was used to assess the

Table 4.3 Dominant polymer types reported in samples of human origin.

Tissue/media	Dominant polymers	Study
Lung	PA, PP, PVC, PE, PU PS, and both PE-PP and PS-PVC copolymers	Amato-Lourenço et al. (2021)
Placenta	PP, PE, PP, PU PVC, PP, PBS, PET, PC, PS, PA, PE, PAM, and PSF	Ragusa et al. (2021), Braun et al. (2021), and Zhu et al. (2023)
Blood	PMMA, PP, PS, and PET	Leslie et al. (2022)
Stool	PP, PET, PS, PE, PVC, PC, PA, and PU	Zhang et al. (2021)

toxicity of plastics of varied polymer and size to human cells. Prediction of the toxicity levels of mixed PS, PVC and ABS plastics revealed that toxicity increased with the level of PS, rather than as a simple function of concentration (Choi et al., 2023). Studies of the effects of micro- and nanoscale PS spheres on lung cells have indicated inflammatory and cytotoxic results as well as the expression of proteins linked with cell death and cycling (Xu et al., 2019). Additionally, coculture models of human intestinal systems using either differentiated Caco-2/HT29 intestinal cells and Caco-2/HT29 + Raji-B cells exposed to PS nanoplastic have been used to determine the impact of nanoplastics entering the body via the gut. While the uptake of plastics was determined in both systems (at dose-dependent levels), no negative effects on cell viability, membrane integrity or apparent oxidative stress were observed (Domenech et al., 2020).

Despite the widespread narrative highlighting the impacts of micro- and nanoplastic on biota, there are a growing number of concerned voices highlighting flaws in these studies which may reduce the reliability of the available literature. Many of the studies use single polymers and shapes not common in the environment, for example studies of the impact of plastics on human lung cells may not be representative of environmentally relevant plastic types and shapes (Amato-Lourenço et al., 2021). Additionally, as has already been indicated, the levels of plastic used in studies are often far above that observed in the environment, and actual risk of the effects outlined therein are very low. In many of the studies discussed above, organisms indicated significant health effects only after a certain threshold was reached, rather than exhibiting a linear effect of plastics on key end points (Fig. 4.3). Indeed in some cases, initial low doses of microplastics have seemed to result in beneficial outcomes for organisms, a state known as hormesis (Agathokleous et al., 2021). Additionally, the abundance of micro- and

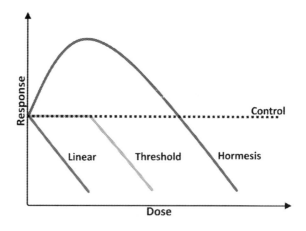

FIGURE 4.3

Dose—response relationships. The varied relationships between microplastic exposure and the change in monitored end-point.

nanoplastics in the environment is much lower than the abundance of other, natural nonnutritive materials in similar size categories (such as silts), that organisms may ingest. As a result, it may not be suitable to conduct such experiments without a suitable particle control in order to estimate relative toxicity (Ogonowski et al., 2018).

4.5 Conclusions

Despite the numerous reports of both micro- and nanoplastic ingestion and its apparent direct effects on organisms, the job of predicting the risk that microplastic poses to the health of individuals and populations remains a challenging one. Many of the laboratory studies which show marked effects utilise concerntrations above those in the environment, restricted numbers of polymers, limited sizes and nonrepresentative shapes, and while many authors have suggested that interactions with microplastics may negatively affect species, others have indicated a lack of, or even positive effect on the organism. Similarly, the mass of plastics accumulated in laboratory studies is often greater than that recorded in wild-caught individuals, and the subsequent impacts inferred from laboratory impact studies may not be representative of that in the environment. Of course, some species demonstrate minimal responses to environmental microplastic exposure. Individual medusae of the moon jellyfish, *Aurelia aurita*, exposed to PS microbeads (100 μm) at concentrations of *c.* 2000 MP/L were seen to ingest very few particles. Those plastics that were ingested after 8 hours and exposed individual demonstrated no significant physiological or histological changes (Sucharitakul et al., 2020). Nevertheless, where many individuals withing a population or community are affected, this may negatively affect energy flows between trophic levels and influence community structure, this will be discussed in greater detail in the next chapter.

References

Agathokleous, E., Iavicoli, I., Barceló, D., & Calabrese, E. J. (2021). Micro/nanoplastics effects on organisms: A review focusing on 'dose. *Journal of Hazardous Materials*, *417*. Available from https://doi.org/10.1016/j.jhazmat.2021.126084, http://www.elsevier.com/locate/jhazmat.

Amaneesh, C., Anna Balan, S., Silpa, P. S., Kim, J. W., Greeshma, K., Aswathi Mohan, A., Robert Antony, A., Grossart, H. P., Kim, H. S., & Ramanan, R. (2023). Gross negligence. Impacts of microplastics and plastic leachates on phytoplankton community and ecosystem dynamics. *Environmental Science and Technology*, *57*(1), 5–24. Available from https://doi.org/10.1021/acs.est.2c05817, http://pubs.acs.org/journal/esthag.

Amariei, G., Rosal, R., Fernández-Piñas, F., & Koelmans, A. A. (2022). Negative food dilution and positive biofilm carrier effects of microplastic ingestion by D. magna cause tipping points at the population level. *Environmental Pollution*, *294*. Available from https://doi.org/10.1016/j.envpol.2021.118622, https://www.journals.elsevier.com/environmental-pollution.

Amato-Lourenço, L. F., Carvalho-Oliveira, R., Júnior, G. R., dos Santos Galvão, L., Ando, R. A., & Mauad, T. (2021). Presence of airborne microplastics in human lung tissue. *Journal of Hazardous Materials*, *416*. Available from https://doi.org/10.1016/j.jhazmat.2021.126124, http://www.elsevier.com/locate/jhazmat.

Ankeny, R., & Leonelli, S. (2020). *Model organisms*. Cambridge University Press. Available from http://doi.org/10.1017/9781108593014.

Austen, K., MacLean, J., Balanzategui, D., & Hölker, F. (2022). Microplastic inclusion in birch tree roots. *Science of the Total Environment*, *808*. Available from https://doi.org/10.1016/j.scitotenv.2021.152085, http://www.elsevier.com/locate/scitotenv.

Basto, M. N., Nicastro, K. R., Tavares, A. I., McQuaid, C. D., Casero, M., Azevedo, F., & Zardi, G. I. (2019). Plastic ingestion in aquatic birds in Portugal. *Marine Pollution Bulletin*, *138*, 19−24. Available from https://doi.org/10.1016/j.marpolbul.2018.11.024, http://www.elsevier.com/locate/marpolbul.

Batel, A., Borchert, F., Reinwald, H., Erdinger, L., & Braunbeck, T. (2018). Microplastic accumulation patterns and transfer of benzo[a]pyrene to adult zebrafish (Danio rerio) gills and zebrafish embryos. *Environmental Pollution*, *235*, 918−930. Available from https://doi.org/10.1016/j.envpol.2018.01.028, http://www.elsevier.com/inca/publications/store/4/0/5/8/5/6.

Bergami, E., Bocci, E., Vannuccini, M. L., Monopoli, M., Salvati, A., Dawson, K. A., & Corsi, I. (2016). Nano-sized polystyrene affects feeding, behavior and physiology of brine shrimp Artemia franciscana larvae. *Ecotoxicology and Environmental Safety*, *123*, 18−25. Available from https://doi.org/10.1016/j.ecoenv.2015.09.021, http://www.elsevier.com/inca/publications/store/6/2/2/8/1/9/index.htt.

Besseling, E., Foekema, E. M., Van Franeker, J. A., Leopold, M. F., Kühn, S., Bravo Rebolledo, E. L., Heße, E., Mielke, L., Ijzer, J., Kamminga, P., & Koelmans, A. A. (2015). Microplastic in a macro filter feeder: Humpback whale Megaptera novaeangliae. *Marine Pollution Bulletin*, *95*(1), 248−252. Available from https://doi.org/10.1016/j.marpolbul.2015.04.007, http://www.elsevier.com/locate/marpolbul.

Besseling, E., Wang, B., Lürling, M., & Koelmans, A. A. (2014). Nanoplastic affects growth of S. obliquus and reproduction of D. magna. *Environmental Science and Technology*, *48*(20), 12336−12343. Available from https://doi.org/10.1021/es503001d, http://pubs.acs.org/journal/esthag.

Bhattacharya, P., Lin, S., Turner, J. P., & Ke, P. C. (2010). Physical adsorption of charged plastic nanoparticles affects algal photosynthesis. *Journal of Physical Chemistry C*, *114*(39), 16556−16561. Available from https://doi.org/10.1021/jp1054759.

Bilbao-Kareaga, A., Menendez, D., Peón, P., Ardura, A., & Garcia-Vazquez, E. (2023). Microplastics in jellifying algae in the Bay of Biscay. Implications for consumers' health. *Algal Research*, *72*. Available from https://doi.org/10.1016/j.algal.2023.103080, http://www.sciencedirect.com/science/journal/aip/22119264.

Booth, A. M., Hansen, B. H., Frenzel, M., Johnsen, H., & Altin, D. (2016). Uptake and toxicity of methylmethacrylate-based nanoplastic particles in aquatic organisms. *Environmental Toxicology and Chemistry*, *35*(7), 1641−1649. Available from https://doi.org/10.1002/etc.3076, http://www.interscience.wiley.com/jpages/0730-7268.

Boots, B., Russell, C. W., & Green, D. S. (2019). Effects of microplastics in soil ecosystems: Above and below ground. *Environmental Science and Technology*, *53*(19), 11496−11506. Available from https://doi.org/10.1021/acs.est.9b03304, http://pubs.acs.org/journal/esthag.

Braun, T., Ehrlich, L., Henrich, W., Koeppel, S., Lomako, I., Schwabl, P., & Liebmann, B. (2021). Detection of microplastic in human placenta and meconium in a clinical setting. *Pharmaceutics*, *13*(7). Available from https://doi.org/10.3390/pharmaceutics13070921, https://www.mdpi.com/1999-4923/13/7/921/pdf.

Campanale., Massarelli., Savino., Locaputo., & Uricchio. (2020). A detailed review study on potential effects of microplastics and additives of concern on human health. *International Journal of Environmental Research and Public Health*, *17*(4), 1212. Available from https://doi.org/10.3390/ijerph17041212.

Cao, J., Xu, R., Wang, F., Geng, Y., Xu, T., Zhu, M., Lv, H., Xu, S., & Guo, My (2023). Polyethylene microplastics trigger cell apoptosis and inflammation via inducing oxidative stress and activation of the NLRP3 inflammasome in carp gills. *Fish and Shellfish Immunology*, *132*. Available from https://doi.org/10.1016/j.fsi.2022.108470, http://www.elsevier.com/inca/publications/store/6/2/2/8/3/2/index.htt.

Carlin, J., Craig, C., Little, S., Donnelly, M., Fox, D., Zhai, L., & Walters, L. (2020). Microplastic accumulation in the gastrointestinal tracts in birds of prey in central Florida, USA. *Environmental Pollution*, *264*. Available from https://doi.org/10.1016/j.envpol.2020.114633, https://www.journals.elsevier.com/environmental-pollution.

Chen, J. C., Chen, M. Y., Fang, C., Zheng, R. H., Jiang, Y. L., Zhang, Y. S., Wang, K. J., Bailey, C., Segner, H., & Bo, J. (2020). Microplastics negatively impact embryogenesis and modulate the immune response of the marine medaka Oryzias melastigma. *Marine Pollution Bulletin*, *158*. Available from https://doi.org/10.1016/j.marpolbul.2020.111349, http://www.elsevier.com/locate/marpolbul.

Choi, D., Kim, C., Kim, T., Park, K., Im, J., & Hong, J. (2023). Potential threat of microplastics to humans: Toxicity prediction modeling by small data analysis. *Environmental Science: Nano.*, *10*(4), 1096−1108. Available from https://doi.org/10.1039/D2EN00192F.

Cole, M., Lindeque, P., Fileman, E., Halsband, C., Goodhead, R., Moger, J., & Galloway, T. S. (2013). Microplastic ingestion by zooplankton. *Environmental Science and Technology*, *47*(12), 6646−6655. Available from https://doi.org/10.1021/es400663f.

Critchell, K., & Hoogenboom, M. O. (2018). Effects of microplastic exposure on the body condition and behaviour of planktivorous reef fish (Acanthochromis polyacanthus). *PLoS One*, *13*(3). Available from https://doi.org/10.1371/journal.pone.0193308, http://journals.plos.org/plosone/article/file?id = 10.1371/journal.pone.0193308&type = printable.

Dantas, N. C. F. M., Duarte, O. S., Ferreira, W. C., Ayala, A. P., Rezende, C. F., & Feitosa, C. V. (2020). Plastic intake does not depend on fish eating habits: Identification of microplastics in the stomach contents of fish on an urban beach in Brazil. *Marine Pollution Bulletin*, *153*. Available from https://doi.org/10.1016/j.marpolbul.2020.110959, http://www.elsevier.com/locate/marpolbul.

Della Torre, C., Bergami, E., Salvati, A., Faleri, C., Cirino, P., Dawson, K. A., & Corsi, I. (2014). Accumulation and embryotoxicity of polystyrene nanoparticles at early stage of development of sea urchin embryos Paracentrotus lividus. *Environmental Science and Technology*, *48*(20), 12302−12311. Available from https://doi.org/10.1021/es502569w, http://pubs.acs.org/journal/esthag.

Devereux, R., Hartl, M. G. J., Bell, M., & Capper, A. (2021). The abundance of microplastics in cnidaria and ctenophora in the North Sea. *Marine Pollution Bulletin*, *173*. Available from https://doi.org/10.1016/j.marpolbul.2021.112992, http://www.elsevier.com/locate/marpolbul.

Domenech, J., Hernández, A., Rubio, L., Marcos, R., & Cortés, C. (2020). Interactions of polystyrene nanoplastics with in vitro models of the human intestinal barrier. *Archives of Toxicology*, *94*(9), 2997−3012. Available from https://doi.org/10.1007/s00204-020-02805-3, http://link.springer.de/link/service/journals/00204/index.htm.

Dong, Y., Gao, M., Song, Z., & Qiu, W. (2020). Microplastic particles increase arsenic toxicity to rice seedlings. *Environmental Pollution*, *259*. Available from https://doi.org/10.1016/j.envpol.2019.113892, https://www.journals.elsevier.com/environmental-pollution.

Duncan, E. M., Broderick, A. C., Fuller, W. J., Galloway, T. S., Godfrey, M. H., Hamann, M., Limpus, C. J., Lindeque, P. K., Mayes, A. G., Omeyer, L. C. M., Santillo, D., Snape, R. T. E., & Godley, B. J. (2019). Microplastic ingestion ubiquitous in marine turtles. *Global Change Biology*, *25*(2), 744−752. Available from https://doi.org/10.1111/gcb.14519, http://onlinelibrary.wiley.com/journal/10.1111/(ISSN)1365-2486.

Farrell, P., & Nelson, K. (2013). Trophic level transfer of microplastic: Mytilus edulis (L.) to Carcinus maenas (L.). *Environmental Pollution*, *177*, 1−3. Available from https://doi.org/10.1016/j.envpol.2013.01.046.

Feng, Z., Zhang, T., Shi, H., Gao, K., Huang, W., Xu, J., Wang, J., Wang, R., Li, J., & Gao, G. (2020). Microplastics in bloom-forming macroalgae: Distribution, characteristics and impacts. *Journal of Hazardous Materials*, *397*. Available from https://doi.org/10.1016/j.jhazmat.2020.122752, http://www.elsevier.com/locate/jhazmat.

Ferreira, C. S. S., Venâncio, C., & Oliveira, M. (2023). Nanoplastics and biota behaviour: Known effects, environmental relevance, and research needs. *TrAC − Trends in Analytical Chemistry*, *165*. Available from https://doi.org/10.1016/j.trac.2023.117129, http://www.elsevier.com/locate/trac.

Van Franeker, J. A., Blaize, C., Danielsen, J., Fairclough, K., Gollan, J., Guse, N., Hansen, P. L., Heubeck, M., Jensen, J. K., Le Guillou, G., Olsen, B., Olsen, K. O., Pedersen, J., Stienen, E. W. M., & Turner, D. M. (2011). Monitoring plastic ingestion by the northern fulmar Fulmarus glacialis in the North Sea. *Environmental Pollution*, *159*(10), 2609−2615. Available from https://doi.org/10.1016/j.envpol.2011.06.008, https://www.journals.elsevier.com/environmental-pollution.

Fudlosid, S., Ritchie, M. W., Muzzatti, M. J., Allison, J. E., Provencher, J., & MacMillan, H. A. (2022). Ingestion of microplastic fibres, but not microplastic beads, impacts growth rates in the tropical house cricket gryllodes sigillatus. *Frontiers in Physiology*, *13*. Available from https://doi.org/10.3389/fphys.2022.871149, http://www.frontiersin.org/Physiology/archive/.

Fu, D., Zhang, Q., Fan, Z., Qi, H., Wang, Z., & Peng, L. (2019). Aged microplastics polyvinyl chloride interact with copper and cause oxidative stress towards microalgae Chlorella vulgaris. *Aquatic Toxicology*, *216*. Available from https://doi.org/10.1016/j.aquatox.2019.105319, http://www.elsevier.com/wps/find/journaldescription.cws_home/505509/description#description.

Gao, M., Liu, Y., & Song, Z. (2019). Effects of polyethylene microplastic on the phytotoxicity of di-n-butyl phthalate in lettuce (Lactuca sativa L. var. ramosa Hort). *Chemosphere*, *237*. Available from https://doi.org/10.1016/j.chemosphere.2019.124482, http://www.elsevier.com/locate/chemosphere.

Gutow, L., Eckerlebe, A., Giménez, L., & Saborowski, R. (2016). Experimental evaluation of seaweeds as a vector for microplastics into marine food webs. *Environmental Science and Technology*, *50*(2), 915−923. Available from https://doi.org/10.1021/acs.est.5b02431, http://pubs.acs.org/journal/esthag.

Gül, S., Karaoğlu, K., Özçifçi, Z., Candan, K., Ilgaz, Ç., & Kumlutaş, Y. (2022). Occurrence of microplastics in herpetological museum collection: Grass snake (Natrix natrix [Linnaeus, 1758]) and dice snake (Natrix tessellata [Laurenti, 1769]) as model organisms. *Water, Air, and Soil Pollution*, *233*(5). Available from https://doi.org/10.1007/s11270-022-05626-5, http://www.kluweronline.com/issn/0049-6979/.

Hall, N. M., Berry, K. L. E., Rintoul, L., & Hoogenboom, M. O. (2015). Microplastic ingestion by scleractinian corals. *Marine Biology*, *162*(3), 725−732. Available from https://doi.org/10.1007/s00227-015-2619-7, http://link.springer.de/link/service/journals/00227/index.htm.

Hernandez-Gonzalez, A., Saavedra, C., Gago, J., Covelo, P., Santos, M. B., & Pierce, G. J. (2018). Microplastics in the stomach contents of common dolphin (Delphinus delphis) stranded on the Galician coasts (NW Spain, 2005−2010). *Marine Pollution Bulletin*, *137*, 526−532. Available from https://doi.org/10.1016/j.marpolbul.2018.10.026, http://www.elsevier.com/locate/marpolbul.

Hoang, T., & Mitten, S. (2022). Microplastic accumulation in the gastrointestinal tracts of nestling and adult migratory birds. *Science of the Total Environment*, *838*. Available from https://doi.org/10.1016/j.scitotenv.2022.155827, http://www.elsevier.com/locate/scitotenv.

Hollerova, A., Hodkovicova, N., Blahova, J., Faldyna, M., Franc, A., Pavlokova, S., Tichy, F., Postulkova, E., Mares, J., Medkova, D., Kyllar, M., & Svobodova, Z. (2023). Polystyrene microparticles can affect the health status of freshwater fish − Threat of oral microplastics intake. *Science of the Total Environment*, *858*. Available from https://doi.org/10.1016/j.scitotenv.2022.159976, http://www.elsevier.com/locate/scitotenv.

Hussain, K. A., Romanova, S., Okur, I., Zhang, D., Kuebler, J., Huang, X., Wang, B., Fernandez-Ballester, L., Lu, Y., Schubert, M., & Li, Y. (2023). Assessing the release of microplastics and nanoplastics from plastic containers and reusable food pouches: Implications for human health. *Environmental Science and Technology*, *57*(26), 9782−9792. Available from https://doi.org/10.1021/acs.est.3c01942, http://pubs.acs.org/journal/esthag.

Ibrahim, Y. S., Tuan Anuar, S., Azmi, A. A., Wan Mohd Khalik, W. M. A., Lehata, S., Hamzah, S. R., Ismail, D., Ma, Z. F., Dzulkarnaen, A., Zakaria, Z., Mustaffa, N., Tuan Sharif, S. E., & Lee, Y. Y. (2021). Detection of microplastics in human colectomy specimens. *JGH Open*, *5*(1), 116−121. Available from https://doi.org/10.1002/jgh3.12457, http://onlinelibrary.wiley.com/journal/10.1002/(ISSN)2397-9070.

Isinibilir, M., Svetlichny, L., Mykitchak, T., Türkeri, E. E., Eryalçın, K. M., Doğan, O., Can, G., Yüksel, E., & Kideys, A. E. (2020). Microplastic consumption and its effect on respiration rate and motility of Calanus helgolandicus from the Marmara Sea. *Frontiers in Marine Science*, *7*. Available from https://doi.org/10.3389/fmars.2020.603321, https://www.frontiersin.org/journals/marine-science#.

Jemec Kokalj, A., Horvat, P., Skalar, T., & Kržan, A. (2018). Plastic bag and facial cleanser derived microplastic do not affect feeding behaviour and energy reserves of terrestrial isopods. *Science of the Total Environment*, *615*, 761−766. Available from https://doi.org/10.1016/j.scitotenv.2017.10.020, http://www.elsevier.com/locate/scitotenv.

Khosrovyan, A., Melkonyan, H., Rshtuni, L., Gabrielyan, B., & Kahru, A. (2023). Polylactic acid-based microplastic particles induced oxidative damage in brain and gills of goldfish Carassius auratus. *Water (Switzerland), 15*(11). Available from https://doi.org/10.3390/w15112133, http://www.mdpi.com/journal/water.

Kolandhasamy, P., Su, L., Li, J., Qu, X., Jabeen, K., & Shi, H. (2018). Adherence of microplastics to soft tissue of mussels: A novel way to uptake microplastics beyond ingestion. *Science of the Total Environment, 610–611*, 635–640. Available from https://doi.org/10.1016/j.scitotenv.2017.08.053, http://www.elsevier.com/locate/scitotenv.

Koongolla, J. B., Lin, L., Pan, Y. F., Yang, C. P., Sun, D. R., Liu, S., Xu, X. R., Maharana, D., Huang, J. S., & Li, H. X. (2020). Occurrence of microplastics in gastrointestinal tracts and gills of fish from Beibu Gulf, South China Sea. *Environmental Pollution, 258*. Available from https://doi.org/10.1016/j.envpol.2019.113734, https://www.journals.elsevier.com/environmental-pollution.

Kuhlman, R. L. (2022). Letter to the editor, discovery and quantification of plastic particle pollution in human blood. *Environment International, 167*. Available from https://doi.org/10.1016/j.envint.2022.107400, http://www.elsevier.com/locate/envint.

Kutralam-Muniasamy, G., Shruti, V. C., Pérez-Guevara, F., & Roy, P. D. (2023). Microplastic diagnostics in humans: "The 3Ps" Progress, problems, and prospects. *Science of the Total Environment, 856*. Available from https://doi.org/10.1016/j.scitotenv.2022.159164, http://www.elsevier.com/locate/scitotenv.

Lagarde, F., Olivier, O., Zanella, M., Daniel, P., Hiard, S., & Caruso, A. (2016). Microplastic interactions with freshwater microalgae: Hetero-aggregation and changes in plastic density appear strongly dependent on polymer type. *Environmental Pollution, 215*, 331–339. Available from https://doi.org/10.1016/j.envpol.2016.05.006, http://www.elsevier.com/locate/envpol.

Leslie, H. A., van Velzen, M. J. M., Brandsma, S. H., Vethaak, A. D., Garcia-Vallejo, J. J., & Lamoree, M. H. (2022). Discovery and quantification of plastic particle pollution in human blood. *Environment International, 163*. Available from https://doi.org/10.1016/j.envint.2022.107199, http://www.elsevier.com/locate/envint.

Liu, Y., Guo, R., Zhang, S., Sun, Y., & Wang, F. (2022). Uptake and translocation of nano/microplastics by rice seedlings: Evidence from a hydroponic experiment. *Journal of Hazardous Materials, 421*. Available from https://doi.org/10.1016/j.jhazmat.2021.126700, http://www.elsevier.com/locate/jhazmat.

Li, J., Yu, S., Yu, Y., & Xu, M. (2022). Effects of microplastics on higher plants: A review. *Bulletin of Environmental Contamination and Toxicology, 109*(2), 241–265. Available from https://doi.org/10.1007/s00128-022-03566-8, https://link.springer.com/journal/128.

Lusher, A. L., Hernandez-Milian, G., O'Brien, J., Berrow, S., O'Connor, I., & Officer, R. (2015). Microplastic and macroplastic ingestion by a deep diving, oceanic cetacean: The True's beaked whale Mesoplodon mirus. *Environmental Pollution, 199*, 185–191. Available from https://doi.org/10.1016/j.envpol.2015.01.023, http://www.elsevier.com/inca/publications/store/4/0/5/8/5/6.

Lusher, A. L., McHugh, M., & Thompson, R. C. (2013). Occurrence of microplastics in the gastrointestinal tract of pelagic and demersal fish from the English Channel. *Marine Pollution Bulletin, 67*(1–2), 94–99. Available from https://doi.org/10.1016/j.marpolbul.2012.11.028.

Lusher, A. L., Welden, N. A., Sobral, P., & Cole, M. (2017). Sampling, isolating and identifying microplastics ingested by fish and invertebrates. *Analytical Methods.*, *9*(9), 1346−1360. Available from https://doi.org/10.1039/C6AY02415G.

López, M. D., Toro, M. T., Riveros, G., Illanes, M., Noriega, F., Schoebitz, M., García-Viguera, C., & Moreno, D. A. (2022). Brassica sprouts exposed to microplastics: Effects on phytochemical constituents. *Science of the Total Environment, 823.* Available from https://doi.org/10.1016/j.scitotenv.2022.153796, http://www.elsevier.com/locate/scitotenv.

Maity, S., Chatterjee, A., Guchhait, R., De, S., & Pramanick, K. (2020). Cytogenotoxic potential of a hazardous material, polystyrene microparticles on Allium cepa L. *Journal of Hazardous Materials, 385.* Available from https://doi.org/10.1016/j.jhazmat.2019.121560, http://www.elsevier.com/locate/jhazmat.

Malafaia, G. (2023). A commentary on the paper "identification of microplastics in human placenta using laser direct infrared spectroscopy": Reflections on identification and typing of microplastics in human biological samples. *Science of the Total Environment, 875.* Available from https://doi.org/10.1016/j.scitotenv.2023.162650, http://www.elsevier.com/locate/scitotenv.

De Marchi, L., Renzi, M., Anselmi, S., Pretti, C., Guazzelli, E., Martinelli, E., Cuccaro, A., Oliva, M., Magri, M., & Bulleri, F. (2022). Polyethylene microplastics reduce filtration and respiration rates in the Mediterranean sponge Petrosia ficiformis. *Environmental Research, 211.* Available from https://doi.org/10.1016/j.envres.2022.113094, http://www.elsevier.com/inca/publications/store/6/2/2/8/2/1/index.htt.

Molin, J. M., Groth-Andersen, W. E., Hansen, P. J., Kühl, M., & Brodersen, K. E. (2023). Microplastic pollution associated with reduced respiration in seagrass (Zostera marina L.) and associated epiphytes. *Frontiers in Marine Science, 10.* Available from https://doi.org/10.3389/fmars.2023.1216299, https://www.frontiersin.org/journals/marine-science#.

Monclús, L., McCann Smith, E., Ciesielski, T. M., Wagner, M., & Jaspers, V. L. B. (2022). Microplastic ingestion induces size-specific effects in Japanese Quail. *Environmental Science and Technology, 56*(22), 15902−15911. Available from https://doi.org/10.1021/acs.est.2c03878, http://pubs.acs.org/journal/esthag.

Murphy, F., & Quinn, B. (2018). The effects of microplastic on freshwater Hydra attenuata feeding, morphology & reproduction. *Environmental Pollution, 234,* 487−494. Available from https://doi.org/10.1016/j.envpol.2017.11.029, http://www.elsevier.com/inca/publications/store/4/0/5/8/5/6.

Nava, V., & Leoni, B. (2021). A critical review of interactions between microplastics, microalgae and aquatic ecosystem function. *Water Research, 188.* Available from https://doi.org/10.1016/j.watres.2020.116476, http://www.elsevier.com/locate/watres.

Nelms, S. E., Galloway, T. S., Godley, B. J., Jarvis, D. S., & Lindeque, P. K. (2018). Investigating microplastic trophic transfer in marine top predators. *Environmental Pollution, 238,* 999−1007. Available from https://doi.org/10.1016/j.envpol.2018.02.016, http://www.elsevier.com/inca/publications/store/4/0/5/8/5/6.

Neves, D., Sobral, P., Ferreira, J. L., & Pereira, T. (2015). Ingestion of microplastics by commercial fish off the Portuguese coast. *Marine Pollution Bulletin, 101*(1), 119−126. Available from https://doi.org/10.1016/j.marpolbul.2015.11.008, http://www.elsevier.com/locate/marpolbul.

Ogonowski, M., Gerdes, Z., & Gorokhova, E. (2018). What we know and what we think we know about microplastic effects − A critical perspective. *Current Opinion in*

Environmental Science and Health, *1*, 41−46. Available from https://doi.org/10.1016/j. coesh.2017.09.001, http://www.journals.elsevier.com/current-opinion-in-environmental-science-and-health.

Park, E. J., Han, J. S., Park, E. J., Seong, E., Lee, G. H., Kim, D. W., Son, H. Y., Han, H. Y., & Lee, B. S. (2020). Repeated-oral dose toxicity of polyethylene microplastics and the possible implications on reproduction and development of the next generation. *Toxicology Letters*, *324*, 75−85. Available from https://doi.org/10.1016/j.toxlet.2020.01.008, http://www.elsevier.com/locate/toxlet.

De Pascalis, F., De Felice, B., Parolini, M., Pisu, D., Pala, D., Antonioli, D., Perin, E., Gianotti, V., Ilahiane, L., Masoero, G., Serra, L., Rubolini, D., & Cecere, J. G. (2022). The hidden cost of following currents: Microplastic ingestion in a planktivorous seabird. *Marine Pollution Bulletin*, *182*. Available from https://doi.org/10.1016/j.marpolbul.2022.114030, http://www.elsevier.com/locate/marpolbul.

Peller, J., Nevers, M. B., Byappanahalli, M., Nelson, C., Ganesh Babu, B., Evans, M. A., Kostelnik, E., Keller, M., Johnston, J., & Shidler, S. (2021). Sequestration of microfibers and other microplastics by green algae, Cladophora, in the US Great Lakes. *Environmental Pollution*, *276*. Available from https://doi.org/10.1016/j. envpol.2021.116695, https://www.journals.elsevier.com/environmental-pollution.

Perez-Venegas, D. J., Toro-Valdivieso, C., Ayala, F., Brito, B., Iturra, L., Arriagada, M., Seguel, M., Barrios, C., Sepúlveda, M., Oliva, D., Cárdenas-Alayza, S., Urbina, M. A., Jorquera, A., Castro-Nallar, E., & Galbán-Malagón, C. (2020). Monitoring the occurrence of microplastic ingestion in Otariids along the Peruvian and Chilean coasts. *Marine Pollution Bulletin*, *153*. Available from https://doi.org/10.1016/j.marpolbul.2020.110966, http://www.elsevier.com/locate/marpolbul.

Pinho, I., Amezcua, F., Rivera, J. M., Green-Ruiz, C., Piñón-Colin, Td. J., & Wakida, F. (2022). First report of plastic contamination in batoids: Plastic ingestion by Haller's Round Ray (Urobatis halleri) in the Gulf of California. *Environmental Research*, *211*. Available from https://doi.org/10.1016/j.envres.2022.113077, http://www.elsevier.com/inca/publications/store/6/2/2/8/2/1/index.htt.

Pitt, J. A., Kozal, J. S., Jayasundara, N., Massarsky, A., Trevisan, R., Geitner, N., Wiesner, M., Levin, E. D., & Di Giulio, R. T. (2018). Uptake, tissue distribution, and toxicity of polystyrene nanoparticles in developing zebrafish (Danio rerio). *Aquatic Toxicology*, *194*, 185−194. Available from https://doi.org/10.1016/j.aquatox.2017.11.017, http://www.elsevier.com/wps/find/journaldescription.cws_home/505509/description#description.

Porter, A., Barber, D., Hobbs, C., Love, J., Power, A. L., Bakir, A., Galloway, T. S., & Lewis, C. (2023). Uptake of microplastics by marine worms depends on feeding mode and particle shape but not exposure time. *Science of the Total Environment*, *857*. Available from https://doi.org/10.1016/j.scitotenv.2022.159287, http://www.elsevier.com/locate/scitotenv.

Qiang, L., & Cheng, J. (2021). Exposure to polystyrene microplastics impairs gonads of zebrafish (Danio rerio). *Chemosphere*, *263*. Available from https://doi.org/10.1016/j. chemosphere.2020.128161, http://www.elsevier.com/locate/chemosphere.

Qi, Y., Yang, X., Pelaez, A. M., Huerta Lwanga, E., Beriot, N., Gertsen, H., Garbeva, P., & Geissen, V. (2018). Macro- and micro-plastics in soil-plant system: Effects of plastic mulch film residues on wheat (Triticum aestivum) growth. *Science of the Total Environment*, *645*, 1048−1056. Available from https://doi.org/10.1016/j.scitotenv.2018.07.229, http://www.elsevier.com/locate/scitotenv.

Ragusa, A., Svelato, A., Santacroce, C., Catalano, P., Notarstefano, V., Carnevali, O., Papa, F., Rongioletti, M. C. A., Baiocco, F., Draghi, S., D'Amore, E., Rinaldo, D., Matta, M., & Giorgini, E. (2021). Plasticenta: First evidence of microplastics in human placenta. *Environment International*, *146*. Available from https://doi.org/10.1016/j.envint.2020.106274, http://www.elsevier.com/locate/envint.

Reynolds, C., & Ryan, P. G. (2018). Micro-plastic ingestion by waterbirds from contaminated wetlands in South Africa. *Marine Pollution Bulletin*, *126*, 330–333. Available from https://doi.org/10.1016/j.marpolbul.2017.11.021, http://www.elsevier.com/locate/marpolbul.

Rist, S., Hartmann, N. B., & Welden, N. A. C. (2021). How fast, how far: Diversification and adoption of novel methods in aquatic microplastic monitoring. *Environmental Pollution*, *291*, 118174. Available from https://doi.org/10.1016/j.envpol.2021.118174.

Rivers-Auty, J., Bond, A. L., Grant, M. L., & Lavers, J. L. (2023). The one-two punch of plastic exposure: Macro- and micro-plastics induce multi-organ damage in seabirds. *Journal of Hazardous Materials*, *442*. Available from https://doi.org/10.1016/j.jhazmat.2022.130117, http://www.elsevier.com/locate/jhazmat.

Savoca, M. S., Wohlfeil, M. E., Ebeler, S. E., & Nevitt, G. A. (2016). Marine plastic debris emits a keystone infochemical for olfactory foraging seabirds. *Science Advances*, *2*(11), e1600395. Available from https://doi.org/10.1126/sciadv.1600395.

Scherer, C., Brennholt, N., Reifferscheid, G., & Wagner, M. (2017). Feeding type and development drive the ingestion of microplastics by freshwater invertebrates. *Scientific Reports*, *7*(1). Available from https://doi.org/10.1038/s41598-017-17191-7, http://www.nature.com/srep/index.html.

Schwabl, P., Köppel, S., Königshofer, P., Bucsics, T., Trauner, M., Reiberger, T., & Liebmann, B. (2019). Detection of various microplastics in human stool. *Annals of Internal Medicine*, *171*(7), 453–457. Available from https://doi.org/10.7326/m19-0618.

Seeley, M. E., Hale, R. C., Zwollo, P., Vogelbein, W., Verry, G., & Wargo, A. R. (2023). Microplastics exacerbate virus-mediated mortality in fish. *Science of the Total Environment*, *866*. Available from https://doi.org/10.1016/j.scitotenv.2022.161191, http://www.elsevier.com/locate/scitotenv.

Sendra, M., Carrasco-Braganza, M. I., Yeste, P. M., Vila, M., & Blasco, J. (2020). Immunotoxicity of polystyrene nanoplastics in different hemocyte subpopulations of Mytilus galloprovincialis. *Scientific Reports*, *10*(1). Available from https://doi.org/10.1038/s41598-020-65596-8, http://www.nature.com/srep/index.html.

Seoane, M., González-Fernández, C., Soudant, P., Huvet, A., Esperanza, M., Cid, Á., & Paul-Pont, I. (2019). Polystyrene microbeads modulate the energy metabolism of the marine diatom Chaetoceros neogracile. *Environmental Pollution*, *251*, 363–371. Available from https://doi.org/10.1016/j.envpol.2019.04.142, https://www.journals.elsevier.com/environmental-pollution.

Seuront, L. (2018). Microplastic leachates impair behavioural vigilance and predator avoidance in a temperate intertidal gastropod. *Biology Letters*, *14*(11). Available from https://doi.org/10.1098/rsbl.2018.0453, http://rsbl.royalsocietypublishing.org/.

Shahsavaripour, M., Abbasi, S., Mirzaee, M., & Amiri, H. (2023). Human occupational exposure to microplastics: A cross-sectional study in a plastic products manufacturing plant. *Science of the Total Environment*, *882*. Available from https://doi.org/10.1016/j.scitotenv.2023.163576, http://www.elsevier.com/locate/scitotenv.

Shen, L., Zhang, P., Lin, Y., Huang, X., Zhang, S., Li, Z., Fang, Z., Wen, Y., & Liu, H. (2023). Polystyrene microplastic attenuated the toxic effects of florfenicol on rice (Oryza sativa L.) seedlings in hydroponics: From the perspective of oxidative response, phototoxicity and molecular metabolism. *Journal of Hazardous Materials, 459.* Available from https://doi.org/10.1016/j.jhazmat.2023.132176, http://www.elsevier.com/locate/jhazmat.

Sjollema, S. B., Redondo-Hasselerharm, P., Leslie, H. A., Kraak, M. H. S., & Vethaak, A. D. (2016). Do plastic particles affect microalgal photosynthesis and growth? *Aquatic Toxicology, 170,* 259–261. Available from https://doi.org/10.1016/j.aquatox.2015.12.002, http://www.elsevier.com/wps/find/journaldescription.cws_home/505509/description#description.

Song, Y., Cao, C., Qiu, R., Hu, J., Liu, M., Lu, S., Shi, H., Raley-Susman, K. M., & He, D. (2019). Uptake and adverse effects of polyethylene terephthalate microplastics fibers on terrestrial snails (Achatina fulica) after soil exposure. *Environmental Pollution, 250,* 447–455. Available from https://doi.org/10.1016/j.envpol.2019.04.066, https://www.journals.elsevier.com/environmental-pollution.

Sucharitakul, P., Pitt, K. A., & Welsh, D. T. (2020). Limited ingestion, rapid egestion and no detectable impacts of microbeads on the moon jellyfish, Aurelia aurita. *Marine Pollution Bulletin, 156.* Available from https://doi.org/10.1016/j.marpolbul.2020.111208, http://www.elsevier.com/locate/marpolbul.

Sundbæk, K. B., Koch, I. D. W., Villaro, C. G., Rasmussen, N. S., Holdt, S. L., & Hartmann, N. B. (2018). Sorption of fluorescent polystyrene microplastic particles to edible seaweed Fucus vesiculosus. *Journal of Applied Phycology, 30*(5), 2923–2927. Available from https://doi.org/10.1007/s10811-018-1472-8, http://www.wkap.nl/journalhome.htm/0921-8971.

Sun, R., Xu, K., Yu, L., Pu, Y., Xiong, F., He, Y., Huang, Q., Tang, M., Chen, M., Yin, L., Zhang, J., & Pu, Y. (2021). Preliminary study on impacts of polystyrene microplastics on the hematological system and gene expression in bone marrow cells of mice. *Ecotoxicology and Environmental Safety, 218.* Available from https://doi.org/10.1016/j.ecoenv.2021.112296, http://www.elsevier.com/inca/publications/store/6/2/2/8/1/9/index.htt.

Sussarellu, R., Suquet, M., Thomas, Y., Lambert, C., Fabioux, C., Pernet, M. E. J., Goïc, N. L., Quillien, V., Mingant, C., Epelboin, Y., Corporeau, C., Guyomarch, J., Robbens, J., Paul-Pont, I., Soudant, P., & Huvet, A. (2016). Oyster reproduction is affected by exposure to polystyrene microplastics. *Proceedings of the National Academy of Sciences of the United States of America, 113*(9), 2430–2435. Available from https://doi.org/10.1073/pnas.1519019113, http://www.pnas.org/content/113/9/2430.full.pdf.

Terepocki, A. K., Brush, A. T., Kleine, L. U., Shugart, G. W., & Hodum, P. (2017). Size and dynamics of microplastic in gastrointestinal tracts of Northern Fulmars (Fulmarus glacialis) and Sooty Shearwaters (Ardenna grisea). *Marine Pollution Bulletin, 116* (1–2), 143–150. Available from https://doi.org/10.1016/j.marpolbul.2016.12.064, http://www.elsevier.com/locate/marpolbul.

Thrift, E., Porter, A., Galloway, T. S., Coomber, F. G., & Mathews, F. (2022). Ingestion of plastics by terrestrial small mammals. *Science of the Total Environment, 842.* Available from https://doi.org/10.1016/j.scitotenv.2022.156679, http://www.elsevier.com/locate/scitotenv.

Tokunaga, Y., Okochi, H., Tani, Y., Niida, Y., Tachibana, T., Saigawa, K., Katayama, K., Moriguchi, S., Kato, T., & Hayama, S.-I. (2023). Airborne microplastics detected in the lungs of wild birds in Japan. *Chemosphere, 321,* 138032. Available from https://doi.org/10.1016/j.chemosphere.2023.138032.

Urbina, M. A., Correa, F., Aburto, F., & Ferrio, J. P. (2020). Adsorption of polyethylene microbeads and physiological effects on hydroponic maize. *Science of the Total Environment, 741.* Available from https://doi.org/10.1016/j.scitotenv.2020.140216, http://www.elsevier.com/locate/scitotenv.

Valente, T., Sbrana, A., Scacco, U., Jacomini, C., Bianchi, J., Palazzo, L., de Lucia, G. A., Silvestri, C., & Matiddi, M. (2019). Exploring microplastic ingestion by three deep-water elasmobranch species: A case study from the Tyrrhenian Sea. *Environmental Pollution*, *253*, 342−350. Available from https://doi.org/10.1016/j.envpol.2019.07.001, https://www.journals.elsevier.com/environmental-pollution.

Wang, W., Do, A. T. N., & Kwon, J. H. (2022). Ecotoxicological effects of micro- and nanoplastics on terrestrial food web from plants to human beings. *Science of the Total Environment*, *834*. Available from https://doi.org/10.1016/j.scitotenv.2022.155333, http://www.elsevier.com/locate/scitotenv.

Wang, Y., Mao, Z., Zhang, M., Ding, G., Sun, J., Du, M., Liu, Q., Cong, Y., Jin, F., Zhang, W., & Wang, J. (2019). The uptake and elimination of polystyrene microplastics by the brine shrimp, Artemia parthenogenetica, and its impact on its feeding behavior and intestinal histology. *Chemosphere*, *234*, 123−131. Available from https://doi.org/10.1016/j.chemosphere.2019.05.267, http://www.elsevier.com/locate/chemosphere.

Wang, T., Tong, C., Wu, F., Jiang, S., & Zhang, S. (2023). Distribution characteristics of microplastics and corresponding feeding habits of the dominant shrimps in the rivers of Chongming Island. *Science of the Total Environment*, *888*. Available from https://doi.org/10.1016/j.scitotenv.2023.164041, http://www.elsevier.com/locate/scitotenv.

Ward, J. E., & Kach, D. J. (2009). Marine aggregates facilitate ingestion of nanoparticles by suspension-feeding bivalves. *Marine Environmental Research*, *68*(3), 137−142. Available from https://doi.org/10.1016/j.marenvres.2009.05.002.

Watts, A. J. R., Urbina, M. A., Goodhead, R., Moger, J., Lewis, C., & Galloway, T. S. (2016). Effect of microplastic on the gills of the shore crab Carcinus maenas. *Environmental Science and Technology*, *50*(10), 5364−5369. Available from https://doi.org/10.1021/acs.est.6b01187, http://pubs.acs.org/journal/esthag.

Wegner, A., Besseling, E., Foekema, E. M., Kamermans, P., & Koelmans, A. A. (2012). Effects of nanopolystyrene on the feeding behavior of the blue mussel (Mytilus edulis L. *Environmental Toxicology and Chemistry*, *31*(11), 2490−2497. Available from https://doi.org/10.1002/etc.1984.

Weitzel, S. L., Feura, J. M., Rush, S. A., Iglay, R. B., & Woodrey, M. S. (2021). Availability and assessment of microplastic ingestion by marsh birds in Mississippi Gulf Coast tidal marshes. *Marine Pollution Bulletin*, *166*. Available from https://doi.org/10.1016/j.marpolbul.2021.112187, http://www.elsevier.com/locate/marpolbul.

Welden, N. A. C., & Cowie, P. R. (2016a). Long-term microplastic retention causes reduced body condition in the langoustine, Nephrops norvegicus. *Environmental Pollution*, *218*, 895−900. Available from https://doi.org/10.1016/j.envpol.2016.08.020, http://www.elsevier.com/locate/envpol.

Welden, N. A. C., & Cowie, P. R. (2016b). Environment and gut morphology influence microplastic retention in langoustine, Nephrops norvegicus. *Environmental Pollution*, *214*, 859−865. Available from https://doi.org/10.1016/j.envpol.2016.03.067, http://www.elsevier.com/inca/publications/store/4/0/5/8/5/6.

Wibowo, A. T., Nugrahapraja, H., Wahyuono, R. A., Islami, I., Haekal, M. H., Fardiansyah, Y., Sugiyo, P. W. W., Putro, Y. K., Fauzia, F. N., Santoso, H., Götz, F., Tangahu, B. V., & Luqman, A. (2021). Microplastic contamination in the human gastrointestinal tract and daily consumables associated with an indonesian farming community. *Sustainability (Switzerland)*, *13*(22). Available from https://doi.org/10.3390/su132212840, https://www.mdpi.com/2071-1050/13/22/12840/pdf.

Wright, S. L., Rowe, D., Thompson, R. C., & Galloway, T. S. (2013). Microplastic ingestion decreases energy reserves in marine worms. *Current Biology*, *23*(23), R1031−R1033. Available from https://doi.org/10.1016/j.cub.2013.10.068, http://www.elsevier.com/journals/current-biology/0960-9822.

Wu, B., Wu, X., Liu, S., Wang, Z., & Chen, L. (2019). Size-dependent effects of polystyrene microplastics on cytotoxicity and efflux pump inhibition in human Caco-2 cells. *Chemosphere*, *221*, 333−341. Available from https://doi.org/10.1016/j.chemosphere.2019.01.056, http://www.elsevier.com/locate/chemosphere.

Xie, M., Lin, L., Xu, P., Zhou, W., Zhao, C., Ding, D., & Suo, A. (2021). Effects of microplastic fibers on Lates calcarifer juveniles: Accumulation, oxidative stress, intestine microbiome dysbiosis and histological damage. *Ecological Indicators*, *133*. Available from https://doi.org/10.1016/j.ecolind.2021.108370, http://www.elsevier.com/locate/ecolind.

Xu, M., Halimu, G., Zhang, Q., Song, Y., Fu, X., Li, Y., Li, Y., & Zhang, H. (2019). Internalization and toxicity: A preliminary study of effects of nanoplastic particles on human lung epithelial cell. *Science of the Total Environment*, *694*, 133794. Available from https://doi.org/10.1016/j.scitotenv.2019.133794.

Yadav, H., Khan, M. R. H., Quadir, M., Rusch, K. A., Mondal, P. P., Orr, M., Xu, E. G., & Iskander, S. M. (2023). Cutting boards: An overlooked source of microplastics in human food? *Environmental Science and Technology*, *57*(22), 8225−8235. Available from https://doi.org/10.1021/acs.est.3c00924, http://pubs.acs.org/journal/esthag.

Yin, K., Wang, Y., Zhao, H., Wang, D., Guo, M., Mu, M., Liu, Y., Nie, X., Li, B., Li, J., & Xing, M. (2021). A comparative review of microplastics and nanoplastics: Toxicity hazards on digestive, reproductive and nervous system. *Science of the Total Environment*, *774*. Available from https://doi.org/10.1016/j.scitotenv.2021.145758, http://www.elsevier.com/locate/scitotenv.

Yokota, K., Waterfield, H., Hastings, C., Davidson, E., Kwietniewski, E., & Wells, B. (2017). Finding the missing piece of the aquatic plastic pollution puzzle: Interaction between primary producers and microplastics. *Limnology and Oceanography Letters*, *2* (4), 91−104. Available from https://doi.org/10.1002/lol2.10040, https://aslopubs.onlinelibrary.wiley.com/loi/23782242.

Young, L. C., Vanderlip, C., Duffy, D. C., Afanasyev, V., & Shaffer, S. A. (2009). Bringing home the trash: Do colony-based differences in foraging distribution lead to increased plastic ingestion in Laysan albatrosses? *PLoS One*, *4*(10). Available from http://www.plosone.org/article/fetchObjectAttachment.action?uri = info%3Adoi%2F10.1371%2Fjournal.pone.0007623&representation = PDF, https://doi.org/10.1371/journal.pone.0007623, United States.

Zhang, N., Li, Y. B., He, H. R., Zhang, J. F., & Ma, G. S. (2021). You are what you eat: Microplastics in the feces of young men living in Beijing. *Science of the Total Environment*, *767*. Available from https://doi.org/10.1016/j.scitotenv.2020.144345, http://www.elsevier.com/locate/scitotenv.

Zhao, Q., Zhu, L., Weng, J., Jin, Z., Cao, Y., Jiang, H., & Zhang, Z. (2023). Detection and characterization of microplastics in the human testis and semen. *Science of the Total Environment*, *877*. Available from https://doi.org/10.1016/j.scitotenv.2023.162713, http://www.elsevier.com/locate/scitotenv.

Zhu, L., Zhu, J., Zuo, R., Xu, Q., Qian, Y., & An, L. (2023). Identification of microplastics in human placenta using laser direct infrared spectroscopy. *The Science of the Total Environment*, *856*(Pt 1), 159060. Available from https://doi.org/10.1016/j.scitotenv.2022.159060.

Ecosystem-level effects

5

5.1 Introduction

While direct interaction with microplastics, their additives and secondary contaminants may have a range of acute and chronic effects on biota, there may also be a range of indirect and secondary effects on populations, communities and the wider environment. These impacts range from alteration of the abiotic properties of substrates, to secondary effects on population health and community dynamics. Such interactions broaden the potential sphere of influence of plastics. As with impacts of microplastics on the individual, population- and ecosystem-level effects are greatly influenced by the size of the plastic particle, which dictates whether it may be ingested, egested and taken up into the tissues, or translocate to various organs of structures. As in the previous chapter, many of the studies we will consider below used plastics across the micro-nanoplastic boundaries.

5.2 Abiotic changes

We have already addressed the unique properties of plastics and their polymers and how their physical and mechanical properties differ from natural materials earlier in this text, thus it may be expected that their widespread distribution and increasing abundance may affect many key environmental processes. Many of the proposed impacts of microplastics on abiotic conditions are associated with their incorporation into sediments and soils. First, the size of microplastics may alter the granulometry (grain size distribution of sediments). The distribution of grain sizes in a sediment can have significant impacts on the conditions within the sediment itself. For example, the tessellation of individual particles within a sediment, in addition to their comparative angularity, will affect the presence and size of pores. Pore size is a key factor in the movement of water and (in intertidal areas) air. Sediments with larger particles typically have larger interstitial spaces (the gaps between sediment grains), which allow an increased area and movement of both liquids and gasses. This movement subsequently influences the formation of sediment zones, such as anoxic regions. The introduction of plastics with particle sizes larger or smaller than the local sediment range may thus influence the relative permeability of sediments (Carson et al., 2011). In soils, the addition of microplastic fibres, beads or fragments has been seen to reduce soil bulk density and increase water saturation and evapotranspiration, as well as somewhat altering microbial activity. However, the scale of these affects are variable, microplastics with similar morphologies to soil particles showed lower impacts on soil health than those of a differing morphology (De Souza Machado et al., 2019). Similar changes to the properties of soil were observed in pot experiments of plant communities of seven species. Here microfibre soil amendments were seen to result in an increased mass of shoots and roots, while preferentially benefiting some species, with implications for ecosystem function (Lozano & Rillig, 2020).

Microplastics may also have significantly different thermal conductivity than that of the surrounding sediments. Plastics typically have lower thermal conductivity (below 0.52) than natural sediments, particularly when wet. Table 5.1 indicates the common conductivity values for common plastics and natural sediments. Here, the substrates (polymer and sediment) have been ranked from lowest to highest conductivity to indicate potential interactions between the two. However, the difference in thermal capacity between wet and dry sediments suggest that the difference in heat transfer may be highest in recently exposed sediments in the intertidal zone, which have not yet had the opportunity to dry out (Hamdhan & Clarke, 2010). Humidity may also play a notable effect in these cases, with the thermal conductivity of many soils ("sands," "loamy sand" and "loams and clays") seen to increase with rising

Table 5.1 The thermal properties of plastics in relation to those of environmental media.

Plastic type	Thermal conductivity (W/mK)
Polystyrene	0.1–0.13
Polypropylene	0.1–0.22
Polyvinylchloride	0.12–0.25
Acrylonitrile-butadiene-styrene	0.14–0.21
Fine sand (dry)	*0.15*
Polyethylene terephthalate	0.15–0.4
Polymethylmethacrylate	0.17–0.19
Polycarbonate	0.19–0.22
Polyamide	0.24–0.3
Coarse sand (dry)	*0.25*
Medium sand (dry)	*0.27*
Polyurethane	0.29
Polyethylene L	0.33
Polyethylene HD	0.45–0.52
Fine sand (saturated)	*2.75*
Medium sand (saturated)	*3.34*
Coarse sand (saturated)	*3.72*
Silt	*4.26*
Gravel	*4.44*

humidity (Nikiforova et al., 2013). Observations of increasing microplastic concentrations in beach sand have revealed increased insulating properties as the percentage microplastic concentration increases. In this case, microplastic contaminated sediment may be characterised as slower warming, with a lower maximum heat capacity (Carson et al., 2011).

In addition, to their impacts in sediments and soils, microplastics have the potential to influence carbon sequestration, particularly in marine systems. Recent observations of microplastics' effects on phytoplankton and algal growth, as well as on zooplankton health and diversity, suggest that there may be a subsequent alteration in the marine biological pump. This pump drives the movement of carbon between the sea surface and deep-sea sediments as a result of carbon dioxide and water uptake during photosynthesis, the use of the resulting energy stores to drive metabolism, growth and reproduction, and the subsequent mortality and sinking of biomass, thus influencing carbon storage (Shen et al., 2020). Meta-analysis of the impacts of plastic debris on planktonic communities has highlighted the potential effects of micro- and nanoplastics on marine autotrophs such as microalgae. Impacts on these species will indirectly affect both carbon drawdown. Phytoplankton may then either be eaten by primary consumers or die, sinking to the seafloor, where the carbon remaining in their bodies is stored (Casabianca et al., 2021). The same review goes on to highlight the dearth of information regarding key areas in the assessment of ecosystem and biome scale impacts, such as transfer and impacts within food webs. However, these potential effects are difficult to predict using presently available data.

5.3 **Population-level effects**

Population-level effects of microplastics may be the result of many factors, for example small populations are more susceptible to the negative effects of environmental change due to the reduced genetic diversity shared by the small pool of individual organisms. Where plastics increase local mortality via reduced individual fitness, the remaining population may be unable to adapt to changing conditions or may experience genetic drift (a change in the frequency of genes, potentially resulting in either the loss of desirable traits or an increase in undesirable traits), increasing the likelihood of localised extinctions (Fig. 5.1). Using experimental data, the impacts of microplastic on populations of copepods, *Temora longicornis,* were determined. It was found that the concentration required to reduce population size by half was 593 ± 376 particles/L. Importantly, while the apparently impactful modelled concentrations were approximately four times less than the effective concentrations at the individual level, these are still much higher than those already experienced in the environment (Everaert et al., 2022). Where populations are already at increased extinction risk, these may experience a greater threat as a result of interactions with microplastic. Such an impact has already been suggested in relation to the floating

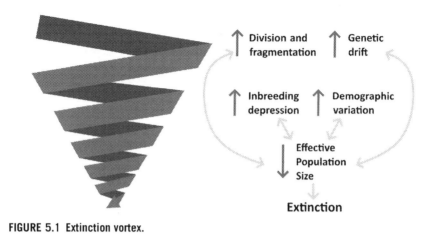

FIGURE 5.1 Extinction vortex.

The interacting factors behind population decline and extinctions.

antlerfern, *Ceratopteris pteridoides*, an aquatic species native in China and Vietnam as well as the tropical and subtropical Americas. However, its range is greatly reduced in China due to anthropogenic activity. Observations on the effects of PS nanoparticles on *C. pteridoides* show that adsorption and accumulation on the spore surface reduced water uptake (imbibition), reducing spore size by as much as 23% and germination by up to 88%, suggesting a risk to reproductive capacity in an already endangered species (Yuan et al., 2019).

However, microplastic may not affect a population evenly, having more significant impacts on key demographics. For example, Welden and Cowie (2016) indicate high plastic aggregation in small langoustine, who appear less able to egest plastics. Conversely, comparisons between plastic exposed (10^2 to 10^5 particles/mL) and control populations of *Daphnia magna* stocked at environmental carrying capacity demonstrate significant reductions in population of up to 21%, a result of the decline of adults rather than eggs (Bosker et al., 2019), although the observed impacts of plastics on *Daphnia* are complex and sometimes contradictory, as we will see below.

Where sublethal effects on the individual are observed, these may influence population viability by decreasing fecundity (the number of offspring per individual) or by bringing about intergenerational effects. Microplastic exposure may limit fecundity by reducing the fitness and body condition in the adult. Transgenerational impacts of microplastics in *D. magna* were explored over four generations exposed to fluorescently label microspheres at 0.1 mg/L in addition to the plastic-free control, experimental condition included populations fed plastic for two successive generations, and populations fed plastic at the first generation (F_0) and observed over a further three. The impact of plastics on individual growth, mortality, reproduction and population growth was determined. Microplastic exposure was seen to result in 10% mortality at the first generation (F_0), additionally, where offspring were successively fed microplastics, growth and brood number were reduced, days until first brood was increased and mortality was between 20% and 100% (Fig. 5.2). In the group where only the first generation was exposed to plastic, there was no mortality at subsequent generations, apparent impacts on growth and brood number were observed in the second generation (F_1), and reduced brood number in the third generation (F_2) (Martins & Guilhermino, 2018). Conversely, a closely related species, *Daphia pulex*, fed nanoplastics and 1 µg/L demonstrated no additional mortality; however, growth and reproduction were affected at the third generation, and recovery in populations where microplastics are removed may take several generations (Liu et al., 2020). Nevetherless, Heinlaan et al. (2023) saw no significant negative effects in *D. magna* exposed to nanoplastics at 0.1 and 1 mg/L over 21 days. The only significant impact being an apparent increase in fertility in the fourth generation (F_3).

In addition to the effects of microplastics on female fitness and the number and health of offspring, exposure of one or both parents may result in altered epigenetic inheritance (changes in the ways genes work or are expressed between generations). In this case, changes may be equally observed via the paternal inheritance line, driven by, for example, oxidative stress in the testes or chromosomal assembly abnormality (Sun et al., 2023). Reduction in breeding success may also be the result of germline apoptosis, the death of cells containing genetic material, as observed in the roundworm, *Caenorhabditis elegans*. Here, parental exposure to PS nanoplastics and amino acid-modified PS nanoplastics was linked to damage during gonad development in the F_1 and F_2 generations (Sun et al., 2021). A similar study of parental exposure to nanoplastics in *C. elegans* has also indicated reduced brood size in the subsequent four generations, in addition to chromosomal aberrations and germline cell apoptosis in the parental generation (Yu et al., 2021).

In addition, plastics may be passed from one generation to another. Parental exposure to PS nanoplastic in the model organism, *Danio rerio*, has been seen to result in the presence of plastics in both embryos and yolk sac. This transfer was

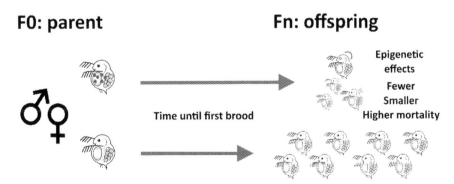

FO: parent   **Fn: offspring**

Time until first brood

Epigenetic effects
Fewer
Smaller
Higher mortality

FIGURE 5.2 Intergenerational effects of plastic exposure.

Impacts on the offspring of *Daphnia magna*—fed plastics.

apparent in maternal (only female exposed) exposure and coparent (both male and female exposed) exposures, but not in paternal (only males exposed) groups. Embryos in these groups also demonstrated bradycardia, a lower than normal heartrate (Pitt et al., 2018). In addition, exposure of *D. rerio* adults to nanoplastics in combination with Bisphenol AF (BPAF) has been seen to cause reduced brood size and lowered hatching rates, with these effects higher in the plastic-mediated group than those exposed to BPAF alone (Wang et al., 2023).

Nevertheless, as with individual-level effects, population-scale impacts are not always observed. For example, in the marine microalgae *Tetraselmis chuii*, experiments examined the impact of microplastic exposure on population growth indicate no significant effects of MP on the specific average of growth rate of *T. chuii* were found up to a concentration of 1.472 mg/L (Davarpanah & Guilhermino, 2015).

5.3.1 Microplastics in the food chain

Since our earliest observations of microplastic ingestion, retention, and bioaccumulation, concerns have been raised regarding the movement of microplastics within the food chain, primarily as a result of active predation, but also the action of detritivores (Fig. 5.3). This movement may take the form of simple trophic transfer, by which the ingestion of microplastic contaminated prey leads to the uptake of microplastics, or biomagnification, during which the ingested plastic load from multiple prey animals results in an increasing microplastic burden at successive stages of the food chain. Additionally, adherence of plastics to algae and animals suggests that the consumed prey need not itself have ingested microplastic for it to be passed to a predator or scavenger (Au et al., 2017).

Trophic transfer, the movement of material from prey to predator, has previously been observed in captive seals fed on wild-caught fish. A subsample of the wild-caught herring was dissected in order to establish the presence and abundance of microplastics within the wider sample. Subsequently, analysis of scat samples from seals fed the remaining herring revealed the presence of microplastics (Nelms et al., 2018). We may also identify trophic transfer in action. For example, in wild-caught fish species. Dissection of whole, sand eels, *Ammodytes tobianus*, found in the stomach content of plaice, *Pleuronectes plastessa*, revealed the presence of microplastics in both predator and prey (Welden et al., 2018).

Less understood is the potential for the biomagnification of microplastics. Biomagnification is the process by which material is retained by each successive stage of the trophic web. This differs from bioaccumulation discussed in Chapter 4, which considered the increase in contamination levels in a single individual over time Fig. 5.4. Previous notorious examples of bioaccumulation include that of DDT in North America, and Mercury poisoning in Minamata Japan. Were biomagnification a regular occurrence in the environment, it would be reasonable to expect to routinely observe elevated abundance of microplastics in top predators. Conversely, it has been observed that smaller organisms and those at the lower levels of the food chain typically demonstrate increased

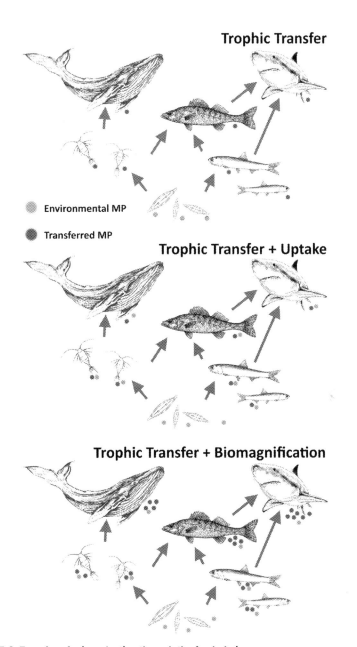

FIGURE 5.3 Transfer of microplastics through the food chain.

Differences in plastic uptake via environmental uptake, trophic transfer, and biomagnification.

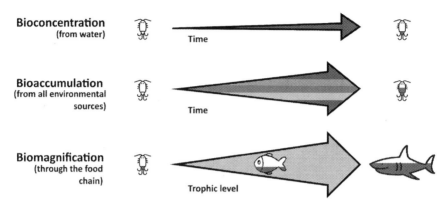

FIGURE 5.4 Bioaccumulation, trophic transfer, biomagnification.

Bioconcentration, bioremediation and biomagnification of plastics.

frequencies of contaminated individuals, and a higher average microplastic abundance. Indeed, in Welden et al.'s (2018) paper exploring uptake of plastic by wild-caught plaice and their consumed prey, sand eels were observed to have elevated levels of ingested plastics over that of their predators. It is probable that this difference between trophic levels is driven by differences in the size and morphology of the digestive system, with small plastics more easily egested by larger individuals.

At smaller scales, the trophic transfer of nanoplastics has been observed in a number of species, which is perhaps not surprisingly considering the ability of these particles to translocate into the tissues of species at lower trophic levels. Here, individual bioaccumulation at each trophic level may result in more substantive biomagnification within the food web. Indeed, PS nanoplastics (24 and 28 nm) have been observed to transfer from algae, *Scenedesmus*, to the primary consumer, *D. manga*, to crucian carp, *Carassius carassius* (Cedervall et al., 2012), a process seen to result in changes to the behaviour and metabolism of fish (Mattsson et al., 2015).

At the bottom of the food chain, adherence or internalisation of micro- and nanoplastics by seaweeds and algae may drive transfer to primary consumers, and observations of microplastics on seagrasses have highlighted potential uptake by grazers (Goss et al., 2018). For example, common periwinkles, *Littorina littorea* exposed to bladderwrack, *Fucus vesiculosus*, contaminated with microbeads and fragments subsequently demonstrated the presence of microplastic within the gut (Gutow et al., 2016). Similarly, crustaceans have been seen to ingest micro- and nanoplastics following exposure to contaminated algae, with nano-polystyrene adhered to alga, *Chlamydomonas reinhardtii*, taken up by the water flea, *D. magna* (Chae et al., 2018), and polyethylene microplastics from *Lemna minor* transferred to *Gammarus duebeni* (Mateos-Cárdenas et al., 2019). Similarly, the transfer of metal-doped polystyrene nanoplastics revealed transfer from periphytic communities (those attached to underwater surfaces) to the bladder snail, *Physa acuta*, by way of the presence of lead in snail faeces (Holzer et al., 2022).

Environmental studies frequently show lower levels of contamination than are utilised in laboratory trials of both uptake and transfer. However, whilst the uptake of microplastics in the environment may be low, representing less than 1 per individual, the high incidence of ingestion of some species and groups — particularly in relation to planktivorous species — may result in increased microplastic ingestion and potential biomagnification. When exposed to plastic via the water or via previously exposed Mysid shrimps, the sculpin, *Myoxocephalus brandtii*, was seen to take up between three and 11 times the microplastic from the proffered prey than the water column (Hasegawa & Nakaoka, 2021).

As indicated in the previous chapter, the potential for microplastic retention (and thus biomagnification) varies between species, with some organisms more able to egest microplastics. In organisms where the rate of egestion exceeds that of ingestion, there may be a reduction in the degree of trophic tranfer and, thus, a break in the biomagnification chain. The resulting patterns of transfer and magnification may then vary significantly depending on the structure of the local food web (Au et al., 2017). A similar study to that of Welden et al. (2018) examined the gut content of hake, *Merluccius merluccius*, for the presence of microplastics in ingested prey. Here it was observed that microplastics were found both in the predator and the ingested species, including blue whiting, *Micromesistius poutassou*, and northern krill, *Meganyctiphanes norvegica* (Cabanilles et al., 2022). As with this earlier study, it was seen that biomagnification of microplastics was not present at higher trophic levels. Of course, these transfers of plastics may also result in human uptake, we will consider the confirmed routes in Box 5.1.

Box 5.1 Microplastic in the human food chain

Objective

As researchers attempt to identify the risks of microplastics to humans multiple potential sources of microplastic uptake have been identified. Below we will consider sources of exposure to microplastics including ambient plastics, air and water-borne particles, and dietary exposure. In comparing the multiple sources of microplastic and the composition and size of this microliter, we may determine the likelihood of translocation. We will compare this information with existing observations of microplastics in human samples, from which the potential for various impacts observed in animal models will be extrapolated.

Micoplastic sources

Microplastics in the air

Sampling of indoor air has revealed MP at concentrations of between 1.0 and 60.0 fibres/m^3 ranging between 50 and 3250 μm, and 2.5–35.4 particles/m^3 (Gaston et al., 2020).

Microplastics in drinking water and other beverages

Microplastics have been observed in piped drinking water from numerous countries (Eerkes-Medrano et al., 2019), with most particles between 100 and 5000 μm. Bottled water may be a substantial source of plastic, with between 28 and 241 particles recorded per litre. Similarly, MPs have

(Continued)

Box 5.1 Microplastic in the human food chain (Continued)

also been found in milk, here at levels of 1−14 particles per litre (Basaran et al., 2023; Kutralam-Muniasamy et al., 2020) and (alarmingly) in infant formula, in which concentrations of 4−7 particles for 100 g were observed (Zhang et al., 2023). Finally, for those of us who enjoy a beer, consumption of 3 pints a week could equate to between 21 and 192 MPs (12−109 fragments/L).

Microplastics in foodstuffs

The most widely studied food source is, of course, seafood, in which average microplastic levels have been observed at between 0.2 and 10 particles (5−5000 μm) per gram in bivalves, and bet1een 1 and 7 (130 to >5,000 μm) per individuals in fish. Analysis of larger microplastics in fish fillets may indicate the potential for contamination during processing and packaging, as well as via dietary exposure. Increasing numbers of studies have also reported the uptake of microplastics in plants, such as carrots (Dong et al., 2021), for human consumption, transferred from contaminated soils.

Condiments and seasonings may be less impactful. See salt has been seen to contain between 550 to 681 particles/kg, however, consumption may be just a few grams. Similarly, honey and sugar (32 fragments/kg) represent equally small sources based on weekly intake. Happily, at this point, the author has been unable to find reference to microplastics in cheese but fears that it is only a matter of time.

Food storage and handling

As indicated above, the plastics in foodstuffs may originate from packing and handling processes. Take-away containers constructed of PP, PS, PE, and PET were examined for their release of microplastics and were observed to release up to 29 items per day (Du et al., 2020). The level of shedding was variable between polymers, with greatest releases from PS. A similar pattern was observed in assessments of differently shaped containers, with average MP loss of up to 38 mg (Fadare et al., 2020). MP contamination may also occur in the home, for example, the loss of MPs from chopping boards was seen to be between 7.4 and 50.7 g/year (Yadav et al., 2023).

Learning and knowledge outcomes

As a result of these above routes and many more besides, plastics have thus far been observed in samples of stool as well as in numerous tissue samples. The range of plastic sizes reported across these samples was between 1.6 and 1600 μm. This range is representative of the lower end of those reported in the above studies. The size of microplastics recovered in tissue samples is below that routinely reported in the human food chain and the surrounding environment. The apparent exposure and subsequent biomagnification as a result of dietary uptake may therefore be limited, resulting in minimal retention without prior reduction in mass. More studies of small-scale MPs are required in order to identify routes of exposure and the apparent risk posed by ambient and dietary MP sources.

5.3.2 Community-level effects

5.3.2.1 Damage to communities

By affecting the health of keystone or habitat forming species, microplastics may also threaten community function. In soils, meta-analysis of 1980 observations of microplastic's effect on microbial activity indicated that the inclusion of microplastics at concentrations over 5% increased soil respiration and that inclusion of microplastics at concentrations over 10% reduced soil enzyme activity, with the scale of both of these effects varying in relation to polymer (Liu et al., 2023). In addition to respiration, microplastics may impact nutrient storage and cycling (Kumar et al., 2023).

The type and extent of any impact, and the species affected, is dependent on the polymer type, shape and concentration, with fibres and films being seen to have greater impacts than particles (Sun et al., 2023). These changes to key abiotic conditions in soils may influence whole plant communities (Rillig et al., 2021).

As with the effects of micro- and nanoplastics in individual species, communities may be affected by the combined impacts of multiple stressors. In a pot experiment, researchers sought to determine the effects of combined drought and microfibre exposure. It was observed that, under drought conditions, soil changes as a result of microplastic pollution improve conditions for some species while disadvantaging others (Lozano & Rillig, 2020). The impact of microplastics on soil communities may also have implications for carbon drawdown and our changing climate, resulting in impacts far beyond the local environment (Kumar et al., 2023), and in agroecosystems, changes in the rhizosphere may have long-term effects on food production (Yadav et al., 2022).

Similar changes may be seen in both nutrient cycling and community composition in aquatic systems. For example, preferential ingestion of microplastics has been seen to negatively affect the health and bacterial assemblages associated with red corals, which may have long-term implications for reef formation (Corinaldesi et al., 2021). Modelling of the apparent resilience of aquatic systems exposed to microplastics also suggests apparent ecosystem-scale effects. In their analysis of secondary effects in shallow lakes, Kong and Koelmans (2019) suggest that the presence of microplastics may influence phosphorous loading, a key factor in the eutrophication of these habitats.

Comparisons have also been made regarding the impact of micro- and nanoplastics on aquatic biofilms. The results of exposure to plastics were seen to differ greatly in relation to size class, with apparent oxidative stress demonstrated by communities exposed to nanoPS (Miao et al., 2019). The impact of microplastics and their leachates on primary production by phytoplankton is a complex one, with a number of studies indicating increased primary production and oxygen output (Amaneesh et al., 2023). As above, community-level effects may also take the form of altered nutrient cycling. For example, PS nanoplastics at 1 and $100\,\mu g/L$ were seen to alter fungal community structure and reduce enzyme activity (despite increases in fungal biomass) in microcosm experiments of decomposition in stream systems. The change significantly decreased the breakdown of poplar, *Populus nigra*, leaf litter, as well as the bioavailability of carbon and nitrogen (Du et al., 2022). Mesocosm studies of the impacts of nanoplastics on communities dominate by submerged macrophytes explored the effects of exposure to PS nanoplastics at a concentration of $20\,mg/L$ on the composition both of periphytic comminities and of planktonic algae and bacteria. It was seen that both algal and bacterial communities changed in response to nanoplastic presence; however, these changes were only significant in bacteria, which demonstrated lower diversity and altered community structure (Hao et al., 2023). In another study of the periphyton, a simplified stream foodchain was studied in relation to its response to metal-doped polystyrene nanoplastics ($c.$ 0.5, 5 and 50 mg plastics/L). Here, nanoplastic exposure resulted in reduced algal biomass and photosynthetic efficiency and reduced growth in grazing bladder snail, *Physa acuta* (Holzer et al., 2022).

These effects are not limited to natural environments and may also occur in managed systems. PS nanoplastics have been identified as a driver of change in the microbiome structure in anaerobic sludge, inhibiting acidogenesis and reducing the production of methane (Wang et al., 2022). Similar subsequent observations of 100 nm PS nanoplastics at 2 μg/L in at sequencing batch reactor revealed no change to process performance, nitrification, or organic matter removal. However, once again, there were changes in the microbiome, with an increasing relative abundance of *Patescibacteria* and reduced growth of *Nitrotoga*, indicating potential for nitrification inhibition in the long term (Alvim et al., 2023).

5.3.2.2 Development of novel communities

In addition to changing the processes in endemic populations, plastics may also facilitate colonisation by new organsisms. For example, the presence of microplastics in aquatic sediments may influence the settlement of planktonic larvae. Macroplastic as debris has previously been identified as a potential site for the colonisation of encrusting sessile organisms in otherwise mobile substrates, resulting in the potential development of new communities (Fig. 5.5). Similarly, rafts of man-made debris may result in the distribution of a range of alien species between countries and continents and are subsequently closely observed (Carlton et al., 2018; Hansen et al., 2018). To microbial organisms, the presence of plastics in the environment may represent a beneficial source of carbon, a harmful substance or vector of harmful chemicals, or a dispersal route. It may also affect the hospitability of the surrounding environment, for example by decreasing the fitness of any host, increasing their apparent immune response, or disrupting chemical processes (Lear et al., 2021). Indeed, some observations have shown preferential settlement on such artificial substrates (Pinochet et al., 2020). In a study of just 68 pieces of microplastics recovered from the waters off Australia, researchers identified 14 genera of diatoms,

FIGURE 5.5 Gooseneck barnacles on a beached bottle.

Colonisation of floating plastics results in the redistribution of species.

Image credit: Bryce Stewart.

and 7 genera of coccolithophores, as well as bryozoans, barnacles, dinoflagellates, isopods, worms and hempitera (Reisser et al., 2014). The presence of microplastics may also influence the development of symbiotic communities of the gut microbiome (Evariste et al., 2019); ingestion of polystyrene by zebrafish was seen the decrease the relative abundance of Proteobacteria from 81% of the microbiome, to as low as 39% of the microbiome, depending on the size of microplastics, with apparent benefit to Firimicutes, Fusobacteria, and fish pathogens (Jin et al., 2018).

5.3.2.3 Wider implications

As indicated in Chapter 4, interactions between micro- and nanoplastics and biota may result in a range of effects on individual organisms, including potential negative implications for human health. If we explore the cumulative impacts of these individual effects, we observe potential damage at population, community and ecosystem scales. At this level, environmental micro- and nanoplastic pollution may represent a significant threat to our collective wellbeing. For example, as indicated above, the presence of microplastics in industrial soils has been predicted to negatively affect food production in addition to having potential direct effects on human health (Okeke et al., 2022). Indeed, the cascade effects of microplastics in the environment may influence a range of ecosystem services (Fig. 5.6), including provisioning services (such as the supply of food and water,

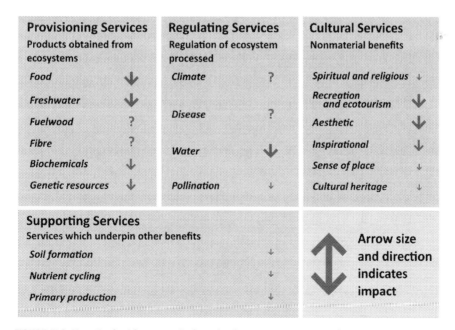

FIGURE 5.6 Hypothetical impacts of microplastics on ecosystem services.

Changes in ecosystems services which may arise as a result of microplastic pollution.

natural fuels and materials), and supporting services (such as carbon drawdown and primary production by plants and algae) (Sridharan et al., 2021). For example, impacts of micro- and nanoplastics on pollinators, such as Heller's stingless bee, *Partamona helleri*, suggested reduced fitness, with individuals exposed to microplastics as larvae needing longer rests and exhibiting unusual walking behaviour (Viana et al., 2023). Such reductions in fitness at a population level may have a substantial effect on plant communities and agricultural production. However, these secondary impacts are equally as hard to predict as direct effects. For example, earthworms, key species involved in bioturbation, nutrient cycling and aeration of soils, showed minimal impacts and even lowered arsenic accumulation when exposed to microplastics (Wang et al., 2019). On the otherhand, simulated bivalve-dominated habitats exposed to microplastics have revealed differences in filtering. Predicted outcomes suggest that mussels may filter less when dosed with high levels of microplastic, whereas oysters filter more. These changes were seen to affect the composition of porewater in sediments, the biomass of cyanobacteria and invertebrate assemblages (Green et al., 2017).

5.4 Methods: determining cumulative effects

Determination of the ecosystem-level impacts of microplastics may be much more challenging to achieve than individual or population-level effects. While combined species trials and mesocosm studies may assist in exploring such imapcts, these studies have only limited ecological validity. Additionally, as mentioned in the previous chapter, existing literature on the single species effects of both micro- and nanoplastics uses forms and levels of plastic that are not environmentally relevant, and vary greatly between studies, this leads to problems when combining the results into wider environmental risk assessments. Moreover, there is a paucity of data regarding the uptake, retention and transfer of data for many taxonomic groups which limits our ability to model potential effects in similar communities. Nevertheless, sensitively designed models including microplastics elements have been seen to accurately explain variation in various measures of ecosystem function (Ladewig et al., 2023).

Studies to date have sought to include observations of whole or partial species assemblages over successive generations. Exploration of the impacts of PS microspheres on growth in a bacterial-feeding nematode community have been explored over multiple generations. The impact on the carrying capacity of each species varied significantly in the first instance, with the number of ingested beads differing between species. In this case, species which typically reproduce quickly and have high energetic costs were the most readily effected and these species showed low population growth rates (Mueller et al., 2020). Again, effects of this kind are typically concentration dependent, with studies of nematode communities from maize-growing regions exposed to PP microplastics showing changes in abundance and diversity. Here plant parasites, bacterivores, fungivores, and omnivores in soils containing the highest levels of microplastics (2% by weight) were reduced by between 76 and 100% (Yang et al., 2022).

Other studies have attempted to determine the impacts of plastics on single relationships within the wider community. For example, the impacts of 0.07, 0.7 and 7 μm PS plastics on population life history of the rotifer, *Brachionus plicatilis*, and their role as a controller of harmful algae, *Phaeocystis*, was explored by Sun et al (2019). In this study, individual rotifers were exposed to plastics at concentrations of 0, 1, 5, 10, and 20 μg/mL and time until first reproduction (eggs and brood), offspring number, and survival time. Exploration of survival times revealed a dose-dependent reduction only in those individuals exposed to 0.07 μm microspheres. Similar patterns were also observed in time until first reproduction, and number of offspring. Subsequently, the effect of 3 concentrations of 0.07 μm microspheres (at 0, 2.04×10^9, and 1.02×10^{10} particles/mL) was explored in relation to the population dynamics of rotifers and *P. globosa*. Results suggested that decreased grazing pressure resulted in higher algal population growth than in the control. Similar observations of the impacts of small size plastics (70 nm) on *B. plicatilis* have also been reported by Li et al. (2023). Here, 70 nm particles were more readily accumulated, resulting in both reduced food consumption and fecundity, as well as subsequent metabolic impairment. Modelling of the reactions of theoretical predator-prey populations in the presence of microplastics using the Lotka—Volterra model suggests that predators may be more sensitive to environmental microplastics than prey (although the response to plastic in each species may remain weak) and that the density of prey is key in maintaining that stability of the affected system (Huang et al., 2020).

5.5 Conclusion

Our understanding of the ecosystem level effects of small plastics is still in its infancy, impacted by the complexity of ecosystems as well remaining gaps in our understanding of individual-level impacts discussed in the previous chapter. Nonetheless, habitat changes and intergenerational effects may threaten populations of single species, and transfer through the food change microplastics may affect organisms at higher trophic levels. Alternatively, impacts on keystone species and those animals higher up the food chain may negatively impact wider food webs. As a result, the impacts on ecosystems may be both top-down and bottom-up, influencing the functioning of many communities (Prata et al., 2019).

References

Alvim, C. B., Ferrer-Polonio, E., Bes-Piá, M. A., Mendoza-Roca, J. A., Fernández-Navarro, J., Alonso-Molina, J. L., & Amorós-Muñoz, I. (2023). Effect of polystyrene nanoplastics on the activated sludge process performance and biomass characteristics. A laboratory study with a sequencing batch reactor. *Journal of Environmental*

Management, *329*. Available from https://doi.org/10.1016/j.jenvman.2022.117131, https://www.sciencedirect.com/journal/journal-of-environmental-management.

Amaneesh, C., Anna Balan, S., Silpa, P. S., Kim, J. W., Greeshma, K., Aswathi Mohan, A., Robert Antony, A., Grossart, H. P., Kim, H. S., & Ramanan, R. (2023). Gross negligence: Impacts of microplastics and plastic leachates on phytoplankton community and ecosystem dynamics. *Environmental Science and Technology*, *57*(1), 5−24. Available from https://doi.org/10.1021/acs.est.2c05817, http://pubs.acs.org/journal/esthag.

Au, S. Y., Lee, C. M., Weinstein, J. E., van den Hurk, P., & Klaine, S. J. (2017). Trophic transfer of microplastics in aquatic ecosystems: Identifying critical research needs. *Integrated Environmental Assessment and Management*, *13*(3), 505−509. Available from https://doi.org/10.1002/ieam.1907, http://www.interscience.wiley.com/jpages/1551-3777.

Basaran, B., Özçifçi, Z., Akcay, H. T., & Aytan, Ü. (2023). Microplastics in branded milk: Dietary exposure and risk assessment. *Journal of Food Composition and Analysis*, *123*. Available from https://doi.org/10.1016/j.jfca.2023.105611, http://www.elsevier.com/inca/publications/store/6/2/2/8/7/8/index.htt.

Bosker, T., Olthof, G., Vijver, M. G., Baas, J., & Barmentlo, S. H. (2019). Significant decline of *Daphnia magna* population biomass due to microplastic exposure. *Environmental Pollution*, *250*, 669−675. Available from https://doi.org/10.1016/j.envpol.2019.04.067, https://www.journals.elsevier.com/environmental-pollution.

Cabanilles, P., Acle, S., Arias, A., Masiá, P., Ardura, A., & Garcia-Vazquez, E. (2022). Microplastics risk into a three-link food chain inside European hake. *Diversity*, *14*(5). Available from https://doi.org/10.3390/d14050308, https://www.mdpi.com/1424-2818/14/5/308/pdf.

Carlton, J. T., Chapman, J. W., Geller, J. B., Miller, J. A., Ruiz, G. M., Carlton, D. A., McCuller, M. I., Treneman, N. C., Steves, B. P., Breitenstein, R. A., Lewis, R., Bilderback, D., Bilderback, D., Haga, T., & Harris, L. H. (2018). Ecological and biological studies of ocean rafting: Japanese tsunami marine debris in North America and the Hawaiian islands. *Aquatic Invasions*, *13*(1), 1−9. Available from https://doi.org/10.3391/ai.2018.13.1.01, http://www.aquaticinvasions.net/2018/AI_2018_JTMD_Carlton_etal.pdf.

Carson, H. S., Colbert, S. L., Kaylor, M. J., & McDermid, K. J. (2011). Small plastic debris changes water movement and heat transfer through beach sediments. *Marine Pollution Bulletin*, *62*(8), 1708−1713. Available from https://doi.org/10.1016/j.marpolbul.2011.05.032.

Casabianca, S., Bellingeri, A., Capellacci, S., Sbrana, A., Russo, T., Corsi, I., & Penna, A. (2021). Ecological implications beyond the ecotoxicity of plastic debris on marine phytoplankton assemblage structure and functioning. *Environmental Pollution*, *290*. Available from https://doi.org/10.1016/j.envpol.2021.118101, https://www.journals.elsevier.com/environmental-pollution.

Cedervall, T., Hansson, L.-A., Lard, M., Frohm, B., Linse, S., & Bansal, V. (2012). Food chain transport of nanoparticles affects behaviour and fat metabolism in fish. *PLoS One*, *7*(2), e32254. Available from https://doi.org/10.1371/journal.pone.0032254.

Chae, Y., Kim, D., Kim, S. W., & An, Y. J. (2018). Trophic transfer and individual impact of nano-sized polystyrene in a four-species freshwater food chain. *Scientific Reports*, *8*(1). Available from https://doi.org/10.1038/s41598-017-18849-y, http://www.nature.com/srep/index.html.

Corinaldesi, C., Canensi, S., Dell'Anno, A., Tangherlini, M., Di Capua, I., Varrella, S., Willis, T. J., Cerrano, C., & Danovaro, R. (2021). Multiple impacts of microplastics can threaten marine habitat-forming species. *Communications Biology, 4*(1). Available from https://doi.org/10.1038/s42003-021-01961-1, https://www.springer.com/journal/11852.

Davarpanah, E., & Guilhermino, L. (2015). Single and combined effects of microplastics and copper on the population growth of the marine microalgae Tetraselmis chuii. *Estuarine, Coastal and Shelf Science, 167,* 269−275. Available from https://doi.org/10.1016/j.ecss.2015.07.023, http://www.elsevier.com/inca/publications/store/6/2/2/8/2/3/index.htt.

De Souza Machado, A. A., Lau, C. W., Kloas, W., Bergmann, J., Bachelier, J. B., Faltin, E., Becker, R., Görlich, A. S., & Rillig, M. C. (2019). Microplastics can change soil properties and affect plant performance. *Environmental Science and Technology, 53* (10), 6044−6052. Available from https://doi.org/10.1021/acs.est.9b01339, http://pubs.acs.org/journal/esthag.

Dong, Y., Gao, M., Qiu, W., & Song, Z. (2021). Uptake of microplastics by carrots in presence of As (III): Combined toxic effects. *Journal of Hazardous Materials, 411.* Available from https://doi.org/10.1016/j.jhazmat.2021.125055, http://www.elsevier.com/locate/jhazmat.

Du, F., Cai, H., Zhang, Q., Chen, Q., & Shi, H. (2020). Microplastics in take-out food containers. *Journal of Hazardous Materials, 399.* Available from https://doi.org/10.1016/j.jhazmat.2020.122969, http://www.elsevier.com/locate/jhazmat.

Du, J., Qv, W., Niu, Y., Qv, M., Jin, K., Xie, J., & Li, Z. (2022). Nanoplastic pollution inhibits stream leaf decomposition through modulating microbial metabolic activity and fungal community structure. *Journal of Hazardous Materials, 424.* Available from https://doi.org/10.1016/j.jhazmat.2021.127392, http://www.elsevier.com/locate/jhazmat.

Eerkes-Medrano, D., Leslie, H. A., & Quinn, B. (2019). Microplastics in drinking water: A review and assessment. *Current Opinion in Environmental Science and Health, 7,* 69−75. Available from https://doi.org/10.1016/j.coesh.2018.12.001, http://www.journals.elsevier.com/current-opinion-in-environmental-science-and-health.

Evariste, L., Barret, M., Mottier, A., Mouchet, F., Gauthier, L., & Pinelli, E. (2019). Gut microbiota of aquatic organisms: A key endpoint for ecotoxicological studies. *Environmental Pollution, 248,* 989−999. Available from https://doi.org/10.1016/j.envpol.2019.02.101, https://www.journals.elsevier.com/environmental-pollution.

Everaert, G., Vlaeminck, K., Vandegehuchte, M. B., & Janssen, C. R. (2022). Effects of microplastic on the population dynamics of a marine copepod: Insights from a laboratory experiment and a mechanistic model. *Environmental Toxicology and Chemistry, 41*(7), 1663−1674. Available from https://doi.org/10.1002/etc0.5336, http://onlinelibrary.wiley.com/journal/10.1002/(ISSN)1552-8618.

Fadare, O. O., Wan, B., Guo, L. H., & Zhao, L. (2020). Microplastics from consumer plastic food containers: Are we consuming it? *Chemosphere, 253.* Available from https://doi.org/10.1016/j.chemosphere.2020.126787, http://www.elsevier.com/locate/chemosphere.

Gaston, E., Woo, M., Steele, C., Sukumaran, S., & Anderson, S. (2020). Microplastics differ between indoor and outdoor air masses: Insights from multiple microscopy methodologies. *Applied Spectroscopy, 74*(9), 1079−1098. Available from https://doi.org/10.1177/0003702820920652, https://journals.sagepub.com/loi/asp.

Goss, H., Jaskiel, J., & Rotjan, R. (2018). Thalassia testudinum as a potential vector for incorporating microplastics into benthic marine food webs. *Marine Pollution Bulletin, 135,* 1085−1089. Available from https://doi.org/10.1016/j.marpolbul.2018.08.024, http://www.elsevier.com/locate/marpolbul.

Green, D. S., Boots, B., O'Connor, N. E., & Thompson, R. (2017). Microplastics affect the ecological functioning of an important biogenic habitat. *Environmental Science and Technology*, *51*(1), 68−77. Available from https://doi.org/10.1021/acs.est.6b04496, http://pubs.acs.org/journal/esthag.

Gutow, L., Eckerlebe, A., Giménez, L., & Saborowski, R. (2016). Experimental evaluation of seaweeds as a vector for microplastics into marine food webs. *Environmental Science and Technology*, *50*(2), 915−923. Available from https://doi.org/10.1021/acs.est.5b02431, http://pubs.acs.org/journal/esthag.

Hamdhan, I.N., & Clarke, B.G. (2010). *Proceedings of world geothermal congress Determination of thermal conductivity of coarse and fine sand soils.*

Hansen, G. I., Hanyuda, T., & Kawai, H. (2018). Invasion threat of benthic marine algae arriving on Japanese tsunami marine debris in Oregon and Washington, USA. *Phycologia*, *57*(6), 641−658. Available from https://doi.org/10.2216/18-58.1, http://www.phycologia.org/doi/pdf/10.2216/18-58.1.

Hao, B., Wu, H., You, Y., Liang, Y., Huang, L., Sun, Y., Zhang, S., & He, B. (2023). Bacterial community are more susceptible to nanoplastics than algae community in aquatic ecosystems dominated by submerged macrophytes. *Water Research*, *232*. Available from https://doi.org/10.1016/j.watres.2023.119717, http://www.elsevier.com/locate/watres.

Hasegawa, T., & Nakaoka, M. (2021). Trophic transfer of microplastics from mysids to fish greatly exceeds direct ingestion from the water column. *Environmental Pollution*, *273*. Available from https://doi.org/10.1016/j.envpol.2021.116468, https://www.journals.elsevier.com/environmental-pollution.

Heinlaan, M., Viljalo, K., Richter, J., Ingwersen, A., Vija, H., & Mitrano, D. M. (2023). Multi-generation exposure to polystyrene nanoplastics showed no major adverse effects in *Daphnia magna. Environmental Pollution*, *323*. Available from https://doi.org/10.1016/j.envpol.2023.121213, https://www.journals.elsevier.com/environmental-pollution.

Holzer, M., Mitrano, D. M., Carles, L., Wagner, B., & Tlili, A. (2022). Important ecological processes are affected by the accumulation and trophic transfer of nanoplastics in a freshwater periphyton-grazer food chain. *Environmental Science: Nano*, *9*(8), 2990−3003. Available from https://doi.org/10.1039/d2en00101b, http://pubs.rsc.org/en/journals/journal/en.

Huang, Q., Lin, Y., Zhong, Q., Ma, F., & Zhang, Y. (2020). The impact of microplastic particles on population dynamics of predator and prey: Implication of the Lotka-Volterra model. *Scientific Reports*, *10*(1). Available from https://doi.org/10.1038/s41598-020-61414-3, http://www.nature.com/srep/index.html.

Jin, Y., Xia, J., Pan, Z., Yang, J., Wang, W., & Fu, Z. (2018). Polystyrene microplastics induce microbiota dysbiosis and inflammation in the gut of adult zebrafish. *Environmental Pollution*, *235*, 322−329. Available from https://doi.org/10.1016/j.envpol.2017.12.088, http://www.elsevier.com/inca/publications/store/4/0/5/8/5/6.

Kong, X., & Koelmans, A. A. (2019). Modeling decreased resilience of shallow lake ecosystems toward eutrophication due to microplastic ingestion across the food web. *Environmental Science and Technology*, *53*(23), 13822−13831. Available from https://doi.org/10.1021/acs.est.9b03905, http://pubs.acs.org/journal/esthag.

Kumar, A., Mishra, S., Pandey, R., Yu, Z. G., Kumar, M., Khoo, K. S., Thakur, T. K., & Show, P. L. (2023). Microplastics in terrestrial ecosystems: Un-ignorable impacts on soil characterises, nutrient storage and its cycling. *TrAC - Trends in Analytical*

Chemistry, 158. Available from https://doi.org/10.1016/j.trac.2022.116869, http://www.elsevier.com/locate/trac.

Kutralam-Muniasamy, G., Pérez-Guevara, F., Elizalde-Martínez, I., & Shruti, V. C. (2020). Branded milks — Are they immune from microplastics contamination? *Science of the Total Environment, 714.* Available from https://doi.org/10.1016/j.scitotenv.2020.136823, http://www.elsevier.com/locate/scitotenv.

Ladewig, S. M., Coco, G., Hope, J. A., Vieillard, A. M., & Thrush, S. F. (2023). Real-world impacts of microplastic pollution on seafloor ecosystem function. *Science of the Total Environment, 858.* Available from https://doi.org/10.1016/j.scitotenv.2022.160114, http://www.elsevier.com/locate/scitotenv.

Lear, G., Kingsbury, J. M., Franchini, S., Gambarini, V., Maday, S. D. M., Wallbank, J. A., Weaver, L., & Pantos, O. (2021). Plastics and the microbiome: Impacts and solutions. *Environmental Microbiomes, 16*(1). Available from https://doi.org/10.1186/s40793-020-00371-w, https://standardsingenomics.biomedcentral.com/.

Li, X., Lu, L., Ru, S., Eom, J., Wang, D., Samreen., & Wang, J. (2023). Nanoplastics induce more severe multigenerational life-history trait changes and metabolic responses in marine rotifer Brachionus plicatilis: Comparison with microplastics. *Journal of Hazardous Materials, 449.* Available from https://doi.org/10.1016/j.jhazmat.2023.131070, http://www.elsevier.com/locate/jhazmat.

Liu, X., Li, Y., Yu, Y., & Yao, H. (2023). Effect of nonbiodegradable microplastics on soil respiration and enzyme activity: A meta-analysis. *Applied Soil Ecology, 184.* Available from https://doi.org/10.1016/j.apsoil.2022.104770, http://www.elsevier.com/inca/publications/store/5/2/4/5/1/8/index.htt.

Liu, Z., Cai, M., Wu, D., Yu, P., Jiao, Y., Jiang, Q., & Zhao, Y. (2020). Effects of nanoplastics at predicted environmental concentration on Daphnia pulex after exposure through multiple generations. *Environmental Pollution, 256.* Available from https://doi.org/10.1016/j.envpol.2019.113506, https://www.journals.elsevier.com/environmental-pollution.

Lozano, Y. M., & Rillig, M. C. (2020). Effects of microplastic fibers and drought on plant communities. *Environmental Science & Technology, 54*(10), 6166−6173. Available from https://doi.org/10.1021/acs.est.0c01051.

Martins, A., & Guilhermino, L. (2018). Transgenerational effects and recovery of microplastics exposure in model populations of the freshwater cladoceran *Daphnia magna* Straus. *Science of the Total Environment, 631-632*, 421−428. Available from https://doi.org/10.1016/j.scitotenv.2018.03.054, http://www.elsevier.com/locate/scitotenv.

Mateos-Cárdenas, A., Scott, D. T., Seitmaganbetova, G., van, V. P., John, O. H., & Marcel, J. A. K. (2019). Polyethylene microplastics adhere to Lemna minor (L.), yet have no effects on plant growth or feeding by Gammarus duebeni (Lillj.). *Science of the Total Environment, 689,* 413−421. Available from https://doi.org/10.1016/j.scitotenv.2019.06.359, http://www.elsevier.com/locate/scitotenv.

Mattsson, K., Ekvall, M. T., Hansson, L. A., Linse, S., Malmendal, A., & Cedervall, T. (2015). Altered behavior, physiology, and metabolism in fish exposed to polystyrene nanoparticles. *Environmental Science and Technology, 49*(1), 553−561. Available from https://doi.org/10.1021/es5053655, http://pubs.acs.org/journal/esthag.

Miao, L., Hou, J., You, G., Liu, Z., Liu, S., Li, T., Mo, Y., Guo, S., & Qu, H. (2019). Acute effects of nanoplastics and microplastics on periphytic biofilms depending on particle size, concentration and surface modification. *Environmental Pollution, 255.*

Available from https://doi.org/10.1016/j.envpol.2019.113300, https://www.journals.elsevier.com/environmental-pollution.

Mueller, M. T., Fueser, H., Höss, S., & Traunspurger, W. (2020). Species-specific effects of long-term microplastic exposure on the population growth of nematodes, with a focus on microplastic ingestion. *Ecological Indicators*, *118*. Available from https://doi.org/10.1016/j.ecolind.2020.106698, http://www.elsevier.com/locate/ecolind.

Nelms, S. E., Galloway, T. S., Godley, B. J., Jarvis, D. S., & Lindeque, P. K. (2018). Investigating microplastic trophic transfer in marine top predators. *Environmental Pollution*, *238*, 999−1007. Available from https://doi.org/10.1016/j.envpol.2018.02.016, http://www.elsevier.com/inca/publications/store/4/0/5/8/5/6.

Nikiforova, T., Savytskyi, M., Limam, K., Bosschaerts, W., & Belarbi, R. (2013). Methods and results of experimental researches of thermal conductivity of soils. *Energy Procedia*, *42*, 775−783. Available from https://doi.org/10.1016/j.egypro.2013.12.034, 18766102, http://www.sciencedirect.com/science/journal/18766102.

Okeke, E. S., Okoye, C. O., Atakpa, E. O., Ita, R. E., Nyaruaba, R., Mgbechidinma, C. L., & Akan, O. D. (2022). Microplastics in agroecosystems—Impacts on ecosystem functions and food chain. *Resources, Conservation and Recycling*, *177*. Available from https://doi.org/10.1016/j.resconrec.2021.105961, http://www.elsevier.com/locate/resconrec.

Pinochet, J., Urbina, M. A., & Lagos, M. E. (2020). Marine invertebrate larvae love plastics: Habitat selection and settlement on artificial substrates. *Environmental Pollution*, *257*. Available from https://doi.org/10.1016/j.envpol.2019.113571, https://www.journals.elsevier.com/environmental-pollution.

Pitt, J. A., Trevisan, R., Massarsky, A., Kozal, J. S., Levin, E. D., & Di Giulio, R. T. (2018). Maternal transfer of nanoplastics to offspring in zebrafish (Danio rerio): A case study with nanopolystyrene. *Science of the Total Environment*, *643*, 324−334. Available from https://doi.org/10.1016/j.scitotenv.2018.06.186, http://www.elsevier.com/locate/scitotenv.

Prata, J. C., da Costa, J. P., Lopes, I., Duarte, A. C., & Rocha-Santos, T. (2019). Effects of microplastics on microalgae populations: A critical review. *Science of the Total Environment*, *665*, 400−405. Available from https://doi.org/10.1016/j.scitotenv.2019.02.132, http://www.elsevier.com/locate/scitotenv.

Reisser, J., Shaw, J., Hallegraeff, G., Proietti, M., Barnes, D. K. A., Thums, M., Wilcox, C., Hardesty, B. D., & Pattiaratchi, C. (2014). Millimeter-sized marine plastics: A new pelagic habitat for microorganisms and invertebrates. *PLoS One*, *9*(6). Available from https://doi.org/10.1371/journal.pone.0100289, http://www.plosone.org/article/fetchObject.action?uri = info%3Adoi%2F10.1371%2Fjournal.pone.0100289&representation = PDF.

Rillig, M. C., Leifheit, E., & Lehmann, J. (2021). Microplastic effects on carbon cycling processes in soils. *PLoS Biology*, *19*(3), e3001130. Available from https://doi.org/10.1371/journal.pbio.3001130.

Shen, M., Ye, S., Zeng, G., Zhang, Y., Xing, L., Tang, W., Wen, X., & Liu, S. (2020). Can microplastics pose a threat to ocean carbon sequestration? *Marine Pollution Bulletin*, *150*. Available from https://doi.org/10.1016/j.marpolbul.2019.110712, http://www.elsevier.com/locate/marpolbul.

Sridharan, S., Kumar, M., Bolan, N. S., Singh, L., Kumar, S., Kumar, R., & You, S. (2021). Are microplastics destabilizing the global network of terrestrial and aquatic ecosystem services? *Environmental Research*, *198*. Available from https://doi.org/10.1016/j.envres.2021.111243, http://www.elsevier.com/inca/publications/store/6/2/2/8/2/1/index.htt.

Sun, J., Teng, M., Zhu, W., Leung, K. M. Y., & Wu, F. (2023). Paternal inheritance is an important, but overlooked, factor affecting the adverse effects of microplastics and nanoplastics on subsequent generations. *Journal of Agricultural and Food Chemistry*, *71*(2), 991−993. Available from https://doi.org/10.1021/acs.jafc.2c08408, http://pubs.acs.org/journal/jafcau.

Sun, L., Liao, K., & Wang, D. (2021). Comparison of transgenerational reproductive toxicity induced by pristine and amino modified nanoplastics in Caenorhabditis elegans. *Science of the Total Environment*, *768*. Available from https://doi.org/10.1016/j.scitotenv.2020.144362, http://www.elsevier.com/locate/scitotenv.

Sun, Y., Xu, W., Gu, Q., Chen, Y., Zhou, Q., Zhang, L., Gu, L., Huang, Y., Lyu, K., & Yang, Z. (2019). Small-sized microplastics negatively affect rotifers: Changes in the key life-history traits and rotifer-phaeocystis population dynamics. *Environmental Science and Technology*, *53*(15), 9241−9251. Available from https://doi.org/10.1021/acs.est.9b02893, http://pubs.acs.org/journal/esthag.

Viana, T. A., Botina, L. L., Bernardes, R. C., Barbosa, W. F., Xavier, T. K. D., Lima, M. A. P., Araújo, Rd. S., & Martins, G. F. (2023). Ingesting microplastics or nanometals during development harms the tropical pollinator Partamona helleri (Apinae: Meliponini). *Science of the Total Environment*, *893*. Available from https://doi.org/10.1016/j.scitotenv.2023.164790, http://www.elsevier.com/locate/scitotenv.

Wang, J., Coffin, S., Sun, C., Schlenk, D., & Gan, J. (2019). Negligible effects of microplastics on animal fitness and HOC bioaccumulation in earthworm Eisenia fetida in soil. *Environmental Pollution*, *249*, 776−784. Available from https://doi.org/10.1016/j.envpol.2019.03.102, https://www.journals.elsevier.com/environmental-pollution.

Wang, J., Ma, D., Feng, K., Lou, Y., Zhou, H., Liu, B., Xie, G., Ren, N., & Xing, D. (2022). Polystyrene nanoplastics shape microbiome and functional metabolism in anaerobic digestion. *Water Research*, *219*. Available from https://doi.org/10.1016/j.watres.2022.118606, http://www.elsevier.com/locate/watres.

Wang, L., Zhu, Y., Gu, J., Yin, X., Guo, L., Qian, L., Shi, L., Guo, M., & Ji, G. (2023). The toxic effect of bisphenol AF and nanoplastic coexposure in parental and offspring generation zebrafish. *Ecotoxicology and Environmental Safety*, *251*. Available from https://doi.org/10.1016/j.ecoenv.2023.114565, http://www.elsevier.com/inca/publications/store/6/2/2/8/1/9/index.htt.

Welden, N. A., Abylkhani, B., & Howarth, L. M. (2018). The effects of trophic transfer and environmental factors on microplastic uptake by plaice, Pleuronectes plastessa, and spider crab, Maja squinado. *Environmental Pollution*, *239*, 351−358. Available from https://doi.org/10.1016/j.envpol.2018.03.110, http://www.elsevier.com/inca/publications/store/4/0/5/8/5/6.

Welden, N. A. C., & Cowie, P. R. (2016). Long-term microplastic retention causes reduced body condition in the langoustine, Nephrops norvegicus. *Environmental Pollution*, *218*, 895−900. Available from https://doi.org/10.1016/j.envpol.2016.08.020, http://www.elsevier.com/locate/envpol.

Yadav, H., Khan, M. R. H., Quadir, M., Rusch, K. A., Mondal, P. P., Orr, M., Xu, E. G., & Iskander, S. M. (2023). Cutting boards: An overlooked source of microplastics in human food? *Environmental Science and Technology*, *57*(22), 8225−8235. Available from https://doi.org/10.1021/acs.est.3c00924, http://pubs.acs.org/journal/esthag.

Yadav, S., Gupta, E., Patel, A., Srivastava, S., Mishra, V. K., Singh, P. C., Srivastava, P. K., & Barik, S. K. (2022). Unravelling the emerging threats of microplastics to agroecosystems. *Reviews in Environmental Science and Biotechnology, 21*(3), 771–798. Available from https://doi.org/10.1007/s11157-022-09621-4. http://www.kluweronline.com/issn/1569-1705.

Yang, B., Li, P., Entemake, W., Guo, Z., & Xue, S. (2022). Concentration-dependent impacts of microplastics on soil nematode community in bulk soils of maize: Evidence from a pot experiment. *Frontiers in Environmental Science, 10*. Available from https://doi.org/10.3389/fenvs.2022.872898, http://journal.frontiersin.org/journal/environmental-science.

Yu, C. W., Luk, T. C., & Liao, V. H. C. (2021). Long-term nanoplastics exposure results in multi and trans-generational reproduction decline associated with germline toxicity and epigenetic regulation in Caenorhabditis elegans. *Journal of Hazardous Materials, 412*. Available from https://doi.org/10.1016/j.jhazmat.2021.125173, http://www.elsevier.com/locate/jhazmat.

Yuan, W., Zhou, Y., Liu, X., & Wang, J. (2019). New perspective on the nanoplastics disrupting the reproduction of an endangered fern in artificial freshwater. *Environmental Science and Technology, 53*(21), 12715–12724. Available from https://doi.org/10.1021/acs.est.9b02882, http://pubs.acs.org/journal/esthag.

Zhang, Q., Liu, L., Jiang, Y., Zhang, Y., Fan, Y., Rao, W., & Qian, X. (2023). Microplastics in infant milk powder. *Environmental Pollution, 323*. Available from https://doi.org/10.1016/j.envpol.2023.121225, https://www.journals.elsevier.com/environmental-pollution.

Associated pollutants and secondary effects

6.1 Introduction

Over the past few chapters, we have observed the manner in which microplastics spread throughout ecosystems and interact with, or otherwise affect, biota. In addition to these direct and indirect effects of uptake and entanglement, there are also a variety of associated secondary impacts on organisms and ecosystems as a result of plastic-mediated chemicals. Thus the movement of plastics within and between environments may act as a driver or enabler of a variety of additional threats, further complicating our understanding of the risks associated with plastic pollution and increasing the importance of understanding its movement both within habitats and around the globe.

6.2 Intentional additions and accidental uptake

We have already established that the term 'plastic' is a catchall for a variety of synthetic polymers of varied physical and mechanical properties, also that these plastics may be amended using a long list of fillers and additives by which desirable characteristics may be conferred on the finished product (Turner & Filella, 2021). Plastics also vary in their degree of crystallinity, functional groups, specific gravities and forms, plus a range of other attributes which alter their fate within the environment. As a result, both primary and secondary microplastics may act in a variety of unpredictable ways. Unfortunately, a number of the additives historically employed to improve the properties of plastics have subsequently been found to have negative effects on both humans and the environment with many now banned or restricted in use. Such substances include bisphenol A (BPA), antimony trioxide, various phthalates, polybrominated diphenyl esters and more, some of the uses of which are listed in Table 6.1.

The impacts of these additives are vaired, for example, BPA, which is a common constituent of polycarbonate plastics, mimics oestrogen and acts as an androgen receptor antagonist. As will be discussed in greater detail in Chapter 9, the ester bonds found in polycarbonate plastics are more susceptible to degradation than those in other plastics, resulting in the leaching of BPA and exposure in both humans and animals (vom Saal & Hughes, 2005). Exposure is known to cause

Microplastics. DOI: https://doi.org/10.1016/B978-0-443-13324-4.00006-6

Table 6.1 Examples of historic and contemporary plastic additives and their known effects.

Compound	Use	Impact
Antimony trioxide	Flame retardants and PET production	Carcinogen
Bisphenol A	Plasticiser	Endocrine disruptor
Cadmium	Pigments	Carcinogen
	Stabiliser	Toxic in aquatic systems
Chlorinated paraffins	Plasticizer	Carcinogen
	Flame retardant	Toxic in aquatic systems
Lead	Plasticizer	Diverse impacts on health and cognitive function
	Pigment	
	Stabiliser	Toxic in aquatic systems
Perfluorinated compounds	Reduced adherence material to food containers	Diverse impacts on health
Phthalates	Plasticizers	Endocrine disruptors
Polybrominated diphenyl esters	Flame retardants	Endocrine disruptors
		Reduced liver function
		Impaired brain development

damage to the reproductive, immune and neuroendocrine systems, in addition to potentially carcinogenic effects (Ma et al., 2019). Such contact may also result in multigenerational impacts, for example *Caenorhabditis elegans* exposed to BPA exhibited dose-dependent impacts on growth, movement and fecundity in subsequent generations (Zhou et al., 2016). The doses of BPA to which an organism is exposed do not have to be high to induce negative effects, occurring at levels below 10^{-12} M or 0.23 parts per thousand (vom Saal & Hughes, 2005).

Another problematic compound, antimony trioxide, is a flame retardant sometimes used as a catalyst in the production of PET. Depending on its use, the concentration of antimony in plastics ranges from a few milligrams per kilogram to up to 10% of the product weight when used as a flame retardant. Antimony trioxide, which may migrate out of plastics when heated (such as in a microwave), has been identified as a possible carcinogen (Filella et al., 2020) and is recognised as hazardous under the Basel Convention (Fu et al., 2023). Antimony may also affect the fate of some plastics at end of life, with the high leachability of the metalloid from bottom incineration ashes, considered a high risk in the reuse of this material (Filella et al., 2020).

Some plasticisers, such as chlorinated paraffins (CPs), have also been seen to have negative environmental effects. CPs are a group of nonbiodegradable, synthetic organic compounds produced by chlorination of particular carbon feedstocks known as *n*-alkanes. They are insoluble, and known to bioaccumulate (Fiedler, 2010), and have been identified as having toxic effects on the liver, kidneys and the thyroid and parathyroid glands, with toxicity inversely proportional to chain length (El-Sayed Ali & Legler, 2010). Before widespread bans, short-chain CPs were frequently used as flame retardants and plasticisers in the production of PVC, these have subsequently been replaced by medium and long-chain CPs (Kobetičová & Černý, 2018). Both historic, short-chain CP containing plastics and modern medium- and long-chain CP containing plastics may release CPs via leaching from landfills, fires and recycling processes (Fiedler, 2010). As can be seen in Table 6.1, many more additives still in use have potentially harmful impacts on biota.

In addition to the introduction of harmful additives during manufacturing, the properties of polymers may also result in unintentional contamination by other substances, both during use and following loss to the environment. Due to their relative inexpensiveness, low weight and resistance to attack by many chemicals, plastics are used in the transport and storage of many substances, which may be sorbed to the polymer surface. Alternatively, when present together in the water column, pollutants will 'preferentially' adhere to the lithophilic plastic structure in the same way in which pasta sauce or fats from cooking may stick to a plastic washing-up bowl.

The uptake of numerous persistent organic pollutants (POPs) to plastics in marine settings has been recorded as part of the Pellet Watch programme, which received samples of plastic pellets from participants around the globe (discussed in greater detail in Box 6.1). POPs are long-lived, frequently hydrophobic

---

Box 6.1 Case study: international pellet watch in monitoring marine pollutants

Objective

To determine the changing composition and distribution of persistent organic pollutants (POPs) in the global oceans. In addition to their relevance to researchers studying sources of pollutants and their distributions, as well as ecotoxicological impacts, the inclusive methods and transparent reporting of IPW results are beneficial to policymakers, regulators and NGOs engaged in the management of toxicological risks, in addition to having significant value as an educational tool (Yeo et al., 2015).

Scope

Instigated following early observations of pollutants on preproduction plastic pellets (nurdles) around the coasts of Japan (Mato et al., 2001), the International Pellet Watch (IPW) was launched in 2005. The programme, which aims to determine the global concentrations of POPs adsorbed to pellets, relies upon the hydrophobic nature of plastics and their resulting affinity to POPs in aquatic systems. Here plastic pellets are employed as randomly distributed passive collectors of potentially harmful chemicals, which are gathered on an *ad hoc* basis by volunteers. Pellets were collected by researchers and volunteers from beaches worldwide and posted to the IPW at the laboratory of Professor Takada in Japan, where the concentration of plastics was determined using a gas chromatograph equipped with a mass spectrometer (FOCUS-Polaris Q GC/MS) (Takada, 2006).

Results

To date, pellets have been received from over 50 countries, from Svalbard in the North to Macquarie Island in the South. Globally, concentrations of Chlordanes up to 151.6 ng/g (Sydney Harbour), PCBs up to 7131.2 ng/g (Robben Island), DDTs up to 1060.5 ng/g (Durres), HCH up to 36.4 ng/g (Portuguese Island) and PAHS up to 119,465.0 ng/g (Waikoura) have been reported.

(Continued)

> **Box 6.1 Case study: international pellet watch in monitoring marine pollutants (Continued)**
>
> Learning and knowledge outcomes
>
> The observations above have been used to create a global map of pollutant distribution, which highlights at-risk areas, as well as underpinning numerous research papers on both regional distribution and specific pollutants. This data has been applied in local and global risk assessments, in studies of potential bioaccumulation and trophic transfer, and as an indicator of the impacts of human activities. For example, the dominance of DDT in relation to its metabolites, which highlights the environmental implications of the use of DDT as a control of malaria-carrying mosquitoes (Ogata et al., 2009), the identification of PAHs and Hopanes as indicators of oil pollution (Yeo et al., 2015) and the identification of e-waste recycling facilities as probable sources of PCBs (Hosoda et al., 2014).
>
> More information regarding the outcomes of the IPW may be found at http://pelletwatch.org.

compounds, either produced intentionally for applications in agriculture, manufacturing or disease control or as waste products from industrial processes (such as the production of dioxin). They represent a significant threat to both people and environments globally and, as a result, most are highly regulated. Testing of recovered plastics has revealed variable concentrations of hydrophobic pollutants, many of which have an organic structure, including oil, pesticides and many pharmaceutical products, frequently observed to have negative effects on one or numerous taxa (Table 6.2).

Many environmental pollutants have an affinity to solids, manmade or otherwise, in aquatic systems. The sorption of materials into plastic substances may occur via a number of routes; *adsorption*, which occurs due to attractive forces between the surface of the plastic and contaminants, and *absorption*, where contaminants enter the plastic matrix. The adsorption of secondary contaminants may be *physical*, the result of van der Waals forces, or *chemical*, in which permanent or semipermanent bonds are formed between the contaminant and the plastic. Alternately the absorption into the plastic matrix by secondary substances is often driven by the hydrophobic nature of many plastics.

The lipophilicity (hydrophobicity) and hydrophilicity of a substance are measured using the n-octanol-water partition coefficient (K_{ow}), with substances with a K_{ow} greater than one classed as hydrophobic. Early thermodynamic modelling efforts to determine the contaminants most likely to be transported by small plastics in aquatic systems indicate that those with a log K_{OW} value greater than 5 have the potential to be transported on PE. However, in the gut, the presence of plastics was calculated to again result in preferential adherence onto plastics for chemicals with a log K_{OW} value between 5.5 and 6.5, this may potentially reduce the level of said chemicals within the affected organism (Gouin et al., 2011). In addition to the variable affinity of different pollutants, the ability of different polymers to transport secondary pollutants is not equal. For example, observations of the transport of phenanthrene and 4,4'-DDT onto PE and PVC revealed that

Table 6.2 Examples of persistent organic pollutants and other restricted chemicals and their occurrence on recovered environmental plastics (* with exemptions).

Pollutant	Use class: pesticide (P), industrial chemical (I) or unintentional by-product (U)	Classified as Annex 1: prohibited*, Annex 2: restricted, Annex 3: reduced, or Annex 4: managed at EoL under EU REACH Regulations	Classified as Annex A: elimination, Annex B: restriction, or Annex C: unintentional Production under the Stockholm Convention	Known effects	Recorded?
Aldrin	P	1, 4	A		Pellets on Chinese beaches
Cadmium	I				Microplastics on English beaches
Chlordane	P	1, 4	A		Pellets on Chinese beaches
Chlordecone	P	1, 4	A		Microplastic on Guadeloupe beaches
DDT	P	1	B		Pellets on Chinese beaches
					San Diego, US
Endosulfan	P	1	A		Pellets on Chinese beaches
					Microplastics on Gulf of Guinea Beaches
Heptachlor	P, I	1, 4	A		Pellets on Chinese beaches
Hexachlorobenzene (HCB)	P, I, U	1, 3, 4	A, C		Pellets on Chilean beaches
Lead	P, I, U	1			Microplastics on English beaches

(Continued)

Table 6.2 Examples of persistent organic pollutants and other restricted chemicals and their occurrence on recovered environmental plastics (* with exemptions). *Continued*

Pollutant	Use class: pesticide (P), industrial chemical (I) or unintentional by-product (U)	Classified as Annex 1: prohibited*, Annex 2: restricted, Annex 3: reduced, or Annex 4: managed at EoL under EU REACH Regulations	Classified as Annex A: elimination, Annex B: restriction, or Annex C: unintentional Production under the Stockholm Convention	Known effects	Recorded?
Lindane	P	1, 4	A		Microplastics on Gulf of Guinea Beaches
Polychlorinated biphenyls (PCBs)	I, U	1, 3, 4	A, C		Pellets on Chinese beaches
					Japanese Beaches
					Mexican Beaches
					San Diego, US

Phenanthrene/PE complexes resulted in the greatest transport, and phenanthrene/PVC the lowest (Bakir et al., 2014).

It has been suggested that hydrophobicity is too simple a measure by which to determine the role of micro- and nanoplastics in the transfer of chemicals to organisms (Marchant et al., 2022). The longer a piece of plastic is lost in the environment, the more extended the period over which it may accumulate secondary contaminants. However, the time over which micro- and nanoplastics persist in the environment may involve the impacts of temperature, UV radiation, mechanical wear, the action of organisms and other factors on both the polymer and its additives. As such, changes to the plastics' structure during environmental exposure may increase or decreasing apparent affinity. Nevertheless, it has previously been noted that the age of the pellet, determined using yellowing as a proxy for time in the environment, may be an indicator of the concentration of adsorbed POPs and other fatty pollutants (Chen et al., 2019).

In addition to the affinity between the polymer and the pollutant, when comparing macro-, micro- and nanoplastics, we must consider both the shape and the surface area to volume ratio (SA:V) of the material. Firstly, let us consider two spheres of equal size, made of marzipan, play-doh or other malleable object, then imagine that a hand has squashed one of them flat (changing the shape). Both the remaining sphere and the flattened sphere have the same volume, but the flattened sphere has a smaller diffusion distance. The larger the diffusion distance, the further a chemical must travel to move between the centre of the plastic and the environment. However, surface area is important too. To help us to visualise this ratio, we can imagine another two perfect spheres. In this case, one has a volume of 20 cm^3, the other is 10 cm^3, half the volume. However, the surface area of the first sphere is 35.63 cm^2, while the second is 22.45 cm^2. The first sphere has an SA:V of 1.782:1 while the smaller sphere has an SA:V of 2.245:1. The smaller the spheres become, the greater the ratio becomes, a sphere with a volume of 5 cm cubed would have and SA:V of 2.828:1 and so on. Of course, not all microplastics are perfect spheres (except in some of the exposure trials that we have previously been somewhat critical of!), and size is not a clear comparative guide to the available surface for diffusion or adsorption.

The presence of harmful additives and contaminants on or within the plastic may be determined via a number of methods, such as spectroscopic techniques following an extraction step (Verla et al., 2019), however, care must be taken to match the extraction method to the polymer under scrutiny, as some are less effective than others (Mato et al., 2001). Options for the extraction step include aqua regia (Rochman et al., 2014), nitric acid (Gao et al., 2019), HF + HNO_3 + H2SO4 (J. Wang et al., 2017), methanol (Teuten et al., 2007), hexane (Carpenter et al., 1972) and dicholoromethane (Rios et al., 2007), which may then be followed by spectroscopic approaches to determine the concentration of a contaminant, including GCMS, CGC-ECD-MD, GC-IMTS, GC-ECD, LSC, ICP-Ms, AAS, FAAS, ICP-OES and FP-XRF (Verla et al., 2019). Whatever the method, suggested quality control measures when establishing the presence of contaminant on or within plastics include a procedural blank to establish background contamination, a surrogate standard and positive controls of both solvent and sample to check recovery rates, and sufficient duplicates to ensure variation is sufficiently captured (Hong et al., 2017).

6.3 Transport within ecosystems

As mentioned above, observations have revealed the presence of POPs and other pollutants on pellets recovered globally. Some pollutants will be taken up directly from the surrounding environment, where the sorption of pollutants to polymers may result in a reduction in the concentration of these contaminants in the immediate locale, however, questions have arisen as to how such adsorption may affect the transport of contaminants and their uptake by organisms which consume the pollutant loaded plastics.

Whether pollutants on recovered plastics were accumulated in situ, or were transported, along with the plastics, to the recovery site is not always apparent; however, it is feasible that the transport of plastic also results in the transport of contaminants within and between habitats (Fig. 6.1). As indicated in Chapter 2, plastics may be moved horizontally and vertically within environments, being transported through soils, rivers and oceans. In addition to transporting secondary pollutants out of habitats, contaminants may also be moved from surface waters into sediments, or from upper to lower soil horizons.

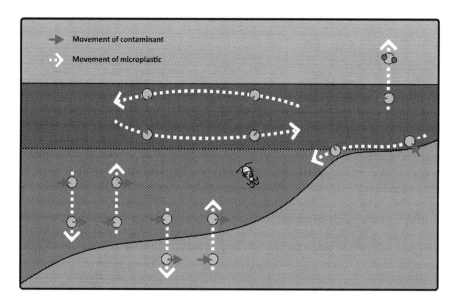

FIGURE 6.1 Potential plastic-mediated movement of contaminants.

The movement of microplastic in the environment and the resulting routes of pollutant redistribution.

6.3.1 **Aquatic transport and accumulation**

Globally, resin pellets have been used to establish the levels of persistent organic pollutants in coastal zones, revealing high levels of dpichlorodiphenyltrichloroethane DDT, hexachlorocyclohexane (HCH), polychlorinated biphenyls (PCBs) (Verla et al., 2019) and polybrominated diphenyl ethers (PBDEs) in locations including the coasts of Australia (Yeo et al., 2015), Chile (Pozo et al., 2020) and remote islands in the Pacific, Atlantic and Indian Oceans (Heskett et al., 2012). The abundance of contaminants associated with beached has shown increases in concentration over time, with the presence of PCBs and DDE were apparently absorbed from the environment, while nonylphenols (NP) were believed to be the result of the presence of additives and their breakdown products (Mato et al., 2001).

The use of many of these identified contaminants has been subject to widespread restrictions due to their known impacts. DDT, for example, has been seen to have significant impacts to insects and songbird populations, as well as range of health effects in humans, such as inducing tremors, vomiting and seizures. Although now heavily restricted, DDT is still used in rare cases for the control of mosquitos in malaria-prone areas. HCH's are another class of insecticides, seen to be neurotoxic, sometimes resulting in dizziness, headaches and convulsions. As with DDT, one form of HCH, Lindane, is still in use for the treatment of serious public health concerns.

A category of chemicals with a wide range of industrial uses, PCBs have been the subject to widespread regulation, including the Stockholm Convention on Persistent Organic Pollutants. Nevertheless, these carcinogenic compounds are still found in legacy plastics and the wider environment. As well as promoting the formation of cancers, PCBs are recognised as neurotoxic and endocrine disrupting, as are PBDEs, which are predominantly used as flame retardants in polyurethane foams for furniture construction. In their Fourth National Report on Human Exposure to Environmental Chemicals (Fourth Report), CDC scientists looked for the occurrence and levels of 10 PBDEs in the blood serum of 1985 participants, revealing the presence of PBDEs in 60% of participants.

In addition to the pollutants described above, plastics are also able to transport a variety of other contaminants such as bromine, heavy metals including aluminium, arsenic, cadmium, chromium, cobalt, copper, iron, manganese, nickel, lead, titanium and zinc (Verla et al., 2019), polycyclic aromatic hydrocarbons (PAHs) and other POPs such as dioxins and furans (Conesa, 2022). PAHs arise from the burning of fossil fuels, wood and other materials, although the extent of their impacts is not fully understood, whereas dioxins, like other POPs, act as carcinogens as well as affecting reproduction and development, causing damage to the immune system, and can interfer with hormones. Furans, on the other hand, are formed during the heating of food and have been shown to have the potential to cause eye, skin and mucus membrane irritation, in addition to pulmonary oedema and bronchiolar necrosis on inhalation. Once in the system, furans may affect the central nervous system as well as driving negative impacts on the liver at high levels (Jackson & Al-Taher, 2010).

Microplastics in aquatic systems have also been associated with the accumulation and transport of other emerging contaminants, such as antibiotics, which may enter the environment from homes, hospitals, and care facilities, as well as a range of agricultural uses, and are not easily removed in typical wastewater treatment (Zhuang & Wang, 2023). Antibiotics have been reported as interacting with numerous polymer types in wastewater, their apparent affinity driven by a combination of factors including hydrogen bonds, electrostatic charges, van der Waals forces and diffusion (Syranidou & Kalogerakis, 2022).

Similarly, experimental observations of the uptake of perfluoroalkylated substances (PFAS) both in the field and the laboratory have also shown the potential for these chemicals to be transported by microplastics (Barhoumi et al., 2022). When comparing PFAS uptake at lake locations and under controlled experimental conditions, levels on LDPE, PP and PET microplastics exposed in lacustrine environments for up to 3 months were from 24 to 259 times the background observed in lake water. Interestingly, there was no clear increase in adsorbed substances between months 1 and 3 suggesting the potential role of initial weathering in uptake (Scott et al., 2021).

6.3.2 Sediment Transport and Accumulation

This movement of microplastics between environmental compartments has significant impacts on both the rate of movement and the final concentration of pollutants in each zone. For example, sediment cores taken from Lake Qianhu in China were used to determine the impacts of microplastics on the input and retention of PAHs, comparing plastics-driven accumulation to that of sediment alone. This study found that the levels of PAHs were much higher on plastics than surrounding sediments, however, their lower total weight resulted in their having a less significant impact on accumulation. Nevertheless, significant variations were noted relating to the polymer and the hydrophobicity of the PAH (L. Zhang & Tao, 2022). Depending on the prevailing conditions, aquatic sediments may represent short or long-term sinks for plastics and associated contaminants, for example in regularly trawled areas, sediment may be resuspended on a frequent basis, whereas sediments in deeper water may be disturbed only in the event of extreme storm conditions. Unlike the transport of contaminants released with a typical sediment plume, microplastic-mediated secondary contaminants may travel further and — depending on polymer density, local conditions such as bathymetry, stratification and prevailing wind —in a different direction.

While the impact of plastics as a carrier of pollutants is most widely studied in aquatic settings, those present in soils also have the capacity to come into contact with and transport secondary pollutants. For example, agri-plastics, such as mulch films, irrigation tubing and containers may come into contact with pesticides, fertilisers and other chemicals. In terrestrial sediments, the migration of pre-aged PP microplastics during rain events has resulted in the redistribution of pollutants between soil layers (Yan et al., 2020). Here, the microplastics were associated with various antibiotic resistance genes, however, these processes are equally as impactful in relation to chemical contaminants. Also in soils, the presence of microplastics derived from agricultural mulch films has been seen to influence the absorption rates of pesticide residues, enhancing adsorption strength. This effect was found to be particularly apparent in the case of aged microplastics. Here, virgin microplastic was seen to decrease the environmental half-life of two pesticides, imidacloprid and flumioxazin, to 0.93 and 0.85 times the original level. However, aged microplastics extended the apparent half-life by 1.64 times (Wu et al., 2022). In some cases, the transport of microplastics and pollutants in sediments and soils may be human mediated. Samples of compost containing macro-, meso- and microplastics have revealed the presence of phthalates, acetyl tributyl citrate, dodecane and nonanal, found at higher concentrations in compost that were plastic free (Scopetani et al., 2022). Redistribution of municipal compost may therefore transport plastics and associated contaminants into agricultural and garden soils.

6.3.3 Final fractions: the cryosphere and atmosphere

Of course, re-mobilisation of microplastics is also driven by by aeolian processes, wind-blown plastics may also be a source of secondary pollutants into new environmental compartments, for example from agricultural environments treated with digested sewage sludge (Borthakur et al., 2022). Plastics carried on water and air currents may also be incorporated into glaciers, ice and snow, transporting secondary contaminants with them (Hamilton et al., 2023). These environments may act as temporary sinks from microplastics in the same way as shallow intertidal sediments, with the different that many areas of the cryosphere have low densities of biota that may otherwise come into contact with these plastics. Subsequent seasonal thaw and flow and the process of deglaciation may serve to relocate plastics (Windsor et al., 2019). Indeed, the effect of darker coloured plastics may reduce the apparent albedo effect and speed the melting of ice due to their light-absorbing properties (Zhang et al., 2022).

6.4 Transfer and effects

As well as carrying secondary contaminants within and between habitats, ingested microplastics may result in the contamination of numerous taxa, with the uptake of contaminated plastics identified or inferred in a variety of species (Fig. 6.2). For example, foraging seabirds around the Aleutian Islands have been see to contain elevated levels of phthalates, with levels differing between feeding strategies. Here plastic ingestion being identified as a key uptake route (Padula et al., 2020). Phthalates have also been observed in corals and water samples around the Maldives, with great levels of contamination being associated with proximity to highly populated islands (Saliu et al., 2019). Clear effects of plastic-mediated pollutants have also been

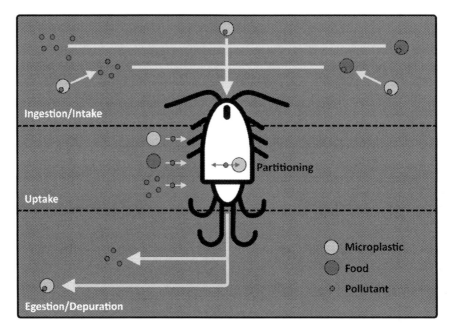

FIGURE 6.2 Movement of contaminants between environmental media, microplastics and biota.

The partitioning of pollutants between the environment and biota, some of which is mediated by plastics.

observed in the cyanobacterium *Anabaena* sp. PCC7120, in which antibiotic-loaded microplastics were seen to affect growth and chlorophyll content, whereas clean microplastics alone did not (González-Pleiter et al., 2020).

There are many routes by which microplastics may influence contaminant uptake by an organism. Direct ingestion of microplastics may result in the movement of contaminants from the plastic into the organisms, or from the organism into the plastic. Those plastics in the water column may also raise or lower the contaminant concentration in the surrounding environment by adsorption or desorption, affecting uptake by individuals in contact with that media, including prey species (Koelmans et al., 2016). Additionally, conditions within the digestive tract may result in the desorption and subsequent uptake of contaminants by the individual and, similarly, the distribution of contaminants within an affected organism may depend on the composition of the tissue and organs, in particular their water and fat content and apparent lithophilicity (Fig. 6.2).

Plastic-mediated pollutant uptake may influence humans too, with digestive system models which mimic the human gut being employed to determine the movement of contaminants associated with plastics. These have indicated that the conditions which develop in the gut may increase the movement of contaminants onto plastics by a factor of between 10 and 20. Such a change may serve to limit bioavailability of PCBs, reducing their uptake across the gut (Mohamed Nor et al., 2023).

The uptake of contaminants is commonly determined in laboratory studies similar to those outlined in Chapter 4, exposing organisms to microplastics treated with individual or mixed contaminants, compared to 'clean' plastic and blank controls. These studies have indicated mixed patterns of uptake. One of the first studies to explore these impacts, undertaken by Besseling et al. (2017), sought to establish the impact of microplastics on the uptake of chemicals in lugworm, *Arenciola marina*, by comparing the level of PCB uptake to that of 'natural' exposure routes. The picture obtained was complex, however, it was determined that the presence of PE spheres had minimal impact on contamination in these worms. Similarly, Browne et al. (2013) exposed *A. marina* to NP, phenanthrene, triclosan and PBDE-47 via both PVC and sediments, with their study demonstrating uptake in addition to subsequent negative health effects.

Nevertheless, while the studies above may indicate the apparent potential for the transfer of pollutants from plastic to biota, we must consider these findings in relation to typical level of microplastic uptake and retention of microplastics in wild caught organisms discussed in Chapter 4. In many species, particularly those from pelagic and deep-sea habitats far from sources of microplastic input, the average abundance of plastics per individual may be below 1, and the apparent retention time less than a day. Thus the potential for transfer to individuals of these species is significantly lower than those species living in areas with elevated levels of both plastics and pollutants. Indeed, not all organisms show clear patterns of increase or decrease in contaminant loads following contact with microplastics. Indeed, in their study of the uptake of PCBs associated with either PE or PS by the Norway Lobtser, *Nephrops norvegicus*, Devriese et al. (2017) observed minimal uptake. Alternatively, microplastics may also reduce the impacts of some pollutants. For example, in rice, *Oryza sativa*, exposed to the PS microplastics and florfenicol, a broad-spectrum bacteriostatic antibiotic, the negative metabolic impacts of the antibiotic were lower than in the florfenicol only treatment (Shen et al., 2023). Thus despite these often concerning accounts, meta-analysis of this experimental data has suggested that the level of contaminants transported by plastics is likely to be minimal when compared the role of other environmental media (Koelmans et al., 2016). Additionally, the ingestion of 'clean' plastics by contaminated organisms may result in a reduction in the pollutant loads of the organisms concerned, particularly where the affinity between the contaminant and the polymer is high. For example, the earthworm, *Metaphire californica*, has previously demonstrated reduced arsenic uptake when exposed in combination with microplastics compared to worms exposed to arsenic alone (Wang et al., 2019).

6.5 Biofouling and harmful organisms

In addition to the transport of chemical substances, microplastics may also harbour hitchhiking organisms, otherwise known as biofouling. Biofouling is the colonisation of objects in water by microorganisms, algae, plants and animals. The process of fouling begins with the adherence of microorganisms and the formation of adhere matrices which alter the surface of the plastic, enabling colonisation by larger species.

Such biofouling is regularly observed on natural and manmade floating debris, as well and on marine structures and vessels. Biofouling on macroplastics has been seen to result in the movement of many species globally, while those on microplastics will be limited in size. We have previously discussed the impact that fouling comminates may have on the buoyancy of microplastic material, with sequential phases of accumulation and sloughing resulting on potential 'bouncing' in the water column, as well as increasing the likelihood of ingestion by feeding species. However, this fouling may also result in the transport of species to areas that they may not otherwise be able to reach. While many of these species are comparatively harmless, some of the fouling organisms have been characterised as either invasive or pathogenic. For example, *Vibrio* spp. are pathogenic bacteria responsible for gastroenteritis and sepsis. Nevertheless, while *Vibrio* have previously seen in fouling assemblages (Kirstein et al., 2016), the effect of microplastic-mediated movement of these organisms remains unclear.

6.6 Conclusions

The structure of plastic polymers may result in the magnification and/or transport of hydrophobic pollutants and other harmful factors in the environment, as well as facilitating their uptake by biota. However, the consistency of these patterns and their importance at current environmental concentrations is questionable. There are a great many unexplored factors which obscure our understanding of these issues. For example in determining the role of microplastics in the transport of heavy metals, various forms of microplastics have been seen to have differential rates of uptake and contaminant desorption is dependent on the pH of the surrounding solution (Khalid et al., 2021). Thus the use of a more representative mix of polymer types and plastic shapes in studies of chemical transfer is required to enable us to understand the role of plastics in mediating contaminant exposure. Additionally, many studies are reliant upon high detection limits and low detection frequencies due to the difficulty involved in sampling from remote locations (Hong et al., 2017).

References

Bakir, A., Rowland, S. J., & Thompson, R. C. (2014). Transport of persistent organic pollutants by microplastics in estuarine conditions. *Estuarine, Coastal and Shelf Science*, *140*, 14–21. Available from https://doi.org/10.1016/j.ecss.2014.01.004.

Barhoumi, B., Sander, S. G., & Tolosa, I. (2022). A review on per-and polyfluorinated alkyl substances (PFASs) in microplastic and food-contact materials. *Environmental Research*, *206*, 112595.

Besseling, E., Foekema, E. M., Van Den Heuvel-Greve, M. J., & Koelmans, A. A. (2017). The effect of microplastic on the uptake of chemicals by the lugworm Arenicola marina (L.) under environmentally relevant exposure conditions. *Environmental Science and Technology*, *51*(15), 8795−8804. Available from https://doi.org/10.1021/acs.est.7b02286, http://pubs.acs.org/journal/esthag.

Borthakur, A., Leonard, J., Koutnik, V. S., Ravi, S., & Mohanty, S. K. (2022). Inhalation risks of wind-blown dust from biosolid-applied agricultural lands: Are they enriched with microplastics and PFAS? *Current Opinion in Environmental Science and Health*, *25*, 100309. Available from https://doi.org/10.1016/j.coesh.2021.100309, http://www.journals.elsevier.com/current-opinion-in-environmental-science-and-health.

Browne, M. A., Niven, S. J., Galloway, T. S., Rowland, S. J., & Thompson, R. C. (2013). Microplastic moves pollutants and additives to worms, reducing functions linked to health and biodiversity. *Current Biology*, *23*(23), 2388−2392. Available from https://doi.org/10.1016/j.cub.2013.10.012.

Carpenter, E. J., Anderson, S. J., Harvey, G. R., Miklas, H. P., & Peck, B. B. (1972). Polystyrene spherules in coastal waters. *Science (New York, N.Y.)*, *178*(4062), 749−750. Available from https://doi.org/10.1126/science.178.4062.749.

Chen, Q., Allgeier, A., Yin, D., & Hollert, H. (2019). Leaching of endocrine disrupting chemicals from marine microplastics and mesoplastics under common life stress conditions. *Environment International*, *130*, 104938. Available from https://doi.org/10.1016/j.envint.2019.104938.

Conesa, J. A. (2022). Adsorption of PAHs and PCDD/Fs in microplastics: A review. *Microplastics*, *1*(3), 346−358. Available from https://doi.org/10.3390/microplastics1030026.

Devriese, L. I., De Witte, B., Vethaak, A. D., Hostens, K., & Leslie, H. A. (2017). Bioaccumulation of PCBs from microplastics in Norway lobster (Nephrops norvegicus): An experimental study. *Chemosphere*, *186*, 10−16. Available from https://doi.org/10.1016/j.chemosphere.2017.07.121, http://www.elsevier.com/locate/chemosphere.

El-Sayed Ali, T., & Legler, J. (2010). *Overview of the mammalian and environmental toxicity of chlorinated paraffins* (pp. 135−154). Springer Nature. Available from 10.1007/698_2010_56.

Fiedler, H. (2010). *Short-chain chlorinated paraffins: Production, use and international regulations* (pp. 1−40). Springer Nature. Available from 10.1007/698_2010_58.

Filella, M., Hennebert, P., Okkenhaug, G., & Turner, A. (2020). Occurrence and fate of antimony in plastics. *Journal of Hazardous Materials*, *390*, 121764. Available from https://doi.org/10.1016/j.jhazmat.2019.121764, http://www.elsevier.com/locate/jhazmat.

Fu, X., Xie, X., Charlet, L., & He, J. (2023). A review on distribution, biogeochemistry of antimony in water and its environmental risk. *Journal of Hydrology*, *625*, 130043. Available from https://doi.org/10.1016/j.jhydrol.2023.130043, http://www.elsevier.com/inca/publications/store/5/0/3/3/4/3.

Gao, F., Li, J., Sun, C., Zhang, L., Jiang, F., Cao, W., & Zheng, L. (2019). Study on the capability and characteristics of heavy metals enriched on microplastics in marine environment. *Marine Pollution Bulletin*, *144*, 61−67. Available from https://doi.org/10.1016/j.marpolbul.2019.04.039, http://www.elsevier.com/locate/marpolbul.

González-Pleiter, M., Pedrouzo-Rodríguez, A., Verdú, I., Leganés, F., Marco, E., & Rosal, R. (2020). Fernández-Piñas\nMicroplastics as vectors of the antibiotics azithromycin and clarithromycin: Effects towards freshwater microalgae. *Chemosphere*, *268*, 128824. Available from https://doi.org/10.1016/j.chemosphere.2020.128824.

Gouin, T., Roche, N., Lohmann, R., & Hodges, G. (2011). A thermodynamic approach for assessing the environmental exposure of chemicals absorbed to microplastic. *Environmental Science and Technology*, *45*(4), 1466−1472. Available from https://doi.org/10.1021/es1032025.

Hamilton, B. M., Baak, J. E., Vorkamp, K., Hammer, S., Granberg, M., Herzke, D., & Provencher, J. F. (2023). Plastics as a carrier of chemical additives to the Arctic: Possibilities for strategic monitoring across the circumpolar North. *Arctic Science*, *9*(2), 284−296. Available from https://doi.org/10.1139/as-2021-0055, http://www.nrcresearchpress.com/journal/as.

Heskett, M., Takada, H., Yamashita, R., Yuyama, M., Ito, M., Geok, Y. B., Ogata, Y., Kwan, C., Heckhausen, A., Taylor, H., Powell, T., Morishige, C., Young, D., Patterson, H., Robertson, B., Bailey, E., & Mermoz, J. (2012). Measurement of persistent organic pollutants (POPs) in plastic resin pellets from remote islands: Toward establishment of background concentrations for International Pellet Watch. *Marine Pollution Bulletin*, *64*(2), 445−448. Available from https://doi.org/10.1016/j.marpolbul.2011.11.004.

Hong, S. H., Shim, W. J., & Hong, L. (2017). Methods of analysing chemicals associated with microplastics: A review. *Analytical Methods*, *9*(9), 1361−1368. Available from https://doi.org/10.1039/C6AY02971J.

Hosoda, J., Ofosu-Anim, J., Sabi, E. B., Akita, L. G., Onwona-Agyeman, S., Yamashita, R., & Takada, H. (2014). Monitoring of organic micropollutants in Ghana by combination of pellet watch with sediment analysis: E-waste as a source of PCBs. *Marine Pollution Bulletin*, *86*(1−2), 575−581. Available from https://doi.org/10.1016/j.marpolbul.2014.06.008, http://www.elsevier.com/locate/marpolbul.

Jackson, L. S. and Al-Taher, F. (2022). Processing issues: acrylamide, furan, and trans fatty acids. In Ensuring Global Food Safety (pp. 229−257). Academic Press.

Khalid, N., Aqeel, M., Noman, A., Khan, S. M., & Akhter, N. (2021). Interactions and effects of microplastics with heavy metals in aquatic and terrestrial environments. *Environmental Pollution*, *290*, 118104. Available from https://doi.org/10.1016/j.envpol.2021.118104, https://www.journals.elsevier.com/environmental-pollution.

Kirstein, I. V., Kirmizi, S., Wichels, A., Garin-Fernandez, A., Erler, R., Löder, M., & Gerdts, G. (2016). Dangerous hitchhikers? Evidence for potentially pathogenic Vibrio spp. on microplastic particles. *Marine Environmental Research*, *120*, 1−8. Available from https://doi.org/10.1016/j.marenvres.2016.07.004, http://www.elsevier.com/locate/marenvrev.

Kobetičová, K., & Černý, R. (2018). Ecotoxicity assessment of short- and medium-chain chlorinated paraffins used in polyvinyl-chloride products for construction industry. *Science of the Total Environment*, *640−641*, 523−528. Available from https://doi.org/10.1016/j.scitotenv.2018.05.300, http://www.elsevier.com/locate/scitotenv.

Koelmans, A. A., Bakir, A., Burton, G. A., & Janssen, C. R. (2016). Microplastic as a vector for chemicals in the aquatic environment: Critical review and model-supported reinterpretation of empirical studies. *Environmental Science and Technology*, *50*(7), 3315−3326. Available from https://doi.org/10.1021/acs.est.5b06069, http://pubs.acs.org/journal/esthag.

Ma, Y., Liu, H., Wu, J., Yuan, L., Wang, Y., Du, X., Wang, R., Marwa, P. W., Petlulu, P., Chen, X., & Zhang, H. (2019). The adverse health effects of bisphenol A and related toxicity mechanisms. *Environmental Research*, *176*, 108575. Available from https://doi.org/10.1016/j.envres.2019.108575, http://www.elsevier.com/inca/publications/store/6/2/2/8/2/1/index.htt.

Marchant, D. J., Iwan Jones, J., Zemelka, G., Eyice, O., & Kratina, P. (2022). Do microplastics mediate the effects of chemicals on aquatic organisms? *Aquatic Toxicology*, *242*, 106037. Available from https://doi.org/10.1016/j.aquatox.2021.106037, http://www.elsevier.com/wps/find/journaldescription.cws_home/505509/description#description.

Mato, Y., Isobe, T., Takada, H., Kanehiro, H., Ohtake, C., & Kaminuma, T. (2001). Plastic resin pellets as a transport medium for toxic chemicals in the marine environment. *Environmental Science and Technology*, *35*(2), 318−324. Available from https://doi.org/10.1021/es0010498.

Mohamed Nor, N. H., Niu, Z., Hennebelle, M., & Koelmans, A. A. (2023). How digestive processes can affect the bioavailability of PCBs associated with microplastics: A modeling study supported by empirical data. *Environmental Science and Technology*, *57*(31), 11452−11464. Available from https://doi.org/10.1021/acs.est.3c02129, http://pubs.acs.org/journal/esthag.

Ogata, Y., Takada, H., Mizukawa, K., Hirai, H., Iwasa, S., Endo, S., Mato, Y., Saha, M., Okuda, K., Nakashima, A., Murakami, M., Zurcher, N., Booyatumanondo, R., Zakaria, M. P., Dung, L. Q., Gordon, M., Miguez, C., Suzuki, S., Moore, C., ... Thompson, R. C. (2009). International Pellet Watch: Global monitoring of persistent organic pollutants (POPs) in coastal waters. 1. Initial phase data on PCBs, DDTs, and HCHs. *Marine Pollution Bulletin*, *58*(10), 1437−1446. Available from https://doi.org/10.1016/j.marpolbul.2009.06.014.

Padula, V., Beaudreau, A. H., Hagedorn, B., & Causey, D. (2020). Plastic-derived contaminants in Aleutian Archipelago seabirds with varied foraging strategies. *Marine Pollution Bulletin*, *158*, 111435. Available from https://doi.org/10.1016/j.marpolbul.2020.111435, http://www.elsevier.com/locate/marpolbul.

Pozo, K., Urbina, W., Gómez, V., Torres, M., Nuñez, D., Přibylová, P., Audy, O., Clarke, B., Arias, A., Tombesi, N., Guida, Y., & Klánová, J. (2020). Persistent organic pollutants sorbed in plastic resin pellet — "Nurdles" from coastal areas of Central Chile. *Marine Pollution Bulletin*, *151*, 110786. Available from https://doi.org/10.1016/j.marpolbul.2019.110786, http://www.elsevier.com/locate/marpolbul.

Rios, L. M., Moore, C., & Jones, P. R. (2007). Persistent organic pollutants carried by synthetic polymers in the ocean environment. *Marine Pollution Bulletin*, *54*(8), 1230−1237. Available from https://doi.org/10.1016/j.marpolbul.2007.03.022.

Rochman, C. M., Hentschel, B. T., & Teh, S. J. (2014). Long-term sorption of metals is similar among plastic types: Implications for plastic debris in aquatic environments. *PLoS One*, *9*(1), e85433. Available from https://doi.org/10.1371/journal.pone.0085433, http://www.plosone.org/article/fetchObject.action?uri = info%3Adoi%2F10.1371%2Fjournal.pone.0085433&representation = PDF.

Saliu, F., Montano, S., Leoni, B., Lasagni, M., & Galli, P. (2019). Microplastics as a threat to coral reef environments: Detection of phthalate esters in neuston and scleractinian corals from the Faafu Atoll, Maldives. *Marine Pollution Bulletin*, *142*, 234−241. Available from https://doi.org/10.1016/j.marpolbul.2019.03.043, http://www.elsevier.com/locate/marpolbul.

Scopetani, C., Chelazzi, D., Cincinelli, A., Martellini, T., Leiniö, V., & Pellinen, J. (2022). Hazardous contaminants in plastics contained in compost and agricultural soil.

Chemosphere, *293*, 133645. Available from https://doi.org/10.1016/j.chemosphere.2022.133645, http://www.elsevier.com/locate/chemosphere.

Scott, J. W., Gunderson, K. G., Green, L. A., Rediske, R. R., & Steinman, A. D. (2021). Perfluoroalkylated substances (PFAS) associated with microplastics in a lake environment. *Toxics*, *9*(5), 106. Available from https://doi.org/10.3390/toxics9050106.

Shen, L., Zhang, P., Lin, Y., Huang, X., Zhang, S., Li, Z., Fang, Z., Wen, Y., & Liu, H. (2023). Polystyrene microplastic attenuated the toxic effects of florfenicol on rice (Oryza sativa L.) seedlings in hydroponics: From the perspective of oxidative response, phototoxicity and molecular metabolism. *Journal of Hazardous Materials*, *459*, 132176. Available from https://doi.org/10.1016/j.jhazmat.2023.132176, http://www.elsevier.com/locate/jhazmat.

Syranidou, E., & Kalogerakis, N. (2022). Interactions of microplastics, antibiotics and antibiotic resistant genes within WWTPs. *Science of the Total Environment*, *804*, 150141. Available from https://doi.org/10.1016/j.scitotenv.2021.150141, http://www.elsevier.com/locate/scitotenv.

Takada, H. (2006). Call for pellets! International Pellet Watch Global Monitoring of POPs using beached plastic resin pellets. *Marine Pollution Bulletin*, *52*(12), 1547–1548. Available from https://doi.org/10.1016/j.marpolbul.2006.10.010.

Teuten, E. L., Rowland, S. J., Galloway, T. S., & Thompson, R. C. (2007). Potential for plastics to transport hydrophobic contaminants. *Environmental Science and Technology*, *41*(22), 7759–7764. Available from https://doi.org/10.1021/es071737s.

Turner, A., & Filella, M. (2021). Hazardous metal additives in plastics and their environmental impacts. *Environment International*, *156*, 106622. Available from https://doi.org/10.1016/j.envint.2021.106622, http://www.elsevier.com/locate/envint.

Verla, A. W., Enyoh, C. E., Verla, E. N., & Nwarnorh, K. O. (2019). Microplastic–toxic chemical interaction: a review study on quantified levels, mechanism and implication. *SN Applied Sciences*, *1*(1400). Available from https://doi.org/10.1007/s42452-019-1352-0.

vom Saal, F. S., & Hughes, C. (2005). An extensive new literature concerning low-dose effects of bisphenol A shows the need for a new risk assessment. *Environmental Health Perspectives*, *113*(8), 926–933. Available from https://doi.org/10.1289/ehp.7713.

Wang, H. T., Ding, J., Xiong, C., Zhu, D., Li, G., Jia, X. Y., Zhu, Y. G., & Xue, X. M. (2019). Exposure to microplastics lowers arsenic accumulation and alters gut bacterial communities of earthworm Metaphire californica. *Environmental Pollution*, *251*, 110–116. Available from https://doi.org/10.1016/j.envpol.2019.04.054, https://www.journals.elsevier.com/environmental-pollution.

Wang, J., Peng, J., Tan, Z., Gao, Y., Zhan, Z., Chen, Q., & Cai, L. (2017). Microplastics in the surface sediments from the Beijiang River littoral zone: Composition, abundance, surface textures and interaction with heavy metals. *Chemosphere*, *171*, 248–258. Available from https://doi.org/10.1016/j.chemosphere.2016.12.074, http://www.elsevier.com/locate/chemosphere.

Windsor, F. M., Durance, I., Horton, A. A., Thompson, R. C., Tyler, C. R., & Ormerod, S. J. (2019). A catchment-scale perspective of plastic pollution. *Global Change Biology*, *25*(4), 1207–1221. Available from https://doi.org/10.1111/gcb.14572, http://onlinelibrary.wiley.com/journal/10.1111/(ISSN)1365-2486.

Wu, C., Pan, S., Shan, Y., Ma, Y., Wang, D., Song, X., Hu, H., Ren, X., Ma, X., Cui, J., & Ma, Y. (2022). Microplastics mulch film affects the environmental behavior of adsorption and degradation of pesticide residues in soil. *Environmental Research, 214*, 114133. Available from https://doi.org/10.1016/j.envres.2022.114133, https://www.sciencedirect.com/journal/environmental-research.

Yan, X., Yang, X., Tang, Z., Fu, J., Chen, F., Zhao, Y., Ruan, L., & Yang, Y. (2020). Downward transport of naturally-aged light microplastics in natural loamy sand and the implication to the dissemination of antibiotic resistance genes. *Environmental Pollution, 262*, 114270. Available from https://doi.org/10.1016/j.envpol.2020.114270, https://www.journals.elsevier.com/environmental-pollution.

Yeo, B. G., Takada, H., Taylor, H., Ito, M., Hosoda, J., Allinson, M., Connell, S., Greaves, L., & McGrath, J. (2015). POPs monitoring in Australia and New Zealand using plastic resin pellets, and International Pellet Watch as a tool for education and raising public awareness on plastic debris and POPs. *Marine Pollution Bulletin, 101*(1), 137−145. Available from https://doi.org/10.1016/j.marpolbul.2015.11.006, http://www.elsevier.com/locate/marpolbul.

Zhang, L., & Tao, Y. (2022). Microplastics contributed much less than organic matter to the burial of polycyclic aromatic hydrocarbons by sediments in the past decades: A case study from an urban lake. *Environmental Science: Processes and Impacts, 24*(11), 2100−2107. Available from https://doi.org/10.1039/d2em00309k, http://pubs.rsc.org/en/journals/journal/em.

Zhang, Y. L., Kang, S. C., & Gao, T. G. (2022). Microplastics have light-absorbing ability to enhance cryospheric melting. *Advances in Climate Change Research, 13*(4), 455−458. Available from https://doi.org/10.1016/j.accre.2022.06.005, http://www.sciencedirect.com/science/journal/aip/16749278.

Zhou, D., Yang, J., Li, H., Lu, Q., Liu, Y. D., & Lin, K. F. (2016). Ecotoxicity of bisphenol A to Caenorhabditis elegans by multigenerational exposure and variations of stress response in vivo across generations. *Environmental Pollution, 208*, 767−773. Available from https://doi.org/10.1016/j.envpol.2015.10.057, http://www.elsevier.com/inca/publications/store/4/0/5/8/5/6.

Zhuang, S., & Wang, J. (2023). Interaction between antibiotics and microplastics: Recent advances and perspective. *Science of the Total Environment, 897*, 165414. Available from https://doi.org/10.1016/j.scitotenv.2023.165414, http://www.elsevier.com/locate/scitotenv.

A great many Rs

7

'Anything not saved will be lost' — Nintendo quit screen

7.1 Introduction

The environmental abundance of microplastic — and secondary microplastics in particular — is linked to the global volume of mishandled plastic wastes. As indicated in Chapter 1, plastics may enter the environment via both accidental loss or intentional discards, both of which are linked to apparent failures in national and international waste policy and infrastructure. For example, ships utilising ports with fees associated with waste disposal may be more likely to illegally dispose of materials at sea, and countries with disconnected or unsuitable waste management infrastructure may see increased loss of material to the environment despite the best interests of those involved. Although the small size of microplastics present significant challenges to traditional waste management structures, the correct handling of plastic parent material may susbtantially limit the formation and introduction of secondary particles.

As with other commercial and domestic wastes, plastics are managed in relation to the waste hierarchy. First suggested in the 1970s and further developed during the 1980s, the heirarchy aims to identify an order of preference for waste handling measures, beginning with those with the smallest environmental and human health impacts and ending with disposal to landfill. In its first iteration, the hierarchy included the Three Rs: *Reduction* (or Prevention), referring to the prevention of new wastes being formed, typically by altering product design or changing the behaviour of both producers and consumers; *Reuse* (and preparation for Reuse), referring to the management and extension of a product lifespan, again as a result of either design or behaviour; and the *Recycling* of the materials within a product. Only those items that may not be treated in this way are then sent for disposal via landfilling, incineration or other suitable management method. More recently, this hierarchy has been expanded in order to separate *Recovery*, typically in the form of waste-to-energy programmes, from recycling, due to the destructive nature of the process. Latterly, other Rs have been suggested, *Replace*, which highlights the use of alternative materials to reduce post-disposal impacts, and *Remove*, which seeks to find options to extract wastes from

Microplastics. DOI: https://doi.org/10.1016/B978-0-443-13324-4.00007-8

polluted environments. Thus the hierarchy provides a guideline by which we might maximise the potential conservation of resources and emphasises the importance of reduced consumption (Fig. 7.1). Over this and the following chapters, we will discuss both established and emerging methods for the management of plastic wastes.

7.2 Reduce

As already indicated, the most favourable position in the waste hierarchy is the reduction, or prevention, of plastic waste generation by limiting the demand for plastic products. Previous initiatives have achieved this by many routes. At its simplest, reduction may focus on ways to incite behavioural change in consumers, encouraging the avoidance of wasteful products. These may be achieved on an individual basis, or as a collective action of community groups or similar organisations. For example, the implementation plastic-free initiatives in schools to both raise awareness and promote sustainable practices, or the frequent social media posting of zero-waste influencers. Subsequently, many individuals have taken a visible stand against consumptive fashion practices by posting themselves wearing the same item of clothing on 100's of successive days (often without the apparent notice of friends and coworkers) (Beddington, 2021; O'Connell, 2009), whereas others make an effort to count and catalogue their consumption, frequently ending their accounts by (rather satisfyingly) returning the packaging to the manufacturer or producer. For example, Green Peace's The Big Plastic Count, which enabled the public to audit 1 week of household plastic waste, using this data in their advocacy for behavioural and policy change.

Unfortunately, appeals to the consumer to change their habits may not be successful in isolation, a problem that we will look at in more detail in Chapter 11. In this event, changes in public policy, which we will discuss in Chapter 10, may be necessary to assist in incentivising driving these behaviours (or disincentivising

FIGURE 7.1

Lasink's ladder. Preferred waste management routes.

consumptive practices). Additionally, the adoption of sustainable behaviours concerning the use of plastics is often reliant on the availability of alternative products as well as the services and networks that underpin their use.

Alternatively, the source of plastic wastes may not be controllable by the consumer, instead reflecting waste generated during the supply chain itself, here the onus also must be placed on industry to reduce their plastic output. For example, businesses may limit their plastic waste production by reducing the volume of packaging per product. Previously the has been achieved either by decreasing the dimensions of existing packaging, or by the exclusion of materials such as film coatings. Examples of industry commitments to minimising plastic waste include a range of highly publicised corporate campaigns. For example, Subaru plants claim to have sent no waste to landfill since 2005 and aiming for over 95% reusability/recyclability in their components. Other 'zero landfill' pledgers include Mars, Proctor and Gamble and Unilever. Many of these companies have been supported in achieving these goals by specialist waste management groups, such as TerraCycle, who provide the services that may otherwise be insufficient in underpinning plastic-free and low-plastic behaviours at both individual and corporate scale.

The drive for behavioural change in industry may be a result of many factors, including public pressure, financial benefits or legislative change. For example, many firms moved to remove microbeads from cosmetics and other household goods prior to the implementation of bans in multiple countries, identifying in advance the change in public opinion and inevitability of regulation. Industrial behavioural change may be also expedited by way of product bans, such as those applied to straws, take-away containers and other single-use plastics at locations around the world. Another suggested option for the reduction of plastics wastes is that of Extended Producer Responsibility schemes which seek to hold manufacturers responsible for the environmental impacts of their products, thus encouraging reduced use. The development and effectiveness of these measures will be discussed in more detail in Chapter 10. Otherwise, reduction in the consumption of plastic products may also be achieved in combination with many of the other Rs outlined below, such as the replacement of plastics in products by alternative materials.

7.3 **Reuse**

Our current overreliance on single-use plastics and the resulting output of plastic waste is partially driven by early promotion of plastics as single use and disposable. Disrupting established patterns of resource extraction, manufacture, use and disposal is a key factor in minimising the wastes generated by both individuals and industry. The reuse of goods is essential in encouraging more sustainable exploitation of global resources. The methods given in this section are frequently aligned with those concerning the replacement of materials outlined below and attempt to extend the lifespan of a product. To keep things clear, here we will focus on reusable *plastic* products, rather that reusable products of any alternative material.

By encouraging the reuse of plastic products, extending their useful life, it is possible to limit the negative impacts of resource extraction, manufacture and disposal that arise as a result of frequent repurchasing. This may be achieved by making previously single-use plastic products more robust, or by otherwise changing the design of more complex goods which contain plastics to delay the point of obsolescence (Box 7.1). Informal reuse of plastic products, both for the intended and unintended purposes, has been apparent since their introduction, and upcycling and repurposing of materials are widespread.

However, over the past decade, extensive efforts to limit our use of single use plastics have seen reusable water bottles, bags, coffee cups and other products become commonplace, with their use supported by both governmental policies and commercial practices. For example, refill services for many household consumables are increasingly available, supported by brands such as Ecover. Similarly, many high street café chains support consumers with refillable cups, often additionally providing discounted costs to those customers, and water bottle-refill stations are increasingly available in many public spaces. At more local scales, bulk stores provide shoppers with discounted prices if they bring

---

Box 7.1 Planned obsolescence and the rise of the repair cafe

Planned obsolescence refers to the intentional creation of products with a predetermined useful lifespan, often resulting in a limited number of reuses and increased product sales. This lifespan may be the result of many factors, such as the use of components likely to have a limited wear time, reduced compatibility with newer technologies or software, limited availability of replacement parts as manufacturers introduce newer models, or the introduction of new aesthetics which may be perceived as more desirable by consumers or negatively affect the desirability of the existing product or device. Examples of many of these intentional or incidental approaches may be seen in the personal electronics such as mobile phones, which have become increasingly difficult to maintain by the users (such as in the replacement of screens, batteries and other components) and may become unsupported by service providers. Unsurprisingly, these practices have been widely criticised as unsustainable for promoting increased consumption and the volume or waste generated as a result.

  In partial response to this practice, as well as to changing economic forces, there has been increasing movement towards extending product lifespans through repairability, modularity and durability of components. The associated public interest has resulted in the development of networks of repair cafés and similar local groups of individuals (often led by skilled volunteers) who come together to share the methods, tools, and expertise required to maintain a variety of products, including the aforementioned phones, appliances and clothing, as well as to advocate for more sustainable behaviours. This drive is associated with the push for Right to Repair legislation, such as the UK Statutory Instrument: The Ecodesign for Energy-Related Products and Energy Information Regulations 2021 No. 745, which would force producers to provide consumers with access to the knowledge and components to repair products themselves.

their own reusable containers to be filled with produce. Alternatively, deposit returns schemes may offer discounts to consumers to surrender packaging, bottles or similar for reuse. We will discuss widespread deposit returns schemes in greater detail in Chapter 10.

Outside of the realm of packaging, there is also a rejuvenation in the market for refurbished goods, particularly in relation to personal technology and white goods. Some brands may offer factory refurbished options at discounted prices to the consumer, and numerous online firms now purchase small electronics from the consumer before refurbishing them and placing them back on the market.

While encouraging the reuse of plastic products is essential in minimising our waste output, these behaviours are not without associated environmental implications. For example, bottles returned for reuse must be stored, transported and cleaned, resulting in increased carbon and water footprints. Additionally, reusable products are generally more durable than their single-use counterparts, which may incur the use of more resources in their initial manufacture. For example, the widespread implementation of both bans and charges relating the single-use carrier bags, designed to encourage consumers to utilise reusable alternatives constructed of either heavy duty plastic or plant-fibre-derived alternatives. This may result in a minimum number of uses before the resource costs incurred are offset. Similarly, plastics reused in the home must also be sanitised. As a result, widespread reuse of plastic items may require the introduction of legislation or infrastructure to enable associated behavioural changes.

7.4 Recycle

The most widely recognised method for the handling of plastic waste is recycling. Recycling processes utilise the material in an existing product to create the raw materials for a new product. It is important to recognise, however, that recycling takes on many forms, with differing levels of intervention, resource recovery and secondary environmental impacts. At its most efficient, recycling processes maintain the quality and purity of the resources contained, enabling the material to be used in the same or equivalent products without the need for large volumes of additional virgin material. We may refer to this as primary' recycling. However, recycling often results in a loss of mechanical properties which reduce the value of the recycled resource. Subsequently, the resulting material is consigned to the production of lower-quality goods. This may be referred to as secondary' recycling (Simón et al., 2018). This reduction in value may relate to the polymer itself and the internal changes resulting from processing, the presence of challenging additives or the purpose to which the original product was put. For example, plastics containing high levels of colourants may not be easily processed or repurposed, except in the event that the colour is desired or may be obscured.

The stages of both primary and secondary recycling may include a variety of physical or mechanical steps. The most commonly applied method is *mechanical recycling*, during which products of similar construction are accumulated and sorted by polymer to ensure that they undergo the correct recycling steps. Sorting may be achieved via induction methods which utilised drum separators, which screen materials by size; sink—float methods which remove plastics based on their specific gravity; sensors to determine different waste types passing under a conveyor; eddy current separators, which use magnetic field to remove nonferrous metals; X-ray methods for the determination of density; and near infrared spectra in order to identify their composition.

Sorted plastics are then cleaned to exclude any contaminants accumulated during use and handling prior to shredding (and, potentially, secondary cleaning) and either agglomeration or melting and extrusion into pellets, nibs or fibres (Garcia & Robertson, 2017). During the manufacturing process, recycled plastics may be mixed with virgin plastics and fillers to improve or maintain the mechanical properties of the product. In the case of polyurethanes, plastics may be re-bonded, a process by which mechanically fragmented foams are moulded together in order to produce secondary products such as matting, underlay and insulation. Alternatively, existing polyurethane waste may be grouped and powdered, then either and mixed with virgin plastic at a weight of up to 30% in the creation of new primary products, or compression moulded to create new primary products mead up of entirely recycled material.

If not mechanically treated, plastics may also be chemically recycled. During chemical recycling, plastics may be converted to shorter polymers, monomer units, gasses, liquids and other by-products. These 'tertiary' recycling processes are beneficial in their ability to treat contaminated and hard to recycle materials, the most common processes for which are depolymerisation, pyrolysis, gasification and solvolysis (Vollmer et al., 2020). The process of depolymerisation utilises heat and a catalyst in order to breakdown otherwise recalcitrant long-chain polymers into monomer units. Catalysts used in these processes include solvents (e.g. methanol and ethanol), acids (e.g. sulphuric and hydrochloric acids), bases (e.g. sodium chloride and potassium hydroxide) and metal-based catalysts (e.g. zinc-based catalysts) (Table 7.1). The resulting monomers may then be purified before being used to create new polymers. This step may take place via filtration, distillation with or without pressure, solvent extraction, ion exchange, crystallisation or adsorption. Novel polymers may also be employed to improve the effectiveness of chemical recycling, limiting degradation of the polymer's properties. Häußler et al. (2021) have previously demonstrated that polycarbonates and polyesters made using renewable feedstocks (in this case microalgae and plant oils) contain vulnerable in-chain functional groups may be efficiently depolymerised with limited loss of mechanical properties. However, such solutions are currently a long way from commercial application.

Pyrolysis breaks down plastics by heating them in under oxygen free conditions. Plastic wastes are exposed to temperature in excess of 400 degrees Celsius, which results in the formation of gases, aqueous components that may be used as

Table 7.1 Processes and reagents applied in the recycling of common polymers.

Plastic	Process	Catalyst	Temperature	Time (hours)	Product
HDPE	Gasification-carbonisation	Ferrocene + ammonium carbonate	700	12.0	Fe₃O₄@C core–shell structures
	Hydrocracking	Pt/SrTiO3	300	96	Fuel oil
		mSiO₂/Pt/SiO₂	300	48	Hydrocarbons
	Pyrolysis	CuCO₃	390	0.3	Hydrocarbons
PET	Gasification-carbonisation	Ferrocene	800	24	Fullerene nanocomposites
	Glycolysis	Magnetite	200	6	Bis(2-Hydroxyethyl) terephthalate
		NiO/Co₂O₃	255	1.5	Bis(2-Hydroxyethyl) terephthalate
		ZnMn₂O₄	260	1	Diethyl terephthalate
PP	Gasification-carbonisation	OMMT	700	0.3	Graphene flakes
		OMMT + NiO	600		Multiwalled carbon nanotubes
	Pyrolysis	HY-2.8	320	0.3	Hydrocarbons
		Clay-ZrZrC₆H₄	400	8	Wax
PVC	Dehydrochlorination	Al(10 wt %)-modified graphitic-C3N4	170	6	Dechlorinated PVC

fuels or feedstocks, as well as residues, the composition of which varies dependent on the waste and the conditions of the treatment, but may include ashes and char, carbon black and remaining polymer, fillers and contaminants. Gasification is similar to pyrolysis, however, the temperatures to which the waste is exposed are much higher, in excess of 700 degrees Celsius. This process produces syngas, a mixture of hydrogen and carbon monoxide, used in power generation and other industrial processes.

Solvolysis utilises solvents to breakdown plastics into their original polymers. The monomers produced are then separated from the solvent via evaporation, distillation, membrane filtration, absorption, crystallisation or extraction using a secondary solvent. These methods are not without their drawbacks, for example solvent dissolution of mixed polymer wastes results in lower purity of the resulting polymer (Zhao et al., 2018). This may have a significant impact on the value of the resulting resource, for example, solvolysis of PET must result in a highly pure monomer to result in a useful product (Sinha et al., 2010). The choice of catalysts is also seen to be important in both solvolysis and pyrolysis (Vollmer et al., 2020).

Finally, glycolysis represents a potentially highly important route in the management of PUR waste (Kemona & Piotrowska, 2020). In this process, the polymer is broken down in the presence of a catalyst (e.g. potassium acetate), in a solvent (e.g. dietheylene gycol) (Wu et al., 2003). This process results in the formation or a polyol and an isocyanate (Donadini et al., 2023).

Despite the many approaches, estimates of total plastic waste published in 2017 suggest that only 9% of the all plastics produced up until that point had been recycled (Geyer et al., 2017). Fortunately, these figures include periods during which there was no systemic recycling of plastic and, as our expertise and ability in this area continues to grow, so will the cumulative proportion of this recycled material. Nevertheless, estimates suggest that, in 2021, approximately just 10.1% of new products were formed from recycled plastic waste. This volume is minimal when compared to the 87.6% of fossil fuel-based virgin plastics produced in the same year.

Of course, not all plastics are suitable for recycling. Some plastic products are of limited value due to low demand, contain harmful chemicals or contaminants that may be released during recycling or persist into the subsequent product, or are challenging to handle. For example, the processing of PVC may result in the release of harmful chemicals as a result of the presence of chlorine, and it is therefore not widely recycled. Other plastics may be limited in the number of times that they can be recycled. In addition to its other drawbacks, mechanical recycling may also be a source of microplastics. The formation, handling and transport of shredded waste may result in the accidental release of microplastics into the environment.

Additional barriers to the recycling of plastics include consumer and industry behaviours, the construction of the plastic waste, such as films and foams, which present issues during mechanical handling, the introduction of harmful or challenging coatings or fillers and contamination during the use-phase. Household recycling is variable both within and between countries and may be influenced by a variety of factors, for example the availability of kerbside recycling schemes. In the UK, the percentage of kerbside household recycling within and between regions was influenced by the type of container provided (which may be linked to ease of handling and aesthetic appearance) rather than aspects such as frequency and method of collection (Abbott et al., 2011).

One waste stream which suffers from many of the aforementioned issues is agricultural plastics. Agriplastics may be used in irrigation, the transport of feeds, pharmaceuticals, fertilisers and pesticides, as well as mulches, films applied to the soil to improve growing conditions, inhibit weeds and reduce moisture loss. Many agriplastics are single-use products with a range of applications (Kyrikou & Briassoulis, 2007). This scale and diversity results in large volumes of waste annually, however, the level of contamination and characteristics of these plastics make them of low value to recyclers. In addition, remote farms may not have the same access to recycling collections compared to those in close proximity to urbanised areas.

Rates of recycling, and the stability of the wider recycling industry, are also affected by changes to both global markets and consumer behaviour. Levels of municipal waste collection vary greatly by region, for example while 98% of local authorities in England provide kerbside (Hahladakis et al., 2018), the same provision reaches only 70% of households in Norway (Xevgenos et al., 2015). Where government-managed facilities are not available the same role may be provided by businesses (Dumbili & Henderson, 2020; Hamidul Bari et al., 2012) or individual informal waste collectors. Indeed, many recyclers are part of a highly interconnected network (Ebner & Iacovidou, 2021). For example, as we use fewer single-use plastic items, there is a reduction in the mass of plastic available for recycling, add to this the demand to more recycled products (driven by their increased attractiveness to consumers) and the relative availability of these recycled materials becomes reduced.

Conversely, during the COVID-19 pandemic, the volume, composition and distribution of wastes were notably altered, factory closures reduced the demand for raw materials, and manual sorting was restricted in order to limit the risk of transmission (Fan et al., 2021). Additionally, falling oil prices also reduced the cost of virgin plastics, making recycled plastics comparatively more expensive (Issifu et al., 2021). As the sale of recycled plastic to producers fell and recyclers were left with surplus material they subsequently began to reduce the mass of plastic that they would accept.

7.5 Replace

In some circumstances, such as in clinical settings, it is not possible to reduce or eliminate single-use high-waste goods or services, and the plastic waste arising may have limited routes for recycling. In such cases, the preferred option under the waste hierarchy is to replace the material or materials within the product with those of plant, animal or mineral origin. The use of alternative materials is often associated with apparently lower environmental impact due to their shorter environmental half-life. For example, the use of cotton and jute shopping bags, metal and glass bottles and food packaging, paper in product packaging and natural fibres in clothing. More recently, bioplastics and biodegradable plastics have also been utilised as alternatives to plastic polymers, reducing our apparent reliance on products of the oil industry and limiting the mass of waste sent to landfill as well as the time the waste remains there.

These alternative materials are not without their own concerns. For example, the use of plant-based alternatives has implications for the land and water footprint of many products. Jute, for example is grown primarily in countries including Bangladesh, China, India and Thailand, where it requires large volumes of water as well as fertile and well-drained soils that might otherwise be used for food crops. Once harvested, the plants must be processed, this involves first being submerged in water for several days. After drying, the nonfibrous material is stripped away and the fibres are again cleaned before spinning and weaving. The resulting cloth may then go through a range of wet processes such as bleaching, dyeing or finishing. The actual volume of water used during the production of jute varies between locations, and it may be considered a less water-intensive source of fibre than other plant (such as cotton) or animal fibres; nervertheless, it is estimated that between 2500 and 4000 L of water are required to produce one kilogram of jute fibre. Similarly, plastic products may be replaced by alternatives made of wood and bamboo. Secondary wood-pulp products such as paper and cardboard may also be used as lightweight packaging, or as coated board where waterproofing or food contact is a concern. These also have implications for land and water use, however, if we compare between the two materials other factors (such as the time until harvesting) are very different.

Potentially less impactful plant-based alternatives may use the by-product of other industries. For example, processed tomato stalks have previously been used at artisanal scales as cardboard cartons for the fruits themselves, while others have sought to utilise seaweed, mushroom mycelium and other plant-based materials as alternative packaging sources which should have significantly shorter degradation periods than plastics while avoiding many of the environmental concerns associated with the use of virgin natural materials.

As with plant based plastic alternatives, there are potential environmental concerns linked to the increased use of metals and glass. The mining of metals has long been linked to habitat destruction, erosion and soil and water contamination from wastes and tailings. Additionally, the processing of metal ores produces bother gaseous and particulate pollutants, effluent water and solid wastes (Dudka & Adriano, 1997). Similarly, glass productions is not without its impacts, in particular, the production processes may result in the output of greenhouse gasses and acidification factors[1], as well as higher energy costs during productions and

---

[1] CO, CO_2, CH_4, N_2O, NO_x, SO_2, SO_x, and VOC.

transport (Vellini & Savioli, 2009). Consideration of tradeoffs remains key in balancing the positive and negative environmental outcomes associated with the replacement of materials within a product.

Of course, when selecting alternative materials, the mechanical properties of the product may be the sole focus of the manufacturer. Some materials will not meet requirements of flexibility, low weight or transparency, for example and there are many compromises to be made. On top of these issues, in selecting materials from renewable or mineral resources, manufacturers should seek to minimise the environmental impacts associated with various sourcing options, prioritising sustainable and ethical sources that do not negatively impact local communities. Similarly, the production phase may be associated with secondary environmental effects, as a result, product handling and manufacturing steps must also be carefully considered. Then, following the initial use phase, provisions for reuse, recovery or disposal should be made in order to minimise energy consumption and maximise resource recovery and reclamation.

7.5.1 Bioplastics

Bioplastics are plastic polymers derived from renewable organic materials, identified as beneficial environmentally due to their capacity to reduce our dependence on fossil fuels. These novel materials are suited to a range of uses, for example polylactic acid (PLA) and polyhydroxyalkanoates (PHA) are commonly used in the in the production of packaging materials as well as medical devices and agricultural films. The products are often widely promoted as a 'natural' materials and are thus believed to be more sustainable. Frequently, the term 'bioplastic' is conflated with biodegradable plastics, however, as we will discuss below, one does not necessarily imply the other.

Many bioplastics, such as polybutylene succinate (PBS), PHA and PLA, are derived from plants rich in sugars and oils. PBS and PLA are formed from succinic acid and either 1,4-butanediol (BDO) or lactic acid, respectively, both of which are metabolic bioproducts of fermentation. PHAs, while also formed during bacterial fermentation, are instead stored as granules in the cells of bacteria.

Feedstocks for the fermentation process may be specific to the desired polymer, for example PLA is commonly made from corn or sugarcane, while PHA can be derived from vegetable oils or agricultural waste. In the case of corn or sugarcane, the starch is converted into simple sugars, while vegetable oils may be broken down into fatty acids. These sugars or fatty acids act as the building blocks for the bioplastic.

In the fermentation process, biomass feedstocks are inoculated using preferred microorganism strains (such as *Actinobacillus succinogenes* and *Mannheimia succiniciproducens*). Under carefully controlled levels of pH, temperature, oxygen and nutrients, these microorganisms metabolise the starches and sugars contained within the feedstock. The desired products may be produced as by-products of these metabolic processes or as energy stores, which are the separated and purified. In the case of succinic acid, the resulting product may then be hydrogenated in order to form BDO.

The monomers produced during these processes are then subjected to polymerisation resulting in plastic pellets, fibres or other manufacturing feedstocks, which may be blended with other bioplastics, oil-based plastics or with additives in order to confer the desired properties to the product.

Polyethylene terephthalate (PET) may also be formulated using plant-based ethylene glycol and terephthalic acid (TPA). Feedstocks are converted to either ethanol (sugarcane, corn), glycerol (bio-oil) or glucose (cellulosic biomass). Ethanol is then dehydrated through catalysis to bio-ethylene, whereas catalysis of glucose may be applied to form sorbitol and glycerol (Pang et al., 2016). These are then converted into ethylene glycol (Fig. 7.2).

Much of the TPA used in the formation of plant-based PET remains petroleum based (Tachibana et al., 2015), however, biobased TPA may be produced using either *p*-exylene (Luo et al., 2023) or *p*-Cymene from citrus peel (Kamitsou et al., 2014). Perhaps the most widely recognised use of biobased PET is Coca-Cola's Plantbottle, approximately 30% of the raw material for which is biobased (Smith, 2015).

7.5.1.1 *Bioplastics at end of life*

As with tradition plastic polymers, the capacity of bioplastics to be recycled is variable as a result of both their chemical structure and the capacities of recycling facilities. PLA may be recycled where facilities exist, however, other bioplastics may contaminate waste streams, reducing the quality of the resulting material.

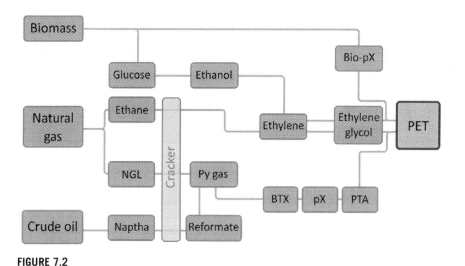

FIGURE 7.2

Example PET production pathways. The routes above represent potential production processes associated with both the hydrocarbon and renewable sources of PET.

While many bioplastics mimic traditional oil-based polymers and are therefore highly recalcitrant, some bioplastics are known to biodegrade. PLA and other starch-based polymers are considered compostable, however, these typically require a specific range of temperature, moisture, pH and oxygen to enable their breakdown. Indeed, observations of many bioplastics in the wider environment have shown that these materials can persist for extended periods. For example PLA exposed to composting conditions at 24 degrees centigrade in tropical climes displayed no apparent degradation after 31 weeks of exposure (Solano et al., 2022).

As with other alternative materials discussed in previous sections, bioplastics may have associated environmental drawbacks that influence their apparent sustainability. Whilst some of the feedstock used in the production of monomers may be waste products, others may have implications for land and wateruse and availability for their growth, resulting in competition with existing industries. The carbon footprint of the products may also be questionable, and many argue that the creation of products that enable our ongoing high consumption and waste production behaviours rather than seeking to change them is not the preferable solution.

7.5.2 Degradable polymers

Whether incidentally or by design, degradable plastics are those which breakdown when exposed to environmental conditions (Table 7.2). This degradation may occur through the direct action of organisms and their enzymes or via abiotic factors and processes, such as the influence of UV or extremes of temperature. Under the catch all term of 'degradable' are biodegradable and oxo-degradation plastics. While biodegradable plastics exhibit breakdown of the polymer chain, even up to mineralisation, oxo-degradable are frequently criticised as being only fragmentable, speeding the formation of micro- and nanoplastics.

Oxo-degradable polymers, by far the least preferable option, include many traditional polymers containing predegradant additives or fillers, such as starch or metal salts, which provide weak spots in otherwise recalcitrant polymers. The apparent problems posed by oxo-degradable polymers have resulted in a variety of restrictions on the use of these polymers, including a ban in the European Union (Directive (EU) 2019/904), and additional controls in Australia (DAWE 2021).

Oxo-degradable plastics should not be confused with oxo-*bio*degradable plastics. Degradation of oxo-biodegradable polymers is 'the result of oxidative and cell-mediated phenomena, either simultaneously or successively' (Chiellini et al., 2011) (Fig. 7.3). This is commonly achieved by way of a prodegradant additive designed to speed the breakdown of the polymer into smaller oligomers at which point it can be more easily accessed by degrading organisms. To prevent this occurring during the useful lifespan of the product, these plastics also commonly include antioxidants which are consumed prior to degradation. Nevertheless, there remains an ongoing debate over the efficacy of these polymers under different

Table 7.2 Degradable plastic polymers.

Class	Degradable polymers
Traditional	Poly(α-amino acid)
	Poly(amide enamine)
	Polybutylene succinate
	Polycaprolactone
	Poly(ethylene succinate)-6-poly(tetramethylene glycol)
	Poly(glycolic acid) (PGA)
	Poly(glycolic acid-co-lactic acid) (PGA/LA)
	Polyester-polyurethane
	Polyether-polyurethane
	Poly(trimethylene carbonate)
Biobased	Cellulosic films
	Polybutylene adipate terephthalate
	Polyhydroxyalkanoates
	Polylactic acid
	Starch blends

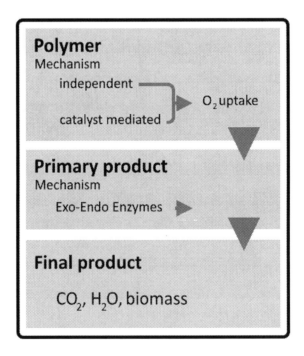

FIGURE 7.3

Degradation routes of oxo-biodegradable plastics. The processes of mineralisation via independent or catalyst mediated oxidation.

Table 7.3 Production of biobased degradable plastics in 2022.

Polymer	Approximate production (million t)	Percentage of all biobased production, including cellulosics
Polylactic acid	0.304	16
Starch blends	0.095	5
Polybutylene adipate terephthalate	0.057	3
Polyhydroxyalkanoates	0.057	3
Cellulosic films	0.114	6

conditions, and it has been suggested that there is a similar threat of microplastic formation.

Biodegradable plastics may be affected by a range of naturally occurring microorganisms such as bacteria, fungi and algae. Unlike traditional polymers, which require prior oxidation and/or a reduction in molecular weight before they become available to microorganisms, many of these polymers are biobased, however, their production compared to traditional plastics is low (Table 7.3), and they frequently require highly specific conditions in order to achieve mineralisation.

A number of standards have been developed in order to determine the effectiveness of biodegradable plastics, for example oxo-biodegradable plastics may be tested to ASTM D6954−04[2]. Accelerated ageing tests measure the period during which antioxidants in the polymer are consumed (an induction period) and the period and extent of prodegradant-mediated oxidation. If a product is to be labelled as compostable, alternative standards are employed, for example ASTM D6400 or EN 13432 for plastics to be aerobically composted in municipal or industrial facilities.

However, while these tests above are highly replicable, many argue that they are not representative of environmental conditions and, as indicated above, degradable polymers may remain in the environment for extended periods. Variation in degradation rate is due to variability in environmental conditions. In soil ecosystems, changability and seasonality in weather, soil conditions and

---

[2] ASTM D6954−04 'Standard guide for exposing and testing plastics that degrade in the environment by a combination of oxidation and biodegradation', which outlines accelerated ageing tests of both thermal- and photooxidation and abiotic degradation, tests of biodegradation and assessment of ecological impacts. Equivalent or comparable test standards include British Standard 8472 Packaging − Method for determining the degradability, oxo-biodegradability and phyto-toxicity of plastics, the French Accord T51−808 Plastics Assessment of oxo-biodegradability of polyolefinic materials in the form of films and French Standard XP_T_54−980__F for oxo-biodegradable plastics in agriculture, Swedish Standard SPCR 141 Polymeric waste degradable by abiotic and subsequent biological degradation − Requirements and test methods, and UAE Standard 5009:2009 Standard & Specification for Oxo-biodegradation of Plastic bags and other disposable plastic objects.

degree of burial (exposure to UV light) may affect both the plastic and the organisms acting upon it (Hoshino et al., 2001; Nishide et al., 1999).

Despite the variability in performance and standards, degradable polymers are frequently posited as an acceptable option for products, such as unavoidable single use items, which may otherwise be sent to landfill or energy-from-waste systems. However, while municipal composting facilities may be the intended destination of the bulk of these novel polymers, it is once again important to consider whether this represents the best management route. Analysis of the relative impacts of biodegradable polymers has suggested that, when comparing incineration, thermophilic and mesophillic anaerobic digestion and industrial and home composting, anaerobic digestion may be the best route for these plastics, enabling the production of both energy and digestate while minimising secondary negative effects (Hermann et al., 2011). Regrettably, under these scenarios, home composting was equal to incineration, however, this may become more favourable as incineration technologies improve.

7.6 **Recover**

In the extended waste hierarchy, Recovery refers to the transformation of wastes to enable their use within other industries, replacing the raw materials that may otherwise have been used. This may include processing at energy-from-waste facilities for the management of municipal solid waste and other refuse not suited to recycling. Energy reclamation may be via combustion, gasification, pyrolisation, anaerobic digestion and landfill gas recovery. In combustion processes, following collection and sorting, wastes are burnt at very high temperatures, releasing heat energy, which may be used to drive steam turbines for the generation of electricity. Noncombustible material may be further treated to recover valuable metals (Brunner & Rechberger, 2015), or used as a filler material in industry, such as the production of ceramic tiles from calcium-rich residues (Yuan et al., 2022). A novel example of recovery of materials from plastics is via the incomplete combustion of polymers, such as polyethylene and polypropylene, and rapid cooling of the combustion products, which results in the formation of shorter molecules which may be used in the production of soap (Xu et al., 2023).

While these processes reduce the volume of waste sent to landfill, the resources remaining in the incinerated wastes are often lost. Additionally, there is the potential for a variety of secondary airborne contaminants and pollutants to be produced. Analysis of residues captured by filters in energy from waste plants in the United Kingdom revealed a combination of metal pollutants, including lead, arsenic, cadmium and copper (Bogush et al., 2015), and anaerobic digestion for the production of biogas is associated with the release of nitrogen oxides (Preble et al., 2020). As a result, scrubbers, filters or other suitable pollution controls must be implemented throughout the relevant processes.

7.6.1 Fibre catchers

Microplastics are a challenging waste stream to manage due to their small size and diffuse sources. However, one accessible and identified source of microplastics is washing machines (Fig. 7.4). Plastic fibres, formed during the washing of clothing, are one of the most prevalent forms of microplastic in the environment. Fibres pass through sewerage systems to wastewater treatment, where up to 98% of them may be removed (Erdle et al., 2021). Unfortunately, these are retained in sewage sludges and may be subsequently transferred into terrestrial ecosystems. One proposed alternative management measure is to capture these fibres either in the washing machine drum or as they exit the machine with effluent water. The technologies employed to retain fibres are diverse. Garment bags may seek to reduce the abrasive action of washing on clothing, as well as to retain fibres release, whereas other in-drum fibre catchers, such as the Coraball, seek to mimic the action of filtering organisms such as coral. External filters have the challenge of retaining fibres whilst allowing the passage of tens of litres of water in a very short period. The design of these filters is a trade-off between energy use (where extra pumps are employed), volume available for the retention of water, and the pore size of the filter surface. The efficiency of these filters may be as high as

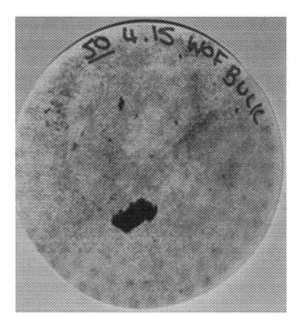

FIGURE 7.4

Fibres released in washing machine effluent. The filter paper pictured contains fibres released during the first machine washing of fleece fabric.

Image Credit: Natalie Welden.

99%, however, they rely on regular consumer intervention and there are few options for the collection and handling of recovered material. At present, there is no specific route to recycling or alternative circular endpoint and the diversity and extent of contamination (such as human and pet hair, fats and greases, dirt and even faecal matter) is substantial. Captured fibre is a potential candidate for recovery, either for energy from waste programmes or in the manufacture of other products (e.g. a stabiliser in concrete or in insulation).

7.7 Limitations and drawbacks

The waste hierarchy and its latter additions are a substantial asset in improving planning for waste management. However, they are not without their drawbacks. The focus on technocentric approaches such as recycling, recovery and removal (as is discussed in Chapter 9) has the benefit of limiting disruption, changes to industrial and economic models or the need for new legal statutes. However, these changes also do little to address either the consumer behaviours or the industrial design and manufacturing decisions behind our current overconsumption. Recovery methods may also lead to the formation and release of secondary pollutants such as lead and arsenic. Additionally, the requirements of infrastructure, investment (both financial and individual), and connectivity vary considerably within and between countries. As a result, the application of the hierarchy's principles may not be achievable in different regions. Conversely, many of the world's wealthiest countries waste management regimes are dependent on the export of waste to low-income countries, transferring many of the environmental and health concerns related to their sorting and management.

As indicated above, the ability of individuals to adopt suitable behaviours is also bound up with regional or national priorities. Even concerned members of the public may be prevented from taking action as a result of lack of access to suitable facilities, or by competing interests or basic needs, such as sourcing of clean water. Where individuals can act to reduce their plastic footprint, adoption of sustainable behaviours may also be influenced by single action bias. Here an individual focusses on one specific action, such as the use of reusable coffee cups. The adoption of which engenders a feeling of having 'done their part', subsequently reducing and individual's guilt when participating in other waste-generating activities. This and other barriers to the widespread uptake of sustainable behaviours will be discussed in Chapter 11.

7.7.1 The role of the waste hierarchy in linear and circular economies

Another criticism of the waste hierarchy is that it does not include the focus on design and manufacture that would enable a more sustainable use of resources.

Box 7.2 Informal waste industries: examining the development of new communal waste management methods

Objective

The relationship between regional plastic waste production and the scale and suitability of local waste facilities is complex, and substantial waste production does not necessarily correlate with the availability of suitable facilities. Here we will explore the development of a Waste for Life programme in Buenos Aires, Argentina, determining its effectiveness in the presence of existing waste challenges.

Scope

Waste for Life brings together students, academics, and professionals across the arts, social sciences, natural sciences and engineering to find solutions the issue of plastic pollution by way of local industries and appropriate technologies. These creative measures foster economic opportunities by adding value to waste products in order to provide livelihood to poorer or marginalised communities. The truly collaborative work of Waste for Life in Buenos Aires began in 2007, with the gathering of a collective of actors from all levels of society to both support informal and semi-informal waste collectors and improve the fate of low value wastes. These efforts continued with iterative interactions in 2008 and 2010, the influences of which (as observed in the groups final site visit in 2014) have spread beyond those locals involved in the initial collective.

Actions Taken

Rather than selling the collected waste for low sums, the collective sought to enable local informal waste collectors (known as cartoneros) to develop new products that could be sold for higher amounts. Waste for Life and their partner organisations collaborated in the development and implementation of novel presses for the creation of composite materials. In order to support this work, individuals were trained in the identification and separation to material types in order to ensure the mechanical properties and marketability of the products created.

The development of this programme was dependent on increasing community engagement and empowerment by way of education and awareness raising, underpinned by research and innovation at both local and global scales, to enable the upcycling of waste materials and increase local production.

In order to achieve its aims, Waste for Life sought to sensitively engage across all levels of society. They did so by actively involving local communities, with a focus on those marginalised groups living in informal settlements, particularly those groups engaged in waste collection, sorting, and transformation. By providing opportunities for skills development and income generation, the Waste for Life programme aimed to empower residents with sustainable livelihoods and improved living conditions. Educational aspects emphasise waste-related issues and the benefits of recycling and upcycling, providing participants with agency over these challenging issues.

The practical aspects of this work emphasise the transformation of waste materials into valuable products, using waste to create building materials, such as tiles and bricks, that can be used for construction projects, reducing the environmental impact of traditional construction materials while creating further business opportunities (Baillie, 2008).

Challenges and solutions

The Waste for Life programme demonstrated several positive outcomes, such as the development of employment opportunities for community members and a change in the relationship with plastic wastes, in addition to influencing living conditions. Nevertheless, apparent conflicts at local and national levels resulted in a series of setbacks to the scheme. In addition to substantial challenges surrounding funding, infrastructure limitations (such as melting wiring and overloading circuits), and regulatory hurdles, the additional issues of competing demands of public and private waste collectors, as well as those informal waste collectors

(Continued)

Box 7.2 Informal waste industries: examining the development of new communal waste management methods (Continued)

working independently and those involved in collectives which have affected acceptance of this new initiative. Competition for waste materials is rife, and it is apparent that independent waste transporters may be selling waste directly rather than delivering it to collection sites. Additionally, ensuring the long-term sustainability of such initiatives requires substantial ongoing community engagement and support.

Learning and knowledge outcomes

Together with their wider international activities, the Waste for Life programme in Buenos Aires showcases how innovative waste management strategies can contribute to sustainable development, poverty reduction, and environmental conservation in developing regions. Upcycling in this manner is not suited to all wastes, and other suitable endpoints and management measures are required, Additionally, successful implementation is reliant on sensitive engagement at numerous levels of society. Nevertheless, similar projects are being explored in various parts of the world to address waste-related challenges while empowering local communities.

For further information regarding Waste for Life projects, visit https://www.wasteforlife.org/projects/.

Much of our current waste management may be considered *linear*, a resource is obtained, transformed into a product, and the resulting waste produced at the product's end of useful life is managed as well as possible. An alternate mode of thinking is that of the Circular Economy, which attempts to maximise efficiency of resource use and minimise waste production. This circularity is represented by an idealised, closed-loop system often referred to as a cradle-to-cradle approach. Here products are designed both for their longevity (reuse, refurbishment and repair) and for effective management at end of life. Design for disassembly at the end of a product's useful life increases the probability that its components may be recycled. Many products are challenging to recycle due to the nature of their components, for example many personal care products, such as toothbrushes, contain numerous elements and are bonded with glues rather than fastenings, making it both costly difficult to separate and recycle the resources contained.

Design for disassembly may take many forms, for example Niche Snowboards utilise a dissolvable adhesive (Recyclamine) to enable the components to be easily isolated for reuse. Alternatively, AIAIAI improves the disassembly of their speaker systems by ensuring that only one tool is required for deconstruction and repair. These principles also apply to the built environment, for example, enabling access for workers involved in disassembly, using standard connectors of a suitable durability to enable reuse, and minimising chemical connectors (such as glues and sealers).

Successful adoption of the principles of the circular economy requires a whole system approach not currently represented by the existing strategies for waste management, an example of which is set out in Box 7.2. Challenges to overcome include the complexity of global supply chains and regional preparedness, the limitation of infrastructure and the cost of transitioning to a new model of design, manufacture and management. Nevertheless, more widespread implementation of these principles has the capacity to significantly increase our chance of attaining sustainability.

7.8 Conclusions

From the above, it is apparent that a wide range of approaches are available to both limit plastic waste production and ensure its management. However, the influence of challenging polymers, polymer blends and potentially harmful additives, as well as the availability of required infrastructure, limit the proportion of plastic wastes that may be 'more sustainably' managed at end-of-life. Where alternative materials exist that may limit the production of traditional fossil fuel-based plastics, the apparent environmental benefits remain largely untested and may result in the contamination of existing waste streams. Additionally, replacing plastics in products with alternative materials may have unintended negative environmental impacts. The use of plant and animal-derived fibres rather than synthetic polymers has implications for the water and land footprints of the product and may result in a product with a more challenging or less favourable waste management root at end of life. Identification of the best options for the reduction of plastic wastes must therefore be the result of substantial consideration, the processes for which we will adress in more detail in Chapter 11, however, one solution will not fit all scenarios.

References

Abbott, A., Nandeibam, S., & O'Shea, L. (2011). Explaining the variation in household recycling rates across the UK. *Ecological Economics*, *70*(11), 2214–2223. Available from https://doi.org/10.1016/j.ecolecon.2011.06.028.

Baillie, C. (2008). Waste for life. *Materials Today*, *11*(10), 6. Available from https://doi.org/10.1016/S1369-7021(08)70187-1, https://www.sciencedirect.com/science/article/pii/S1369702108701871.

Beddington, E. (2021). The Guardian Could you wear a dress for 100 days? https://www.theguardian.com/fashion/2021/may/02/could-you-wear-a-dress-for-100-days.

Bogush, A., Stegemann, J. A., Wood, I., & Roy, A. (2015). Element composition and mineralogical characterisation of air pollution control residue from UK energy-from-waste facilities. *Waste Management*, *36*, 119–129. Available from https://doi.org/10.1016/j.wasman.2014.11.017, http://www.elsevier.com/locate/wasman.

Brunner, P. H., & Rechberger, H. (2015). Waste to energy - Key element for sustainable waste management. *Waste Management*, *37*, 3–12. Available from https://doi.org/10.1016/j.wasman.2014.02.003, http://www.elsevier.com/locate/wasman.

Chiellini, E., Corti, A., D'Antone, S., & Wiles, D. M. K. (2011). *Oxo-biodegradable polymers: Present status and future perspectives. Handbook of biodegradable polymers:*

Isolation, synthesis, characterization and applications (pp. 379−398). Italy: Wiley-VCH. Available from http://onlinelibrary.wiley.com/book/10.1002/9783527635818, https://doi.org/10.1002/9783527635818.ch16.

DAWE. (2021). *National plastics plan in 2021*, Department of Agriculture, Water and the Environment.

Donadini, R., Boaretti, C., Lorenzetti, A., Roso, M., Penzo, D., Dal Lago, E., & Modesti, M. (2023). Chemical recycling of polyurethane waste via a microwave-assisted glycolysis process. *ACS Omega, 8*(5), 4655−4666. Available from https://doi.org/10.1021/acsomega.2c06297, http://pubs.acs.org/journal/acsodf.

Dudka, S., & Adriano, D. C. (1997). Environmental impacts of metal ore mining and processing: A review. *Journal of Environmental Quality, 26*(3), 590−602. Available from https://doi.org/10.2134/jeq1997.00472425002600030003x, https://www.agronomy.org/publications/jeq.

Dumbili, E., & Henderson, L. (2020). *The challenge of plastic pollution in Nigeria. Plastic Waste and Recycling: Environmental Impact, Societal Issues, Prevention, and Solutions* (pp. 569−583). Nigeria: Elsevier. Available from https://www.sciencedirect.com/book/9780128178805, https://doi.org/10.1016/B978-0-12−817880-5.00022-0.

Ebner, N., & Iacovidou, E. (2021). The challenges of Covid-19 pandemic on improving plastic waste recycling rates. *Sustainable Production and Consumption, 28*, 726−735. Available from https://doi.org/10.1016/j.spc.2021.07.001, http://www.journals.elsevier.com/sustainable-production-and-consumption/.

Erdle, L. M., Nouri Parto, D., Sweetnam, D., & Rochman, C. M. (2021). Washing machine filters reduce microfiber emissions: Evidence from a community-scale pilot in Parry Sound, Ontario. *Frontiers in Marine Science, 8.* Available from https://doi.org/10.3389/fmars.2021.777865, https://www.frontiersin.org/journals/marine-science#.

Fan, Y. V., Jiang, P., Hemzal, M., & Klemeš, J. J. (2021). An update of COVID-19 influence on waste management. *Science of the Total Environment, 754.* Available from https://doi.org/10.1016/j.scitotenv.2020.142014, http://www.elsevier.com/locate/scitotenv.

Garcia, J. M., & Robertson, M. L. (2017). The future of plastics recycling. *Science (New York, N.Y.), 358*(6365), 870−872. Available from https://doi.org/10.1126/science.aaq0324, http://science.sciencemag.org/content/sci/358/6365/870.full.pdf.

Geyer, R., Jambeck, J. R., & Law, K. L. (2017). Production, use, and fate of all plastics ever made. *Science Advances, 3*(7). Available from https://doi.org/10.1126/sciadv.1700782.

Hahladakis, J. N., Purnell, P., Iacovidou, E., Velis, C. A., & Atseyinku, M. (2018). Postconsumer plastic packaging waste in England: Assessing the yield of multiple collection-recycling schemes. *Waste Management, 75*, 149−159. Available from https://doi.org/10.1016/j.wasman.2018.02.009, http://www.elsevier.com/locate/wasman.

Hamidul Bari, Q., Mahbub Hassan, K., & Ehsanul Haque, M. (2012). Solid waste recycling in Rajshahi city of Bangladesh. *Waste Management, 32*(11), 2029−2036. Available from https://doi.org/10.1016/j.wasman.2012.05.036.

Häußler, M., Eck, M., Rothauer, D., & Mecking, S. (2021). Closed-loop recycling of polyethylene-like materials. *Nature, 590*(7846), 423−427. Available from https://doi.org/10.1038/s41586-020-03149-9, http://www.nature.com/nature/index.html.

Hermann, B. G., Debeer, L., De Wilde, B., Blok, K., & Patel, M. K. (2011). To compost or not to compost: Carbon and energy footprints of biodegradable materials' waste treatment. *Polymer Degradation and Stability, 96*(6), 1159−1171. Available from https://doi.org/10.1016/j.polymdegradstab.2010.12.026.

Hoshino, A., Sawada, H., Yokota, M., Tsuji, M., Fukuda, K., & Kimura, M. (2001). Influence of weather conditions and soil properties on degradation of biodegradable plastics in soil. *Soil Science and Plant Nutrition*, *47*(1), 35−43. Available from https://doi.org/10.1080/00380768.2001.10408366.

Issifu, I., Deffor, E. W., & Sumaila, U. R. (2021). How COVID-19 could change the economics of the plastic recycling sector. *Recycling*, *6*(4). Available from https://doi.org/10.3390/recycling6040064, https://www.mdpi.com/2313-4321/6/4/64/pdf.

Kamitsou, M., Panagiotou, G. D., Triantafyllidis, K. S., Bourikas, K., Lycourghiotis, A., & Kordulis, C. (2014). Transformation of α-limonene into p-cymene over oxide catalysts: A green chemistry approach. *Applied Catalysis A: General*, *474*, 224−229. Available from https://doi.org/10.1016/j.apcata.2013.06.001.

Kemona, A., & Piotrowska, M. (2020). Polyurethane recycling and disposal: Methods and prospects. *Polymers*, *12*(8). Available from https://doi.org/10.3390/POLYM12081752, https://res.mdpi.com/d_attachment/polymers/polymers-12-01752/article_deploy/polymers-12-01752-v2.pdf.

Kyrikou, I., & Briassoulis, D. (2007). Biodegradation of agricultural plastic films: A critical review. *Journal of Polymers and the Environment*, *15*(2), 125−150. Available from https://doi.org/10.1007/s10924-007-0053-8.

Luo, Z. W., Choi, K. R., & Lee, S. Y. (2023). Improved terephthalic acid production from p-xylene using metabolically engineered Pseudomonas putida. *Metabolic Engineering*, *76*, 75−86. Available from https://doi.org/10.1016/j.ymben.2023.01.007, http://www.elsevier.com/inca/publications/store/6/2/2/9/1/3/index.htt.

Nishide, H., Toyota, K., & Kimura, M. (1999). Effects of soil temperature and anaerobiosis on degradation of biodegradable plastics in soil and their degrading microorganisms. *Soil Science and Plant Nutrition*, *45*(4), 963−972. Available from https://doi.org/10.1080/00380768.1999.10414346.

O'Connell, S. (2009). The Guardian The Uniform Project: One woman, one dress, one year. Available from https://www.theguardian.com/lifeandstyle/2009/jun/24/uniform-project-one-dress-year.

Pang, J., Zheng, M., Sun, R., Wang, A., Wang, X., & Zhang, T. (2016). Synthesis of ethylene glycol and terephthalic acid from biomass for producing PET. *Green Chemistry*, *18*(2), 342−359. Available from https://doi.org/10.1039/c5gc01771h, http://pubs.rsc.org/en/journals/journal/gc.

Preble, C. V., Chen, S. S., Hotchi, T., Sohn, M. D., Maddalena, R. L., Russell, M. L., Brown, N. J., Scown, C. D., & Kirchstetter, T. W. (2020). Air pollutant emission rates for dry anaerobic digestion and composting of organic municipal solid waste. *Environmental Science and Technology*, *54*(24), 16097−16107. Available from https://doi.org/10.1021/acs.est.0c03953, http://pubs.acs.org/journal/esthag.

Simón, D., Borreguero, A. M., de Lucas, A., & Rodríguez, J. F. (2018). Recycling of polyurethanes from laboratory to industry, a journey towards the sustainability. *Waste Management*, *76*, 147−171. Available from https://doi.org/10.1016/j.wasman.2018.03.041, http://www.elsevier.com/locate/wasman.

Sinha, V., Patel, M. R., & Patel, J. V. (2010). Pet waste management by chemical recycling: A review. *Journal of Polymers and the Environment*, *18*(1), 8−25. Available from https://doi.org/10.1007/s10924-008-0106-7.

Smith, P. B. (2015). Bio-based sources for terephthalic acid. *ACS Symposium Series, 1192*, 453–469. Available from https://doi.org/10.1021/bk-2015-1192.ch027, http://global.oup.com/academic/content/series/a/acs-symposium-series-acsss/?cc = in&lang = en.

Solano, G., Rojas-Gätjens, D., Rojas-Jimenez, K., Chavarría, M., & Romero, R. M. (2022). Biodegradation of plastics at home composting conditions. *Environmental Challenges, 7.* Available from https://doi.org/10.1016/j.envc.2022.100500, https://www.journals.elsevier.com/environmental-challenges.

Tachibana, Y., Kimura, S., & Kasuya, K. I. (2015). Synthesis and verification of biobased terephthalic acid from furfural. *Scientific Reports, 5.* Available from https://doi.org/10.1038/srep08249, http://www.nature.com/srep/index.html.

Vellini, M., & Savioli, M. (2009). Energy and environmental analysis of glass container production and recycling. *Energy, 34*(12), 2137–2143. Available from https://doi.org/10.1016/j.energy.2008.09.017, http://www.elsevier.com/inca/publications/store/4/8/3/.

Vollmer, I., Jenks, M. J. F., Roelands, M. C. P., White, R. J., van Harmelen, T., de Wild, P., van der Laan, G. P., Meirer, F., Keurentjes, J. T. F., & Weckhuysen, B. M. (2020). Beyond mechanical recycling: Giving new life to plastic waste. *Angewandte Chemie - International Edition, 59*(36), 15402–15423. Available from https://doi.org/10.1002/anie.201915651, http://onlinelibrary.wiley.com/journal/10.1002/(ISSN)1521-3773.

Wu, C.-H., Chang, C.-Y., Cheng, C.-M., & Huang, H.-C. (2003). Glycolysis of waste flexible polyurethane foam. *Polymer Degradation and Stability, 80*(1), 103–111. Available from https://doi.org/10.1016/S0141-3910(02)00390-7, https://www.sciencedirect.com/science/article/pii/S0141391002003907.

Xevgenos, D., Papadaskalopoulou, C., Panaretou, V., Moustakas, K., & Malamis, D. (2015). Success stories for recycling of MSW at municipal level: A review. *Waste and Biomass Valorization, 6*(5), 657–684. Available from https://doi.org/10.1007/s12649-015-9389-9, http://www.springer.com/engineering/journal/12649.

Xu, Z., Eric Munyaneza, N., Zhang, Q., Sun, M., Posada, C., Venturo, P., Rorrer, N. A., Miscall, J., Sumpter, B. G., & Liu, G. (2023). Chemical upcycling of polyethylene, polypropylene, and mixtures to high-value surfactants. *Science (New York, N.Y.), 381* (6658), 666–671. Available from https://doi.org/10.1126/science.adh0993, https://www.science.org/doi/10.1126/science.adh0993.

Yuan, Q., Robert, D., Mohajerani, A., Tran, P., & Pramanik, B. K. (2022). Utilisation of waste-to-energy fly ash in ceramic tiles. *Construction and Building Materials, 347.* Available from https://doi.org/10.1016/j.conbuildmat.2022.128475, https://www.journals.elsevier.com/construction-and-building-materials.

Zhao, Y. B., Lv, X. D., & Ni, H. G. (2018). Solvent-based separation and recycling of waste plastics: A review. *Chemosphere, 209*, 707–720. Available from https://doi.org/10.1016/j.chemosphere.2018.06.095, http://www.elsevier.com/locate/chemosphere.

Recover and remove: rise of the litter collector

8

8.1 Introduction

As discussed in earlier chapters, plastics and microplastics make their way into the environment via a variety of routes, including intentional discards, unintentional losses and mismanaged wastes. Once in the environment, large plastics may break down, further increasing the proportion of meso- micro- and nanoplastic and, subsequently, their impacts on organisms and communities (Welden & Lusher, 2017). Chapter 7 introduced many of the measures applied in an attempt to reduce the volume of plastic waste produced each year, increasing sustainable management following unsustainable disposal and decreasing the mass of plastics lost to the wider environment. Unfortunately, these methods will not address the mass of plastic already present in the environment or that which will continue to escape proper waste management.

Fortunately, there is a long tradition of recovering litter from the environment, driven by both an appreciation of its apparent aesthetic and environmental effects and of potential economic advantages. It is no surprise then, that these efforts have increased in line with our awareness of plastics abundance and its impacts. As a result, there are now many approaches by which macroplastic and microplastic may be removed from the environment. While many of these are best suited to the recovery of macroplastic debris, the act of reducing the volume of large plastic parent material may limit the in situ formation of secondary microplastics and have a significant part to play in both their prevention and control. In the following sections, we will explore the options for the recovery of plastics from the environment, from the very simple to the highly complex.

8.2 Terrestrial and beached litter

8.2.1 The informal waste collection industry

One of the simplest ways to extract plastics from the environment is to do so by hand, however, the scale, frequency and intensity of these efforts are highly variable globally, as are the motivations of the collectors. While some undertake collections as a result of environmental concerns, to others it is a source of

Microplastics. DOI: https://doi.org/10.1016/B978-0-443-13324-4.00008-X

livelihood. Many countries have formal and informal networks of waste collectors seeking to either benefit their environment or earn money from the materials gathered. These smaller collection networks, run by small businesses or individuals rather than authorities, frequently operate in areas of low waste management provision where civic organisations may struggle to operate (Dumbili & Henderson, 2020). Additionally, these groups may collect and sort a wider range of wastes than are handled in either government-run or commercial networks. They may thus add an additional tier to existing systems, enabling a more complete reclamation of wastes (Gutberlet & Carenzo, 2020; Schenck & Blaauw, 2011). An essential part of these networks are the individual informal waste collectors (IWCs), who independently collect wastes to sell to both independent and municipal facilities.

The scale of this industry may be substantial, with *c.*140 recycling shops in one city alone in Bangladesh in 2012 (Hamidul Bari et al., 2012), and similar enterprises are also found at varying scales globally, including India, Kenya, Nigeria and the United Kingdom (Dunn & Welden, 2023). Depending on location, the reliance on waste as a source of income may be high, and comparisons between the number of reported IWCs and the national population reveal nonnegligible proportions of the total populace earning money from waste. For example, in Argentina, 200,000 people were identified as IWCs, representing approximately 0.4% of the total population, in Brazil, 387,910 IWCs accounted for just under 0.2%, and in South Africa, the range of estimated IWCs (60,000−215,000) was between 0.001 and 0.38% of the total population (Godfrey & Oelofse, 2017; Gutberlet & Carenzo, 2020).

Recovery of plastics on this scale may have a significant impact on the mass of material removed from (or prevented from entering) the environment. For example, in Manila in the Philippines, a mass of between 20.32 and 1.63 kg of plastic per individual per day was collected by IWCs working dump sites, whereas municipal collection workers recovered approximately 9.79 kg of plastic each per day (Environmental Management Bureau Republic of the Philippines Department of Environment and Natural Resources, 2018). Indeed, in many places, the mass of waste gathered by informal collectors may far exceed that of civic operators where the two coexist, and their actions result in improvements in the state of the local environment for the wider population. Nevertheless, the working conditions for those engaged in waste collection are often challenging with significant threats to health and wellbeing (Godfrey & Oelofse, 2017; Nzeadibe, 2013), and groups working as IWCs are frequently stigmatised. More information on IWC networks can be found in Chapter 7.

8.2.2 Beach cleans and litter picks

Other informal (although frequently highly organised) waste collection includes beach cleans and litter picks. These activities serve a dual purpose in the sphere of plastic management, firstly in their ability to remove environmental plastic and secondly (and perhaps more impactfully) in generating awareness that may lead to long-term behavioural change in the use of plastic products. Unlike municipal waste collection and the work of IWCs, the key drivers of participation are not linked to income and are more frequently associated with concern for the environment or a sense of stewardship over a certain location.

Beach cleans and litter picks are variable in their frequency and intensity; litter picks may be undertaken individually, as part of a group, or in relation to national campaigns, such as the Marine Conservation Society's *Great British Beach Clean* and *Source to Sea: Litter Quest*, New Zealand's *Love your coast* (Sustainable Coastlines, 2009), and the European Surfrider *Ocean Initiatives* programme (Surfrider Foundation, 2023) among others.

Collaboration in this manner has been suggested to encourage environmental community building, bringing together participants under a shared purpose and promoting more widespread environmental awareness. These acts may also provide the public with a sense of agency and improved motivation arising from additional secondary impacts that litter picks may have on the wider problem of plastic waste. For example, some litter picks are structured so as to generate citizen science data which may be used to influence changes in legislation, such as the PADI AWARE Foundation's Dive Against Debris and Marine Debris Program (Roman et al., 2020). Similarly, the results of these programmes may also be used to promote alternative plastics management measures, such as the impact of FIDRA's Great Nurdle Hunt on improved practice by plastics manufacturers. Here the concentrations of nurdles observed in coastal habitats have been used to encourage producers to sign up to best practice in the transport and handling of these preproduction plastic pellets.

Of course, the key benefit may be considered to be the impact of sustained effort on local levels of plastics debris. Individual beach clean programmes have previously reported the collection of substantial quantities of litter. For example, participants in the 2021 Great British Beach Clean collected over 5000 kg of litter in a single week, recording reductions in key items such as cotton bud sticks and plastic bags, and localised increases in cigarette filters. Over extended periods and geographic scales, these repeated actions can have a significant effect. The Ocean Conservancy's clean-ups have collected over 136 million kilograms of litter over 30 years. A closer look at this data is provided in Box 8.1.

8.2.3 **Mechanical cleaning**

In addition to the clean-up methods outlined above, natural and manmade debris, including plastics, are removed from sandy beaches on a daily basis by a range of mechanical methods. The focus of these activities is typically less to limit the impacts of solid wastes on coastal environments but rather to maintain the aesthetic value of the beach for residents and visitors.

There are two primary approaches to the removal of unwanted debris, mechanical raking or sifting and screening. Mechanical rakes may be static arrangements of teeth pulled behind a vehicle, or they may be attached to a conveyor, drawn through the sand in the direction of travel, lifting objects larger than the tine

Box 8.1 The effectiveness of volunteer clean-up schemes

Objective

Volunteer beach cleans are internationally recognised tools for the communication of environmental issues, as well as for their ability to providing community agency in their management. These methods can result in large volumes of collected waste (reducing microplastic parent materials in the environment), in addition to the creating valuable data on plastic distribution and abundance. In this case study, we will examine the impact of the Ocean Conservancy's International Coastal Cleanup, both as a communication measure and as a method of reducing marine litter.

Scope

Implemented first in 1986 and focussing, in its initial stages, on the beaches of the United States alone, the International Coastal Cleanup provides volunteers with standardised survey methods and recording sheets that enable the consistent litter collection and reporting of comparable data. In addition to recording the number of items in different debris categories, the mass of removed material is returned to the Ocean Conservancy for collation and reporting.

Results

The Coastal Cleanup has now accumulated over 36 years' worth of data, beginning with the US only counts in 1986 and including international data as of 1989. Over this period approximately 164 thousand tonnes of plastic have been collected, with participation averaging 494,200 individuals per year and covering an average of 16,585.3 miles. Average collection rates are approximately 0.274 tonnes per person or 0.009 tonnes per mile. However, the surveys have changed over the years, increasing the coverage of inland water areas and subaquatic zones, as well as increasing their focus on small debris fractions (sub. 2.5 cm material dubbed 'tiny trash'). If we explore data returned from the period between 2015 and 2022 in detail, we may observe apparent trends in the data, in addition to identifying the difference in land-based, subsurface and floating litter.

Land, subsurface and boat-based surveys

First, we may examine the cumulative mass of collected debris across all survey types. In Fig. 8.1, we can see that the mass of material recovered from land-based surveys greatly exceeds that from subsurface and boat-based collections. The Coastal Cleanup has seen consistent increases in the number of participants since its 1986 inception, however, it has not been exempt from the effect of global challenges. In the mass of debris recovered above, we may note a consistent in the cumulative total until 2019. We may associate this dip in annual recovery with the drop in participants during the COVID-19 period (Fig. 8.2). Thankfully, recovery in participant numbers is apparent and appears to be tracking pre-COVID trends.

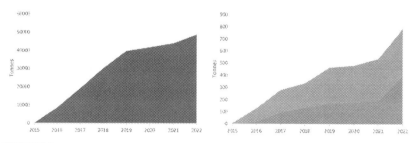

FIGURE 8.1

Mass of stranded, floating and subsurface plastic recovered. The cumulative mass of plastic recovered during clean-up events.

(Continued)

Box 8.1 The effectiveness of volunteer clean-up schemes
(Continued)

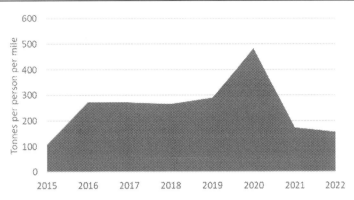

FIGURE 8.2

The annual mass of recovered plastic, normalised by the number of people per mile. The normalised plastic data shows no evidence of plateauing, suggesting that there is plenty more debris to collect, even at sites which are the most intensively cleaned.

Land surveys: effort vs recovery

The link between the mass of recovered debris and the number of participants is not direct. The relationship between the number of people and the total area covered is significant. Up to a certain point, the level of effort (number of people) at a site should result in increased plastic recovery, however, above a threshold level there will be no additional benefit to adding more participants at a single site (there is simply no more debris to collect, no matter how many more people are present). In Fig. 8.3, we can see that, across all surveys, this level is far from being reached, with a clear correlation between the number of people per mile and the recovered tonnage of debris. Interestingly, while mass of items is linked to the density of participants, this does not relate to the number of items recovered, illustrating the substantial range of sizes in the debris recovered each year.

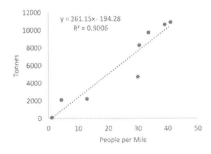
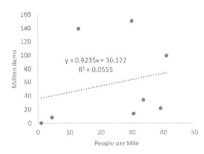

FIGURE 8.3

Plastic recovered, normalised by the number of people for mile. The mass of plastic recovered appears closely correlated with the mass of plastic recovered.

(Continued)

> **Box 8.1** The effectiveness of volunteer clean-up schemes (Continued)
>
> **Learning and knowledge outcomes**
>
> Estimates indicated that around 2.5 billion metric tonnes of waste are produced every year, while much is directed to waste management, large proportions are directed to the environment. Approximately 14 million tonnes of plastic alone are mishandled in this manner, a figure far above the annual average mass of collected debris in this (4,551.3 tonnes) or any collection scheme. Nevertheless, the data gathered is hugely valuable for its understanding of the spatial distribution of debris types, the split between floating, subsurface and stranded debris, and the apparent density of debris types, in addition to providing information regarding the success of policy interventions.
>
> To interrogate this data for yourself, visit https://www.coastalcleanupdata.org/

width. The conveyor then lifts this material into the vehicle hopper. Alternatively, sifters remove material from the sand by way of an oscillating ridged screen. The screen is angled so that it cuts down into the sediment, lifting and separating lighter material. More recently, there have also been proposals for a BeachBot, an AI-controlled robot, trained to identify and remove cigarette butts (Project, 2023).

Cleaning beaches in this manner certainly results in the removal of plastic (Jayasiri et al., 2013), but what is done with this waste (a mixture of differing litter types combined with natural debris) is often undocumented and its management is dependent on the quality of municipal collection and handling facilities. What is more, beach cleaning in this manner may disguise the scale of local issues by burying or 'tidying away' the problem (Shiber, 1987).

Additionally, mechanical beach cleaning has been associated with a range of negative impacts on the health of beach communities. Firstly, assessment of routine beach cleaning by surface sifters suggests that consistent overturning of the surface sediments may result in the removal of sand by aeolian processes. After one cleaning event, observations of reductions in organic matter content have been made (Gheskiere et al., 2006), and long-term cleaning may prevent the formation of plant communities in the backshore that might otherwise support dune formation and the stabilisation of sediments (Dugan & Hubbard, 2010). Similarly, the presence of natural debris on the strandline may limit wave run-up, and their removal may further increase erosional forces.

In addition to changing the abiotic profiles of sediments, mechanical cleaning has the potential to influence the communities therein. The natural material of the strandline often represents a high-humidity refuge for invertebrates and a vital foraging zone for shorebirds (Dugan et al., 2003). Observations of shoreline communities before and after cleaning using a sifter have noted that, after a single beach cleaning event to a depth of 5 cm, there was a reduced abundance of organisms and changes to the community structure, although the long-term impact of one-off events is believed to be minor (Gheskiere et al., 2006). Nevertheless, certain species groups, such as burrowing crabs may be particularly affected (Stelling-Wood et al., 2016). Fortunately, the impact of mechanical beach cleaning events appears to be linked with their intensity and frequency, with only a minimal effect observed on bacterial, meiofaunal and macrofaunal assemblages (Malm et al., 2004; Morton et al., 2015). Following cessation of cleaning, population numbers may recover, or be reintroduced with subsequent inputs of natural beach wrack. Most importantly in our case, while beach cleaning is effective in removing much of the visible macroplastic, microplastics frequently remain and may be added to by the action of machinery on weathered, embrittled litter.

8.3 **Floating debris**

In addition to the collection of terrestrial or stranded litter, numerous measures have been taken to intercept plastics during their transport within and between aquatic systems. In order to achieve these goals a variety of plastic collectors have been developed for use in riverine and marine environments. Many of these make use of the buoyancy of plastics to herd and capture this material in the upper layers of the water column.

8.3.1 **Riverine collectors**

Unlike the movement of plastics in marine and coastal settings, riverine transport is predominantly unidirectional, a factor that has been taken advantaged of in collector designs; however, the form that these collectors take is highly dependent on the river conditions such as flow, depth, presence of river traffic and other factors (Fig. 8.4).

The simplest riverine collectors employ passive plastic aggregation, for example floating booms which direct material at the water's surface into an aggregation area or mechanical collector. These booms are typically constructed of rigid or flexible PE or PVC and are either anchored or tied in place. One such boom, deployed in collaboration with CLAIM[1] and New Naval, is the 'CLEAN TRASH'[2] system. This utilises a barge with sensors that automate the lifting and closing of removable trash cages when full. An alternative approach to passive collection is that of rakes, vertical fixed screens in the water column which trap litter as water passes through them, an example of which is the Thames litter collectors. These systems may be applied within the main river flow or at the sites of inlets and outfalls in order to prevent litter from entering the river. Both are well adapted to trap larger plastic items but struggle to retain smaller meso- and microplastics. However, both approaches to plastic aggregation may fail to capture debris at depth. Aeration systems such as 'The Great Bubble Barrier' seek to float and capture submerged plastics. Instead of sitting at the water's surface, these barriers are made up of tubes deployed across the riverbed. Air is forced

---

[1] CLAIM—Cleaning Litter & Applying Innovative Methods in European Seas.
[2] CLEAN TRASH— CLAIM's Litter Entrapping Autonomous Network Tactical Recovery Accumulation System Hellas.

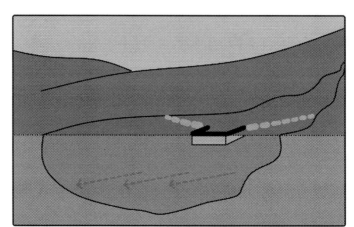

(A) Passive collector (boom)

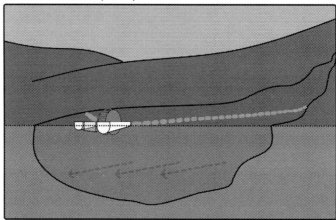

(B) Mechanical collector (trash wheel with boom)

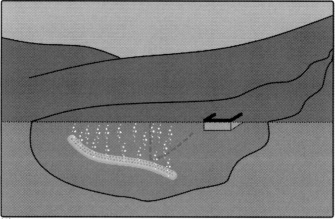

(C) Aeration

FIGURE 8.4

Litter collectors in riverine systems. Options for the design of riverine litter collectors including floating booms, aeration systems and trash wheels.

through these tubes and escapes via small holes, forcing plastics up and across the prevailing current into waiting collectors.

In comparison, fixed active collectors of riverine plastic and other debris include a variety of surface skimmers which utilise conveyors with brushes or panels to extract surface debris. These systems are commonly used in tandem with surface floats and booms which aggregate litter at the conveyor mouth. Examples include the H_2OPE 'River Whale', and The Ocean Cleanup's 'Interceptor', each of which uses renewable solar energy to run their belt systems. The similarly designed 'Mr Trash Wheel' (and later 'Professor' and 'Captain Trash Wheel'), is powered by water currents, with backup solar. Conversely, mobile collectors include a variety of garbage boats, such as those deployed by Collectix and WaterNet, which manage Amsterdam's waterways using five garbage boats (previously observed to fish out 42,000 kg of plastic per year).

The operation of these collectors is not without its challenges, capture efficiency is frequently influenced by environmental conditions and extreme weather and flows may result in damage or loss. For example in 2022, an avalanche net installed by The Ocean Cleanup across a river in Guatemala was damaged after a storm event washed 1000 tonnes of plastic downriver over the space of 20 minutes (Nordling, 2023). Similarly, in early 2023, a newly deployed pilot Trash Interceptor 007 sited in Ballona Creek in Playa del Rey was damaged by storms. Nevertheless, such systems have an essential role in retrieving waste at a key environmental bottleneck.

8.3.2 Coastal and offshore collectors

In marine settings, conditions vary greatly between coasts and offshore, pelagic, zones. Inshore collectors must compete with the multiple challenges of waves, tides and the presence of both biota and natural debris whereas those in offshore environments may be exposed to the significant swells and extreme conditions of the high seas. Macro-, meso- and microplastics in marine settings may be found at elevated levels close river mouths and the locations of key anthropogenic activities, as well as at sites of aggregation. As with riverine collectors, the technologies used to retrieve plastics are diverse, with a number of potential benefits and drawbacks.

8.3.2.1 Seabin

One form of collector frequently deployed in these settings (albeit sheltered ones) is the Seabin. These small, pumped collectors seek to capture floating debris such as plastic and oil by drawing surface water into a bucket-shaped collector and through a mesh screen. Typically positioned in marinas and ports, their 2-year deployment in Sydney harbour reportedly resulted in the capture of 100 tons of marine little from 14 billion filtered litres, however, the global average is 3.9 kg per day or 1.4 tons per year (Winterstetter et al., 2021). However, as with other collectors, the Seabin is unselective in the type of floating debris which it collects, as a result, the devices may become clogged with seaweed, requiring regular emptying. Additionally, deployment in Laucala Bay, Fiji, resulted in frequent by-catch (Paris et al., 2022), as did field trials in the Plymouth Sound, UK, which demonstrated that (alongside an average of 58 litter items weighing just under 0.006 kg), the Seabin caught an average of 13 organisms per day, although it was not possible to determine whether the organisms found in this study were already deceased upon capture. Concerningly, in addition to the problem of by-catch, it was also observed thatthe manual collection of plastics may be more efficient than seabin deployment (Parker-Jurd et al., 2022).

8.3.2.2 Trawls

Also suited to deployment in deeper coastal and offshore waters are surface tows such as the Thomsea trawl and The Ocean Cleanup systems. These trawls are comprised of an inflated boom and frame, supporting a fine mesh, designed for the capture of macroplastics. Due to high variability in the methods applied, the efficacy of trawls is mixed, however, The Ocean Cleanup's recently deployed system 002, 'Jenny', retrieved a reported 282,787 kg (282.8 tonnes) of litter during its July and October 2021 trial.

In addition to these specialist trawls, fishing for litter schemes aim to engage fishers in the retrieval and disposal of plastics and other marine litter retained in nets to be appropriately disposed of at ports. The efficacy of these schemes is dependent on a number of factors, including the quality of port facilities and costs or benefits to the individual. While not explicitly targeted to marine litter, fishing for litter schemes have significant potential to contribute to our understanding of plastic litter distribution. The deployment of research vessels is costly and, as a result, dedicated ship-based sampling is both spatially and temporally limited. Collaboration with fishers may assist in addressing existing data gaps. For example, twelve fishing vessels working out of two Italian ports on the Adriatic Sea collected more than 600 kg of litter over 1 month, revealing the contribution of plastic packaging, fishing gear and mussel nets, as well as identifying areas of apparent aggregation (Pasanisi et al., 2023). Participation in such schemes can also result in secondary environmental benefits by engaging participants with key issues. In their 2019 study of the perceptions of stakeholders regarding the UK Fishing For Litter programme, Wyles and colleagues identified that the participating fishers have higher perceptions of the efficacy of the programme and its ability to raise awareness than nonparticipating fishers (Wyles et al., 2019). While positive, the results of this study also indicate a certain degree of disengagement in some groups, despite widespread awareness of the scheme. The development of new markets for recovered plastics may help to incentivise participation by sceptical stakeholders (Nguyen & Brouwer, 2022). Additionally, transboundary consistency and well-defined responsibilities would increase the effectiveness of existing schemes (Ronchi et al., 2019).

8.3.2.3 Pelagic collectors

We have already touched upon the challenges of actively reclaiming plastics and microplastics from offshore environments, away from the continental shelf. These zones include the much discussed 'Garbage Patches', areas of elevated plastic concentration at the centre of gyres which have previously been identified as a potential focal point for plastic recovery. Approaches to plastic removal at these sites are similar to those in nearshore and riverine settings and include the use of trawling (such as the previously mentioned 'Jenny') in the upper water layers as well as the deployment of passive boom-and-net systems such as The Ocean Cleanup's early technologies. However, these systems must be scaled to withstand long-term offshore deployment and the extreme conditions in these environments, including waves of up to 30 m. Fluid dynamic modelling studies highlight that passive collection devices work best at flows less than 2.5/ms and wave height below 0.4 m, and their efficacy drops significantly at wave heights above 4/ms (Shaw et al., 2019). Significant wave heights[3] above this size are regularly observed (Kumar et al., 2022). Indeed, during its initial 2018–19 trial period, The Ocean Cleanup's 600-m long high-density polyethylene boom 'Wilson' was returned to port due to the loss of an 18-m end section. Following an assessment of the factors influencing the low capture efficiency and mechanical issues of this first system, the second boom (001/B) was more successful; with the addition of a sea anchor to influence drift speed and retain greater amounts of plastics.

As above, potential impacts of extraction measures on biota have led to concern from marine researchers, who highlight that floating booms may attract pelagic species which may accidentally become entangled, and that towed systems risk additional by-catch (although data on the distribution of species in offshore areas targeted for plastic retrieval is limited (Falk-Andersson et al., 2020)). Indeed, even the positioning of these collectors in offshore areas is in dispute, with models suggesting that coastal sites may be the most effective places at which plastics may be retrieved (Sherman & van Sebille, 2016), and source reduction highlighted as a more impactful way of addressing the problem of ocean plastic pollution (Rochman, 2016). Analysis of the cost-effectiveness of these methods in relation to the increase in plastic output driven by population and economic growth has suggested that they are insufficient in isolation and that up to €708 billion (which the authors highlighted as 1% of the world GDP at time of study) would be needed to remove the minimum 135 million tons of plastics required (Cordier & Uehara, 2019).

8.4 Conclusion

The recovery of plastic debris from the environment has been achieved by a variety of means designed to address the needs of a range of commercial, subsistence and sustainability concerns. Consistent efforts by individuals – such as volunteers

---

[3] The average height of the largest third of waves recorded.

and informal waste collectors — may result in the removal of substantial masses of plastic not yet equalled by mechanical means, with fewer apparent risks to the environment (although studies outlining these impacts present a mixed picture). Indeed, the efficacy of most plastic extractors, both riverine and marine is impacted by plastic density and concentration in addition to site characteristics such as flow, wave strength, debris composition and the presence of biota, thus site-specific assessments may be required. However, all approaches to plastic recovery may result in positive effects in terms of engagement and awareness both within and between stakeholder groups. This can result in positive feedback in terms of removal effort, whereby stakeholders not part of the original participant group become engaged, a factor which may further reinforce the commitment of the primary participant group (Page, 2006). Thus significant efforts must be made to remove barriers to participation and to devise a suitable fate for recovered litter.

References

Cordier, M., & Uehara, T. (2019). How much innovation is needed to protect the ocean from plastic contamination? *Science of the Total Environment*, *670*, 789—799. Available from https://doi.org/10.1016/j.scitotenv.2019.03.258, http://www.elsevier.com/locate/scitotenv.

Dugan, J.E., Hubbard, D.M., McCrary, M.D., & Pierson, M.O. (2003), The response of macrofauna communities and shorebirds to macrophyte wrack subsidies on exposed sandy beaches of southern California. *Estuarine, Coastal and Shelf Science*, http://www.elsevier.com/inca/publications/store/6/2/2/8/2/3/index.htt 58

Dugan, J. E., & Hubbard, D. M. (2010). Loss of coastal strand habitat in Southern California: The role of beach grooming. *Estuaries and Coasts*, *33*(1), 67—77. Available from https://doi.org/10.1007/s12237-009-9239-8.

Dumbili, E., & Henderson, L. (2020). *The challenge of plastic pollution in Nigeria. Plastic waste and recycling: Environmental impact, societal issues, prevention, and solutions* (pp. 569—583). Nigeria: Elsevier. Available from https://www.sciencedirect.com/book/9780128178805, 10.1016/B978-0-12-817880-5.00022-0.

Dunn, R. A., & Welden, N. A. (2023). Management of environmental plastic pollution: A comparison of existing strategies and emerging solutions from nature. *Water, Air, and Soil Pollution*, *234*(3). Available from https://doi.org/10.1007/s11270-023-06190-2, https://www.springer.com/journal/11270.

Environmental Management Bureau Republic of the Philippines Department of Environment and Natural Resources (2018). 2021 *National solid waste management status report* https://emb.gov.ph/national-solid-waste-management-status-report/

Falk-Andersson, J., Larsen Haarr, M., & Havas, V. (2020). Basic principles for development and implementation of plastic clean-up technologies: What can we learn from fisheries management? *Science of the Total Environment*, *745*. Available from https://doi.org/10.1016/j.scitotenv.2020.141117, http://www.elsevier.com/locate/scitotenv.

Gheskiere, T., Magda, V., Greet, P., & Steven, D. (2006). Are strandline meiofaunal assemblages affected by a once-only mechanical beach cleaning? Experimental findings. *Marine Environmental Research, 61*(3), 245−264. Available from https://doi.org/10.1016/j.marenvres.2005.10.003.

Godfrey, L., & Oelofse, S. (2017). Historical review of waste management and recycling in South Africa. *Resources, 6*(4), 57. Available from https://doi.org/10.3390/resources6040057.

Gutberlet, J., & Carenzo, S. (2020). Waste pickers at the heart of the circular economy: A perspective of inclusive recycling from the Global South. *Worldwide Waste: Journal of Interdisciplinary Studies, 3*(1). Available from https://doi.org/10.5334/wwwj.50.

Hamidul Bari, Q., Mahbub Hassan, K., & Ehsanul Haque, M. (2012). Solid waste recycling in Rajshahi city of Bangladesh. *Waste Management, 32*(11), 2029−2036. Available from https://doi.org/10.1016/j.wasman.2012.05.036.

Jayasiri, H. B., Purushothaman, C. S., & Vennila, A. (2013). Plastic litter accumulation on high-water strandline of urban beaches in Mumbai, India. *Environmental Monitoring and Assessment, 185*(9), 7709−7719. Available from https://doi.org/10.1007/s10661-013-3129-z, https://link.springer.com/journal/10661.

Kumar, P., Sardana, D., Kaur, S., Remya, P. G., Rajni., & Weller, E. (2022). Influence of climate variability on wind-sea and swell wave height extreme over the Indo-Pacific Ocean. *International Journal of Climatology, 42*(12), 6183−6203. Available from https://doi.org/10.1002/joc.7584, http://onlinelibrary.wiley.com/journal/10.1002/(ISSN)1097-0088.

Malm, T., Råberg, S., Fell, S., & Carlsson, P. (2004). Effects of beach cast cleaning on beach quality, microbial food web, and littoral macrofaunal biodiversity. *Estuarine, Coastal and Shelf Science, 60*(2), 339−347. Available from https://doi.org/10.1016/j.ecss.2004.01.008, http://www.elsevier.com/inca/publications/store/6/2/2/8/2/3/index.htt.

Morton, J. K., Ward, E. J., & de Berg, K. C. (2015). Potential small- and large-scale effects of mechanical beach cleaning on biological assemblages of exposed sandy beaches receiving low inputs of beach-cast macroalgae. *Estuaries and Coasts, 38*(6), 2083−2100. Available from https://doi.org/10.1007/s12237-015-9963-1, http://www.springerlink.com/content/120846/.

Nguyen, L., & Brouwer, R. (2022). Fishing for litter: Creating an economic market for marine plastics in a sustainable fisheries model. *Frontiers in Marine Science, 9.* Available from https://doi.org/10.3389/fmars.2022.722815, https://www.frontiersin.org/journals/marine-science#.

Nordling, L. (2023). Why I'm a garbage collector for the world's oceans. *Nature, 613* (7945), 800. Available from https://doi.org/10.1038/d41586-023-00179-x, https://www.nature.com/nature/.

Nzeadibe, T. C. (2013). Informal waste management in Africa: Perspectives and lessons from Nigerian garbage geographies. *Geography Compass, 7*(10), 729−744. Available from https://doi.org/10.1111/gec3.12072.

Page, S. E. (2006). Path dependence. *Quarterly Journal of Political Science, 1*(1), 87−115. Available from https://doi.org/10.1561/100.00000006.

Paris, A., Kwaoga, A., & Hewavitharane, C. (2022). An assessment of floating marine debris within the breakwaters of the University of the South Pacific, Marine Studies Campus at Laucala Bay. *Marine Pollution Bulletin, 174.* Available from https://doi.org/10.1016/j.marpolbul.2021.113290, http://www.elsevier.com/locate/marpolbul.

Parker-Jurd, F. N. F., Smith, N. S., Gibson, L., Nuojua, S., & Thompson, R. C. (2022). Evaluating the performance of the 'Seabin' − A fixed point mechanical litter removal device for sheltered waters. *Marine Pollution Bulletin, 184*. Available from https://doi.org/10.1016/j.marpolbul.2022.114199, http://www.elsevier.com/locate/marpolbul.

Pasanisi, E., Galasso, G., Panti, C., Baini, M., Galli, M., Giani, D., Limonta, G., Tepsich, P., Delaney, E., Fossi, M. C., & Pojana, G. (2023). Monitoring the composition, sources and spatial distribution of seafloor litter in the Adriatic Sea (Mediterranean Sea) through Fishing for Litter initiatives. *Environmental Science and Pollution Research, 30*(39), 90858–90874. Available from https://doi.org/10.1007/s11356-023-28557-y, https://www.springer.com/journal/11356.

Project BB (2023). https://project.bb/en [Accessed 31 January 2024].

Rochman, Chelsea M. (2016). Strategies for reducing ocean plastic debris should be diverse and guided by science. *Environmental Research Letters, 11*(4), 041001. Available from https://doi.org/10.1088/1748-9326/11/4/041001, https://doi.org/10.1088/1748-9326/11/4/041001.

Roman, L., Hardesty, B. D., Leonard, G. H., Pragnell-Raasch, H., Mallos, N., Campbell, I., & Wilcox, C. (2020). A global assessment of the relationship between anthropogenic debris on land and the seafloor. *Environmental Pollution, 264*. Available from https://doi.org/10.1016/j.envpol.2020.114663, https://www.journals.elsevier.com/environmental-pollution.

Ronchi, F., Galgani, F., Binda, F., Mandić, M., Peterlin, M., Tutman, P., Anastasopoulou, A., & Fortibuoni, T. (2019). Fishing for Litter in the Adriatic-Ionian macroregion (Mediterranean Sea): Strengths, weaknesses, opportunities and threats. *Marine Policy, 100*, 226–237. Available from https://doi.org/10.1016/j.marpol.2018.11.041, http://www.elsevier.com/inca/publications/store/3/0/4/5/3/.

Schenck, R., & Blaauw, P. F. (2011). The work and lives of street waste pickers in Pretoria—A case study of recycling in South Africa's Urban Informal Economy. *Urban Forum, 22*(4), 411–430. Available from https://doi.org/10.1007/s12132-011-9125-x.

Shaw, H. J., Chen, W. L., & Li, Y. H. (2019). A CFD study on the performance of a passive ocean plastic collector under rough sea conditions. *Ocean Engineering, 188*. Available from https://doi.org/10.1016/j.oceaneng.2019.106243, http://www.journals.elsevier.com/ocean-engineering/.

Sherman, P., & van Sebille, E. (2016). Modeling marine surface microplastic transport to assess optimal removal locations. *Environmental Research Letters, 11*(1), 014006. Available from https://doi.org/10.1088/1748-9326/11/1/014006, https://doi.org/10.1088/1748-9326/11/1/014006.

Shiber, J. G. (1987). Plastic pellets and tar on Spain's Mediterranean beaches. *Marine Pollution Bulletin, 18*(2), 84–86. Available from https://doi.org/10.1016/0025-326X(87)90573-X, https://www.sciencedirect.com/science/article/pii/0025326X8790573X.

Stelling-Wood, T. P., Clark, G. F., & Poore, A. G. B. (2016). Responses of ghost crabs to habitat modification of urban sandy beaches. *Marine Environmental Research, 116*, 32–40. Available from https://doi.org/10.1016/j.marenvres.2016.02.009, http://www.elsevier.com/locate/marenvrev.

Surfrider Foundation. (2023). *Ocean initiatives. A Surfrider Foundation Europe Programme*. Europe: Surfrider Foundation. https://www.initiativesoceanes.org/en/

Sustainable Coastlines. (2009). Love your coast. https://sustainablecoastlines.org/about/our-programmes/love-your-coast/

Welden, N. A. C., & Lusher, A. L. (2017). Impacts of changing ocean circulation on the distribution of marine microplastic litter. *Integrated Environmental Assessment and*

Management, 13(3), 483−487. Available from https://doi.org/10.1002/ieam.1911, http://www.interscience.wiley.com/jpages/1551-3777.

Winterstetter, A., Grodent, M., Kini, V., Ragaert, K., & Vrancken, K. C. (2021). A review of technological solutions to prevent or reduce marine plastic litter in developing countries. *Sustainability (Switzerland), 13*(9). Available from https://doi.org/10.3390/su13094894, https://www.mdpi.com/2071-1050/13/9/4894/pdf.

Wyles, K. J., Pahl, S., Carroll, L., & Thompson, R. C. (2019). An evaluation of the Fishing For Litter (FFL) scheme in the UK in terms of attitudes, behavior, barriers and opportunities. *Marine Pollution Bulletin, 144*, 48−60. Available from https://doi.org/10.1016/j.marpolbul.2019.04.035, http://www.elsevier.com/locate/marpolbul.

Consume: biological fragmentation and degradation of polymers

9

9.1 Introduction

Despite the variety of approaches applied to either limit plastic production or manage plastic wastes, large volumes of unavoidable material still end up in either landfill or the wider environment. Although some petroleum-based plastics may be readily biodegradable (e.g. PBAT, formed from adipate and terephthalate, which has been seen to be biodegradable under optimal conditions), during typical environmental exposure, the high molecular weight, recalcitrant bonds and degree of crystallinity (the regular structure and alignment of polymer chains within the plastic) limit degradation through typical pathways. As a result, plastic debris is slow to fragment and even slower to degrade, resulting in the formation, transport and build-up of microplastics.

A complementary approach to the waste management methods outlined in Chapter 7, suited to those wastes that end up in 'less desirable' endpoints, is to limit the lifespan of plastic debris either in the environment or in landfill. This has previously been achieved by replacing traditional plastic polymers with either renewable materials or plastics containing either additives or weaker function groups in order to speed the rate of degradation. However, we may also look to nature to find organisms capable of achieving the same effect, removing problematic plastics in a manner similar to the bioremediation of other contaminants as discussed in Box 9.1. An emerging body of evidence highlights the ability of specialised groups of biota and their associated enzymes to affect polymer structures, inducing mechanical or chemical breakdown of plastics and their constituent polymers. However, these organisms differ in their usefulness in combating the issues of microplastic pollution, for example boring marine invertebrates have been seen to mechanically breakdown macroplastics resulting in the release of microplastics into the environment, while others have been seen to reduce the mass and molecular weight of polymers, speeding the process of complete mineralisation. To date, species capable of breaking down plastics come from a range of taxa, including invertebrates, such as waxworms and mealworms, fungi, algae and bacteria (Table 9.1). The ways in which this degradation has been monitored

Microplastics. DOI: https://doi.org/10.1016/B978-0-443-13324-4.00009-1

Box 9.1 A background in bioremediation: positives and drawbacks.

Bioremediation has long been used as a treatment for potentially harmful environmental contaminants, either by transformation or complete mineralisation. These methods may occur at the site of contamination (in situ) or involve the removal of a contaminated environmental matrix to a secondary setting in order to achieve complete treatment. For example, bacterial have been seen to effectively consume spilled oil, the microorganisms responsible for which are believed to have developed around natural oil seeps (Atlas & Hazen, 2011). Additionally, as in wastewater treatment, bioremediation steps may be used alongside the chemical or physical treatment approaches to increase the rate of either detoxification or destruction. By facilitating the growth of these organisms, the rate of contaminant degradation may be optimised. A variety of plastic degrading microorganisms have been identified, however, their usefulness in both waste treatment and the wider environment remains in question. By reviewing our current understanding of bioremediation methods, we may identify both barriers and avenues in the path of the successful remediation of microplastic wastes.

Options for remediation

In this chapter we touch upon the variety of organisms identified as influencing plastics. Similarly, wider bioremediation approaches may also be mediated by a range of taxa, including, aerobic, anaerobic and methylotrophic bacteria, as well as fungi (Vidali, 2001). Similarly, the identification, application and augmentation of these microorganisms for use in- or ex situ require a range of molecular and genetic measures to enable the researcher to identify the species and enzymes of interest (Singh et al., 2008).

Approaches to bioremediation are variable. In addition to biodeterioration, bioaccumulation and biotransformation, we may consider:

- bioattenuation: the reduction of the impact or magnitude of negative effects
- bioaugmentation: the introduction of additional microorganisms at sites of contamination to facilitate increased detoxification
- biobleaching: the use of suitable organisms for the treatment of paper pulp, for example in moving dyes, inks and metals from paper
- bioleaching: the extraction and reclamation of desirable compounds by biota
- biosorption: the extraction of reclamation of harmful substances from water via metabolic or physio-chemical pathways
- biostimulation: altering environmental conditions in order to facilitate the growth of beneficial microorganisms

Prior to the implementation of the above measures, the characterisation of both contaminants and existing microbial communities is of key importance. The determination of the composition and growth of degrading communities may be assessed by a variety of measures, such as PCR techniques (Iwamoto & Nasu, 2001). If applied in ex situ settings, systems must also be consistently inspected to ensure the maintenance of optimum conditions, apparent effectiveness and changes to the surrounding environment.

Challenges and solutions

Remaining with the use of bioremediation in the management of oil spills at various scales, we may see two of the above strategies in action. Leaks and spills may be addressed using both biostimulation and bioaugmentation: introducing or otherwise increasing the number of degrading microbes or improving the conditions for the growth of microorganisms. In such cases, the rate of oil reduction increases with population growth. Following an initial lag period, there is a period of exponential growth and rapid increase in the rate of oil degradation, followed by a final plateauing as the available substrate or level of secondary toxics restricts the size of the population.

(Continued)

> **Box 9.1 A background in bioremediation: positives and drawbacks. (Continued)**
>
> Alternative approaches to the control of oil spills include the deployment of booms, the use of absorbents or the application of dispersants (many of which have their own negative environmental effects), however, bioremediation measures have the benefit of being both cost-effective and no more time intensive than alternative approaches (Juwarkar et al., 2010).
>
> Unfortunately, bioremediation is not appropriate in all cases. It has already been recognised that high molecular weight compounds are not widely suited to microbial degradation (Megharaj et al., 2014). The degradation of some wastes may also result in the formation of toxic by-products which may influence either the degrading organisms or the wider environment. Similarly, the presence of secondary contaminants in mixed waste settings may inhibit the growth of degrading microbes.
>
> **Learning and knowledge outcomes**
>
> Unlike many contaminants, plastic polymers are relatively inert, thus the route of detoxification is perhaps the least feasible route for management, and destruction is the preferable outcome. This destruction (mineralisation) may take place in composting environments or anaerobic bioreactors (potentially complemented by bioaugmentation with suitable cultures). Many of the current plastic degrading organisms have been identified from communities in environments with a high proportion of plastics and, in the long term, contaminated environments may be subject to biostimulation to promote the growth of these adapted communities. However, wider unknowns such as the behaviour of target bacteria and changes in community structure during the remediation process continue to hamper progress and represent a significant area for ongoing research.

are diverse, typically comparing the number and characteristics of the organism, population or community to observed changes in the mechanical, chemical and physical properties of the plastic (Fig. 9.1).

9.2 Degradation by invertebrates

As discussed in earlier chapters, a variety of animals, including invertebrates, have been seen to ingest plastics. Of these, very few have been identified as *consuming* plastics, that is, using the plastic polymer as a useful substrate. Here we will focus predominantly on those species seen to be involved in the degradation of the polymer structure, rather than the fragmentation of plastic debris alone. Some of the earliest observations of the invertebrate-mediated breakdown of plastics were made by high school students in China working with various species of darkling beetle larvae, who noted their ability to reduce the mass of PS (Mealworms Digest Plastic, 2023). Indeed, observations of the impacts of beetles and other invertebrates apparently able to penetrate the plastic packing of a range of goods have been studied for many years (Hassan et al., 2014). Only as we have focused on the environmental problems caused by plastic waste has the primary focus of these studies shifted from one of preservation to one of degradation.

Table 9.1 Examples of species identified as degraders of plastics.

Taxa	Latin name	Affected plastic
Invertebrates	*Galleria mellonella, Plodia interpunctella, Tenebrio molitor, T. obscurus*	PET PS
Fungi	*Cryptococcus* sp.	PBS
	Papiliotrema laurentii	PBS
	Acremonium kiliense	PE
	Chrysonilia setophila	PE
	Colletotrichum fructicola	PE
	Curvularia lunata	PE
	Diaporthe italiana	PE
	Gliocladium virens	PE
	Aspergillus sp.	PHB
	Candida guilliermondii	PHB
	Cephalosporium sp.	PHB
	Cladosporium sp.	PHB
	Debaryomyces hansenii	PHB
	Eupenicillium sp.	PHB
	Paecilomyces sp.	PHB
	Penicillium sp.	PE, PHB, PVC
	Aspergillus sp.	PBS, PE, PET, PHB, PU, PVC
	Alternaria sp.	PE, PUR
	Bionectria sp.	PUR
	Edenia gomezpompae	PUR
	Guignardia mangiferae	PUR
	Lasiodiplodia sp.	PUR
	Nectria sp.	PUR
	Pestalotiopsis sp.	PUR
	Phaeosphaeria sp.	PUR
	Zopfiella karachiensis	PUR
Bacteria	*Vibrio alginolyticus* and *Vibrio parahaemolyticus*	PVA-LLDPE blend
	Zalerion maritimum	PE
	Acinetobacter sp.	PE
	Chitinophaga sp.	PE
	Enterobacter sp.	PE
	Flavobacterium	PE
	Rhodococcus	PE
	Aspergillus	PET
	Ideonella sakaiensis	PET
	Staphylococcus	PET
	Streptococcus	PET
	Thermobifida	PET

(Continued)

Table 9.1 Examples of species identified as degraders of plastics. *Continued*

Taxa	Latin name	Affected plastic
	Exiguobacterium	PS
	Kocuria	PS
	Aureobasidium	PUR
	Curvularia	PUR
	Fusarium	PUR

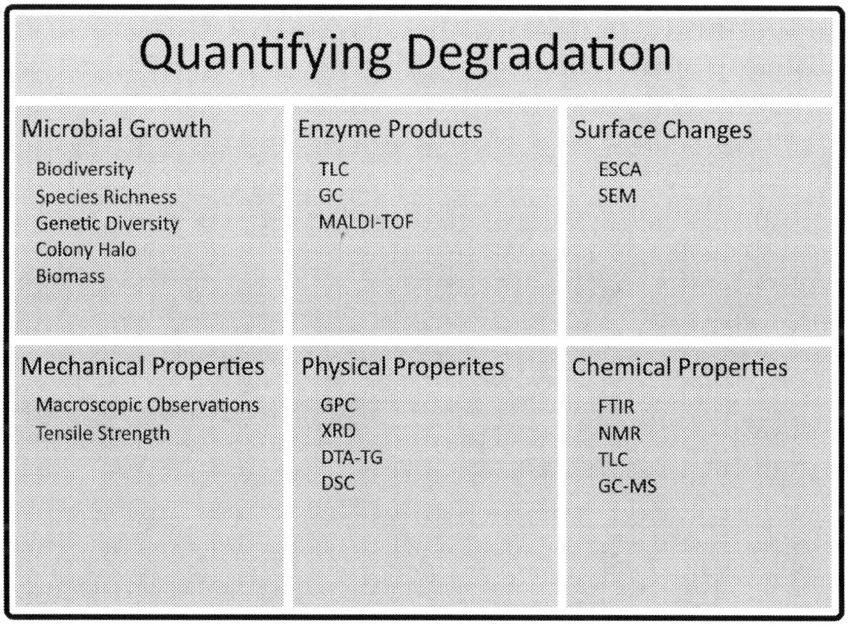

Quantifying Degradation

Microbial Growth

Biodiversity
Species Richness
Genetic Diversity
Colony Halo
Biomass

Enzyme Products

TLC
GC
MALDI-TOF

Surface Changes

ESCA
SEM

Mechanical Properties

Macroscopic Observations
Tensile Strength

Physical Properites

GPC
XRD
DTA-TG
DSC

Chemical Properties

FTIR
NMR
TLC
GC-MS

FIGURE 9.1 Assessment of the degradation of polymers.

The effectiveness of organisms in the breakdown of plastic polymers may be determined chemically, physically, mechanically or biologically.

Subsequently, beetle larvae, such as the mealworm, *Tenebrio molitor*, have been seen to interact with a variety of plastic foams including PS, PUR and PE (Przemieniecki et al., 2020; Wu et al., 2019). Previous observations have indicated that, during 60 days of exposure, mealworms were seen to remove between 46.9% and 69.7% of the initial plastic weight, averaged by polymer type (Bulak et al., 2021). And then there are the so-called superworms, *Zophobas atratus*, apparently capable of consuming four times more PS per day than the average mealworm (Yang et al., 2020). Analysis of the 'frass' (excrement) of plastic fed *Z. atratus* using a combination of gel permeation chromatography (GPC), solid-state 13C cross-polarisation/magic angle spinning nuclear magnetic resonance (CP/MAS NMR) spectroscopy and thermogravimetric interfaced with Fourier transform infrared (TG-FTIR) spectroscopy revealed the depolymerisation of long-chain PS molecules and the presence of low molecular-weight products.

Also widely studied are the caterpillar larvae of wax moth species, such as *Galleria mellonella* and *Plodia interpunctella*, which were amongst the first recorded invertebrate decomposers of plastics. The impact of these larvae on plastics may be remarkably fast, for example *G. mellonella* introduced to samples of polyethylene bags produced holes in as little as 40 minutes, with 100 worms apparently ingesting approximately 92 mg of plastic over a 12-hour period. While this initial fragmentation of the plastic may have been driven by the mechanical grazing action of worms, application of worm homogenate (a preparation of disrupted cells) to plastic samples resulted in a 13% mass loss after 14 hours. Degradation of the plastic polymer was confirmed using high-performance liquid chromatography coupled with mass spectrometry, which showed the development of new absorbance peaks when compared to an untreated sample (Bombelli et al., 2017). Similarly, holes were seen to form after approximately 45 minutes in HDPE samples in contact with the lesser waxworm, *Achroia grisella*. These samples demonstrated a 43% weight loss over an 8-day exposure to 100 waxworms (Kundungal et al., 2019). It was also seen that mixing plastics with nonplastic feed increased the rate of plastic consumption to up to 69.6%. Further, observation of the effectiveness of plastic degradation by *G. mellonella* has indicated that this rate may be further increased by initial solar exposure of the plastics, which may increase the bioavailability of the polymer (Kundungal et al., 2021).

As indicated above, simply ingesting plastics is not the same as consuming them, and many studies of plastic ingestion seek to determine whether these invertebrates are capable of living solely or in part on plastic wastes. A number of authors indicate a significant proportion of plastic is entirely broken down by these organisms, for example mealworms fed PS alone for a period of 1 month were estimated to convert approximately half the ingested carbon to CO_2 (Yang et al., 2015). Interestingly, the offspring of plastic-fed individuals may be more effective at degrading plastics themselves (Yang et al., 2018).

The capacity of invertebrates to degrade plastics is believed to be, primarily, the result of bacterial strains present in the gut. These include *Enterobacter asburiae* YT1, *Bacillus sp.* YP1, *Meyerozyma guilliermondii* and *Serratia marcescens*, isolated from *P. interpunctella*, each of which has been found to be effective in the degradation of PE (Lou et al., 2022; Yang et al., 2014). Similarly, the mealworms *Tribolium castaneum* and *T. molitor* are known to employ *Acinetobacter* sp. AnTc-1 and *Mixta tenebrionis* sp. nov. BIT-26[T] respectively in the breakdown of plastic. The composition of gut bacteria may also change in response to plastic in the diet, for example *Z. atratus* fed on plastic was seen to have polymer-dependent increases in *Enterococcus, Citrobacter, Dysgonomonas, Sphingobacterium* and *Mangrovibacter* (Luo et al., 2021). We will discuss these and similar plastic-degrading bacteria in greater detail below.

While these records are promising, they are not without their drawbacks, for example the current literature base predominantly considers the effectiveness of waxworms and other invertebrates on the breakdown of thin films that may be more susceptible to the action of grazers and similar species. More work is required to determine the wider application of many of these species.

9.3 **Degradation by fungi**

To date, there have been over 200 reports of fungi that can or may be capable of degrading plastics. Many fungi are saprotrophic[1] feeders on wood, plants and other cellulosic material. Such species may produce enzymes like esterases and lipases, capable of breaking down polymer structures. This is particularly likely in the presence of predegraded or weakened plastic material (Yamada-Onodera et al., 2001). Plastic degrading fungi have been isolated from, or are adapted to, a range of environments, from landfill sites to rainforests and, while a variety of polymers have been seen to be affected by fungi, LDPE is by far the most regularly studied substrate (Ekanayaka et al., 2022). Typical observations of the effectiveness of fungi in degrading plastic polymers have been conducted over periods between 14 days and a few months, with reduction observed in size, mass and molecular mass of the product, however, others also include consideration of community growth in situ, establishing the development of communities and the apparent dominance of some fungal strains.

Fungal strains identified as breaking down plastics include: *Penicillium simplicissimum* YK, seen to reduce the molecular weight of PE in liquid cultivation (Yamada-Onodera et al., 2001); *Aspergillus tubingensis* VRKPT1 and *Aspergillus flavus* VRKPT2, isolated from litter in coastal zones, which have been seen to degrade HDPE (Sangeetha Devi et al., 2015); *Aspergillus niger* and *Penicillium pinophilum*, observed to breakdown thermo-oxidised LDPE (Volke-Sepúlveda et al., 2002); and *Pestalotiopsis microspore*, seen to breakdown PUR (Russell et al., 2019). A phylogenetic analysis of fungi previously seen to degrade plastics highlights the importance of eleven biological classes in phyla Ascomycota (Dothideomycetes, Eurotiomycetes, Leotiomycetes, Saccharomycetes and Sordariomycetes), Basidiomycota (Agaricomycetes, Microbotryomycetes, Tremellomycetes, Tritirachiomycetes and Ustilaginomycetes) and Mucoromycota (Mucoromycetes), and the most plastic-degrading fungal records belong to the Eurotiomycetes. These Classes include plant pathogens, saprobes, microparasites and yeasts, as well as some edible fungi (Ekanayaka et al., 2022). In this latter

---

[1] Predominantly living on dead or decaying organic matter.

case, it was the edible oyster mushroom, *Pleurotus ostreatus*, which was tested for its ability to degrade of oxo-biodegradable plastic bags and 'green' polyethylene (Luz et al., 2019).

9.4 Degradation by algae

Algal species have previously been seen to interact with micro- and nanoplastics, colonising the surface of plastic debris either as biofilms or attached macroalgae. However, this group of organisms has only been superficially studied in terms of their impact on plastics. It is hypothesised that it is the process of colonisation of the plastics surface by algae, and the production of ligninolytic[2] and exopolysaccharide enzymes, that damages the polymers at the plastic's surface (Sarmah & Rout, 2018). *Chlorella vulgaris*, for example, is believed to carry enzymes capable of breaking down plastics. Indeed, PET films exposed to *C. vulgaris* over a month were seen to lose approximately 5.5% of their mass in comparison to a 0% loss in control samples (Falah et al., 2020). The production of useful enzymes may be engineered too, with PETase previously identified in the bacteria *Ideonella sakaiensis* successfully synthesised the chloroplasts of *Chlamydomonas reinhardtii*, a single-celled microalga found in temperate soils. This PETase was later found to be active on both PET film and postconsumer products (Di Rocco et al., 2023).

9.5 Degradation by bacteria

As indicated above, bacteria, typically free-living single-celled organisms, have also been seen to act upon plastic polymers. In 2016, the bacteria *Ideonella sakainesis* was isolated at a bottle recycling facility in Osaka, Japan. This strain of bacteria was seen to attach to the surface of PET films, degrading them at a rate of 0.13 mg/cm^2/day at 30°C (Yoshida et al., 2016). This degradation was the result of the action of two enzymes, PETase and MHETase, which acted upon the PET polymer (Yoshida et al., 2016). Subsequently identified plastic-degrading taxa include *Arthrobacter*, *Bacillus*, *Corynebecterium*, *Micrococcus*, *Nocardia*, *Pseudomonas*, *Streptomyces* and *Rhodococcus*. In addition to monitoring the mass loss experienced by plastics, further examination using spectroscopic techniques has been used to determine changes to the polymer structure and the formation of degradation products. For example, the degradation of PE exposed to *Penicillium simplicissimum* and *Nocardia asteroides* over a period of between three and seven months revealed the formation of the degradation product ethylene glycol (Bonhomme et al., 2003; Yamada-Onodera et al., 2001).

---

[2] Able to break down the organic polymer, lignin.

Many of the bacteria capable of degrading polymers have been isolated at sites handling high volumes of plastic or related products, such as the identification of *Bacillus cereus* and three species of *Pseudomonas* at an oilfield in Houstan, Texas (Vague et al., 2019). However, this is not always the case, and studies have seen signs of elevated plastic degradation rates arising from cultures taken from wastewater and sludge, as well as sediments and other environmental media (Auta et al., 2018).

Extremophile taxa, those bacteria adapted to live in particularly extreme conditions of pH (alkaliphiles), temperature (thermophiles), salinity (halophiles) or other environmental factor, may also be particularly useful in breaking down plastic polymers. However, this may be less the result of the adaptation of the bacteria to act upon plastics, but rather that the effect of these extreme conditions may weaken the polymer structure, increasing its bioavailability.

In addition to acting on the plastic polymer, the formation of biofilms may change the surface pH of the plastic, resulting in weathering and cracking or other biodeterioration of the plastic's surface. Surface changes have previously been observed in PP exposed to *Rhodococcus* sp. and *Bacillus* sp. over 40 days (Auta et al., 2018) as well as PE, PET and PP exposed to *Bacillus gottheili* and *B. cereus* (Auta et al., 2017). These changes are partly mediated by the production of extracellular polymeric substances (EPS), which form a matrix around microbial communities. These EPS matrices form microclimates conducive to the growth of the biofilm, improving contact with the plastic substrate, and retaining moisture. Following surface biodeterioration, enzymes then act upon the plastics, conversion of the polymer to oligomers (biofragmentation). If taken into the cell, oligomers (a polymer of few repeating units) may then be mineralised. These latter two processes will be discussed in greater detail below.

9.6 Mechanical breakdown of plastics

The action of animals on plastic waste may result in rapid damage to the plastic surface in addition to the subsequent uptake of the secondary plastics formed. Grazing waxworms and other plastic-consuming organisms chew upon plastics, creating sufficient physical stress to break the material. The rate at which plastics are mechanically degraded will be dependent on the number and feeding rate of the population present.

Other species have also been seen to fragment plastics into smaller pieces or affect the plastic surface, either by impaction, burrowing or chewing or grazing; although the majority of these species only hasten the speed of microplastic formation (e.g. the production of microplastics by organisms burrowing into PS floats in marine habitats) and promote the inclusion of microplastics into the food chain (Zheng et al., 2023). Nevertheless, damage to the surface of plastic debris and the fragmentation of large plastic litter increases the area for enzymatic attack. However, while increasing the surface area to volume ratio of the plastics may give a greater area of substrate for organisms and their enzymes to act upon, this may also enable increased transfer of harmful chemicals between the plastic and the environment or organism.

9.7 Enzymes

As indicated above, bacteria and fungi colonising the surface of plastics may secrete extracellular enzymes capable of breaking down vulnerable bonds in the polymer, the rate and effectiveness of which is affected by polymer crystallinity, temperature and pH (Maurya et al., 2020). Degradation products may then be taken up and, in the case of fungi, some smaller plastic fragments and may be internalised. Examples of plastic degrading enzymes include polyethylene terephthalate hydrolase (PETase), mono(2-hydroxyethyl)terephthalic acid esterase, cutinases and polylactic acid (PLA) hydrolases and polyurethane-degrading enzymes. We may separate these isolated enzymes into two categories, those responsible for the initial breakdown of the polymer (known as depolymerising) and those responsible for the breakdown of the resulting oligomers (Verschoor et al., 2022). In either case, the structure of enzymes, outlined in Box 9.2, enables them to bind to specific chemical bonds within a polymer, breaking the bonds by hydrolysis.

Box 9.2 Enzyme action.

Enzymes are highly specific polypeptides capable of acting as catalysts in chemical reactions, intitiating or speeding the processes without being used up or changed. The specificity of an enzyme is defined by its structure, which is itself determined by nucleotide sequences in an organisms DNA. Following transcription, mRNA travels through the cytoplasm to, in most cases, the ribosomes. Here, amino acids are assembled against the mRNA strand, forming polypaptide chains. Once complete it may undergo further modification, then transported to where it will be active.

The resulting enzyme has a complex three-dimensional structure formed by the position and type bonds within the molecule, comprised of the following: the Primary Structure, the sequence of amino acids in the protein; the Secondary Structure, the helices and sheets stabilised by hydrogen bonds between amino acids within the protein; the Tertiary Structure, the further folding of the protein and, in some cases; Quaternary Structure, which includes multiple proteins (Fig. 9.2). Of particular importance is the 'active site' which is of a specific shape to bind to target substrates to form an enzyme—substrate complex. Here, amino acid residues that interact with the substrate drive the reformation of chemical bonds and the rearrangement of atoms in the substrate, resulting in the formation of the desired products. After the products are formed, they are released from the active site of the enzyme, which is then free to act upon the next substrate. The enzymatic degradation is, in fact, a two-stage process, adsorption of enzymes on the polymer surface, followed by hydro-peroxidation/hydrolysis of the bonds.

The effectiveness of these reactions is influenced by both pH and temperature. Fastest reactions will occur within the optimal range, whereas extremes may lead to denaturing of the proteins. Additionally, the presence of cofactors such as coenzymes or metals can enhance the degradation process.

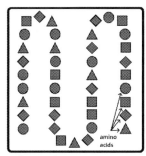
Primary Structure

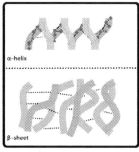
Secondary Structure

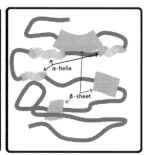
Tertiary Structure

FIGURE 9.2 The structure of enzymes.

The primary, secondary and tertiary structure of enzymes.

Some plastics are more susceptible to the action of enzymes than others. For example, polymers which contain vulnerable functional groups with amides, esters or other hydrolysable bonds (Fig. 9.3). PET, for instance, is comprised of terephthalic acid and ethylene glycol units linked by ester bonds. Ester bonds are also found in many natural polymers, such as those in plant cells. Organisms which produce enzymes capable of breaking down these natural polymers may therefore have a head start in degrading PET and other similar plastics. One example of a potentially effective enzyme group is that of cutinases, such as those isolated from *Thermobifida* spp., which have been seen to act on both plant cutin (a waxy polymer in a plant's cuticle) and PET (Ribitsch et al., 2012). Similarly, lipases, which act on fats, and esterases have also been seen to be effective in breaking down polyesters (Hajighasemi et al., 2016). Some polyesters are already widely recognised as being biodegradable, for example PBS, synthesised from either petroleum-based or biobased polymerising butanediol and succinic acid, and polycaprolactone (PCL), synthesised from caprolactone, both of which contain ester bonds formed through condensation reactions.

During the hydrolysis of ester bonds in PET, carbonyl groups ($C = O$ bonds) are broken to form an alcohol and a carboxylic acid. In this process, a nucleophile (typically a water molecule) donates an electron pair, attacking the carbonyl carbon atom, resulting in the cleavage of the $C = O$ bonds. Following this scission, the oxygen from the carbonyl group becomes part of the acid, and the carbonyl carbon becomes part of the alcohol. Additionally, a further $H+$ ion is released that may go on to further influence the degradation of the plastic.

Similarly, polyurethanes are formed of two main monomer types. An isocyanate compound which contains the isocyanate functional group ($-NCO$), such as Toluene diisocyanate (TDI) or Hexamethylene diisocyanate (HDI), and a polyol compound which contains multiple hydroxyl functional groups ($-OH$), such as polyester polyols (produced from esterification reactions), polyether polyols

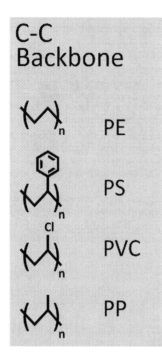

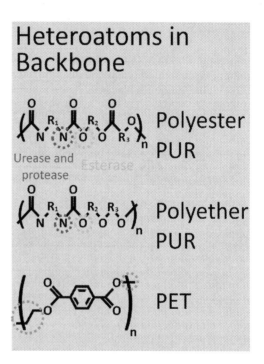

Polymer	Crystalinity (%)	Polymer	Crystalinity (%)
HDPE	80-90	Polyester PUR	40-50
LDPE	45-65	Polyether PUR	40-50
PP	60-70	PET	40-60

FIGURE 9.3 Structural susceptibility of plastics.

Vulnerable functional groups in plastic polymers.

(produced from alkoxide opening reactions) and polycarbonate polyols (produced from carbonate precursors). These monomers are joined by step-growth condensation polymerisation, which results in the formation of urethane bonds (−NHCOO−).

As with esters, urethane bonds may also be more vulnerable to biodegradation due to their similarity to naturally occurring, short-chain polymers. In this case, urethane bonds are similar in structure to those in proteins. While *some* enzymes, such as polyester hydrolases, have been seen to degrade polyurethanes (Álvarez-Barragán et al., 2016; do Canto et al., 2019; Howard et al., 1999; Russell et al., 2011) (Fig. 9.4), as the formation of these bonds (and the forces within them) differs from that of esters, different enzymes are able to break the polymer chain (Kjeldsen and Zubarev, 2011).

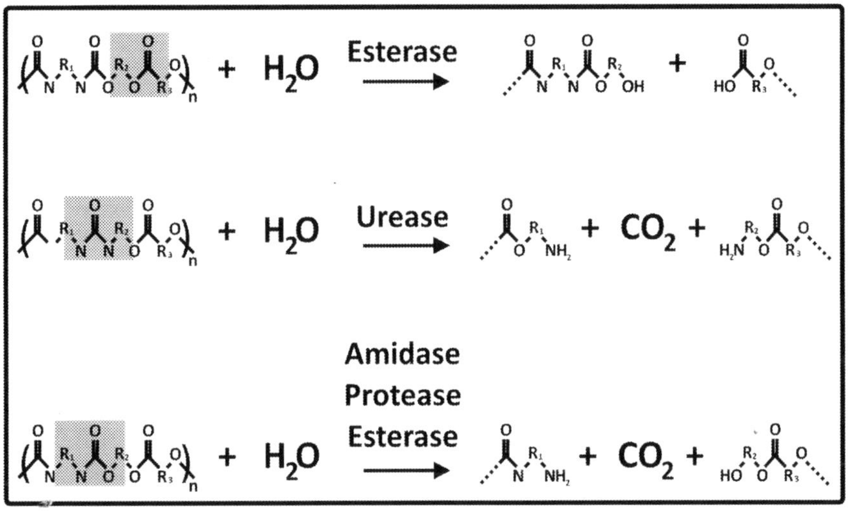

FIGURE 9.4 Example degradation pathways of polyurethanes.

Sites of vulnerability in PUR by enzyme and their associated degradation products.

As with polyesters, the degradation of a urethane bond may be by the action of a nucleophile in the presence of the enzyme (esterases, amidases or urethanases), attacking the carbon atom adjacent to the nitrogen atom. This process forms an alcohol (or phenol) and an amine (or aromatic amine), along with other possible by-products depending on the specific reaction conditions.

Despite their lack of comparatively weak funtional groups, some C-C backbone plastics have been seen to be biodegraded. Bacteria capable of utilising PE have been isolated from terrestrial mulch films, oil contaminated land, as well as in sewage sludge and landfills. These changes were identified based on mass loss, as well as deterioration of the plastic surface and reduction in properties such as tensile strength (Ru et al., 2020). Indeed, over a third of the weight of PE exposed to *Serratia marcescens* was lost over 70 days (Azeko et al., 2015).

Despite previous studies indicating that mealworms and other organisms may break down PS, little is known about the mechanisms by which this occurs (Hou & Majumder, 2021). Unlike the PUR and PET above, PS forms through addition polymerisation, during which styrene monomers with unsaturated double bonds are mixed with a catalyst or initiator. This catalyst drives the reaction by introducing a free radical (an unstable chemical species with an unpaired electron in its outermost shell) which affects the monomer, opening the double bond to form a single bond. This process forms a secondary radical, which in turn affects the next molecule, forming a third free radical. This process repeats, adding monomer units sequentially until a termination step (either the joining of two free radical chains or reaction with another molecule). These reactions result in simple C-C bonds in the polymer backbone.

Analysis of the genomes of species seen to degrade plastic polymers has suggested that the enzymes likely to break down PS include those able to affect the C−C bonds in the polymer backbone (such as cytochrome P4500s, alkane hydroxylases and monooxygenases) and those able to break PS side chains and oxidise aromatic ring compounds (such as ring-hydroxylating dioxygenases) (Hou & Majumder, 2021).

9.8 Other degrading factors

The onset of enzymatic degradation of plastic polymers may be affected by the various metabolites produced by the organism, for example those of fouling organisms. Some microorganisms may produce acids, bases and even hydrogen peroxide which act upon the plastic polymer. Acids and bases may influence the stability of bonds within plastics, weakening the polymer structure and promoting its hydrolysis. For example, the breakdown of PUR may also by catalysed by acids, which donate a proton that interacts with the urethane bond, increasing its susceptibility to attack by a nucleophile. Furthermore, the use of surfactants, additives such as $Ca2+$, $Mg2+$ and hydrophobins (cysteine-rich proteins), has also been reported to improve PET hyrdolysis (Maurya et al., 2020).

However, the action of enzymes on plastics may also be advanced or retarded by other characteristics of the plastic which may increase their susceptibility to enzyme attack. For example, the crystallinity of a polymer is known to affect degradation rates, with low crystallinity being associated with faster and more complete degradation. Fortunately, the crystallinity of polymers may be affected by temperature. For example, highly crystalline PET may be converted to amorphous PET by prolonged exposure to elevated temperatures. Degradation of PET by *I. sakaiensis* may thus be enabled by prior heat treatment of plastic wastes (Wallace et al., 2020).

Additionally, the hydrophobicity of a polymer may influence enzyme attack. Analysis of apparent hydrophobicity indicates that, when presented on a log scale, polymers such as PVA, Nylon 4, PEG, PVP and PLGA are hydrophilic, PLA and PBS show low hydrophobicity, PBAT (known to be degradable) was seen to have mid-range hydrophocibity and PVC, PC, PS, PE and PP, had the highest hydrophobicity. The degree of hydrophobicity has thus been highlighted as inversely proportional to the probability of surface erosion by fouling organisms (Min et al., 2020).

At the most basic level, factors such as temperature and sample depth (in sediment or water), may influence degradation. UV-radiation and oxygen are the most important factors which initiate degradation of polymers with a carbon−carbon backbone, leading to chain scission. Smaller polymer fragments formed by chain scission are more susceptible to biodegradation and therefore abiotic degradation is expected to precede biodegradation. When heteroatoms are present in the main chain of a polymer, degradation may proceed by photo-oxidation, hydrolysis and biodegradation (Gewert et al., 2015), it is thus unsurprising that observations of PHBV, PCL, PLA and Nylon in aquatic settings have demonstrated that increasing water temperature increases degradation of the polymer surface (although this effect varies in relation to the type of plastic).

9.9 Engineering degradation

As indicated in Box 9.1, the structure of enzymes is predetermined by the sequence of amino acids in DNA. However, it is possible to optimise enzymes for a specific purpose. For example, lipases have been historically developed to work effectively at a broader range of both temperature and pH, enabling them to be more effectively utilised in the detergent industry. Similarly, α-amylase enzymes are designed to work at specified temperatures and pHs for the breakdown of starch in the food and beverage industry, and restriction endonucleases are routinely adapted for use in genetic engineering.

In engineering novel or optimised enzymes for the degradation of plastic, researchers must first identify their target polymer and any naturally occurring enzymes suited to the degradation of this target or the functional group(s) it contains. This may occur by way of the screening of microbial communities in environments containing high plastic abundance or in the presence of naturally occurring or introduced material of a similar molecular make-up or in silico screening of enzymes already seen to work on similar functional groups. These enzymes are then characterised in order to understand their structure (via X-ray crystallography or similar) and mode of action. At this point, researchers may seek to optimise the candidate by either laboratory evolution or modelling methods (Verschoor et al., 2022). For example, it has previously been reported that so-called 'FAST-PETases' are able to completely depolymerise PET in just a week at room temperature, and as little as a day if the temperature is raised to 50°C (Lu et al., 2021), and that engineered cytochrome P450 may act as a biocatalyst in the biodegradation of PE (Yeom et al., 2022).

9.10 Real-world applicability

9.10.1 Understanding observed degradation rates

Whilst the discovery of plastic-consuming organisms is highly promising, the overall impact on plastic management is uncertain. The rate of both surface degradation and mass loss of plastics waste will vary depending on the organism acting on the polymer and the availability, structure and composition of the plastic and its additives. Questions currently remain as to the affordability, scalability and timescales involved in these processes.

Additionally, the polymers on which these enzymes are known to act represent only a small fraction of plastics produced annually (Chow et al., 2023). From the above, we might note that PET, which contains potentially vulnerable ester bonds, is one of the most regularly used polymers when exploring the biodegradation of plastics.

At the time of writing, PET represents only 6.2% of the global plastic produced (Plastics Europe, 2022). However, it is commonly used in packaging, textiles and beverage containers, and, as such, is a major contributor to plastic waste. Although recyclable, the energy required in processing PET can be high. Biotechnology firms seek to apply these enzymes in the depolymerisation of polymers, particularly PET in food containers and clothing in order to improve the recycling of plastics. The uses of other polymers containing ester bonds such as polybutylene terephthalate, PLA, polyglycolic acid, polyhydroxyalkanoates and polyethylene terephthalate glycol are listed in Table 9.2.

In addition to considering the structure of the polymer, we must also identify the requirements of the organism. Many of the species that we have discussed have optimal ranges of temperature, salinity, pH and other abiotic factors, outside which their efficacy as plastic consumers may not be guaranteed. Species capable of breaking down plastic may be most useful within waste treatment facilities, where conditions can be controlled. Nevertheless, the action of plastic degrading organisms on plastic litter in the enviornment has been suggested as a route of remediation. However, to date, only a small number of species have been identified from a handful of "natural" environments.

Of course, overtime, more plastivores may emerge; however, if we wish to intervene to increase plastic degradation in overly contaminated settings, we require the identification or development of organisms suited to the habitats that act as plastic sinks such as gyre centres, benthic habitats, beaches or agricultural land. However, the introduction of such species has significant implications regarding the safety of plastics in use in the environment. What might the impact be on the plastics that we wish to preserve in these settings? For example, how would the presence of plastic-degrading soil bacteria affect the use of mulch films? Or those in coastal settings affect coastal infrastructure? Currently, antifouling to protect ships from the growth of biofilms and the settlements of larger organisms is big business. We may see a similar need for the protection of plastics (or, counter intuitively, a move towards more recalcitrant polymers) if such effective plastic degrading organsisms were to become widespread.

We must also understand how these species are affected by changes in environmental conditions over time. Some studies of the bacterially mediated breakdown of plastic report a mass loss of a few percent during the trial periods (Auta et al., 2017), while others may reach 10% (J. Yang et al., 2014) or higher over a typical trial period of up to 3 months. This indicates that not only should environmental conditions be optimised for degradation, but that these conditions must be maintained over long periods for the full mineralisation of plastics to occur.

Table 9.2 Uses of degradable polymers.

Polymer	Example uses
Polyethylene terephthalate (PET)	Beverage bottles
	Food packaging (jars, containers)
	Polyester fabrics (clothing, textiles)
Polybutylene terephthalate (PBT)	Electrical components (connectors, switches)
	Automotive parts (housings, connectors)
	Appliances (enclosures, handles)
Polycaprolactone (PCL)	3D printing materials
	Drug delivery systems (sutures, implants)
	Modelling and sculpting material
Polylactic acid (PLA)	Food packaging (containers, cutlery)
	3D printing filament
Polyglycolic acid (PGA)	Biomedical sutures
	Tissue engineering scaffolds
	Drug delivery devices
Polyhydroxyalkanoates (PHA)	Biodegradable plastics (films, packaging)
	Medical applications (tissue engineering)
	Agricultural films
Polyethylene terephthalate glycol (PETG)	Food containers and bottles
	Display and signage materials
	Medical device components
Polyethylene naphthalate (PEN)	Flexible printed circuit boards
	Packaging films
	Automotive parts
Polytrimethylene terephthalate (PTT)	Textiles and apparel
	Carpets and upholstery
	Fibre applications
Polycyclohexylene dimethylene terephthalate (PCT)	Engineering plastics
	Automotive applications
	Electronics components
Polymethylene terephthalate (PMT)	Electrical insulation
	Automotive components
	Industrial applications
Polyethylene adipate (PEA)	Food packaging
	Industrial films
	Agricultural films
Polycyclohexylenedimethylene terephthalate (PCTG)	Medical equipment
	Consumer goods
	Packaging

(Continued)

Table 9.2 Uses of degradable polymers. *Continued*

Polymer	Example uses
Polybutylene naphthalate (PBN)	Electrical connectors
	Automotive parts
	Engineering applications
Polyethylene sebacate (PES)	Biodegradable films
	Packaging materials
	Textiles
Polysulfone (PSU)	Medical devices
	Aerospace components
	Automotive parts
Polyphthalamide (PPA)	Engineering plastics
	Electrical connectors
	Automotive parts
Polyarylate (PAR)	Electrical insulation
	Transparent containers
	Optical applications
Polyether ester ketone (PEEK)	Aerospace components
	Medical implants
	High-performance industrial parts

9.11 Conclusion

The biodegradation of plastics represents a promising strategy for the control of polymer wastes as well as a potential source of valuable byproducts (Mohanan et al., 2020). For example, the degradation products released during the breakdown of plastic may be suitable for use as feedstock in the plastics industry or related chemical processes, thus reducing reliance on fossil fuel extraction. However, while plastics containing ester, urethane and ether bonds may be broken down by enzymes (and their associated organisms), these polymers do not represent the majority of plastic production. Plastics such as PE and PP, which make up 16.9% and 19% of 2021 plastic production respectively, contain highly recalcitrant carbon-carbon bonds. These plastics must be broken down and may require significant preprocessing before they become widely available to biota. Additionally, while plastic-degrading organisms and enzymes have been seen to be effective under laboratory conditions, their applicability for application in large-scale waste management or plastic control in the wider environment is as yet unclear. Nevertheless, while there are numerous challenges to overcome, biodegradation may become a valuable aspect of the wider management of plastic wastes.

References

Álvarez-Barragán, J., Domínguez-Malfavón, L., Vargas-Suárez, M., González-Hernández, R., Aguilar-Osorio, G., & Loza-Tavera, H. (2016). Biodegradative activities of selected environmental fungi on a polyester polyurethane varnish and polyether polyurethane foams. *Applied and Environmental Microbiology*, *82*(17), 5225−5235. Available from https://doi.org/10.1128/AEM.01344-16, http://aem.asm.org/content/82/17/5225.full.pdf.

Atlas, R. M., & Hazen, T. C. (2011). Oil biodegradation and bioremediation: A tale of the two worst spills in U.S. history. *Environmental Science and Technology*, *45*(16), 6709−6715. Available from https://doi.org/10.1021/es2013227.

Auta, H. S., Emenike, C. U., & Fauziah, S. H. (2017). Screening of Bacillus strains isolated from mangrove ecosystems in Peninsular Malaysia for microplastic degradation. *Environmental Pollution*, *231*, 1552−1559. Available from https://doi.org/10.1016/j.envpol.2017.09.043, http://www.elsevier.com/inca/publications/store/4/0/5/8/5/6.

Auta, H. S., Emenike, C. U., Jayanthi, B., & Fauziah, S. H. (2018). Growth kinetics and biodeterioration of polypropylene microplastics by Bacillus sp. and Rhodococcus sp. isolated from mangrove sediment. *Marine Pollution Bulletin*, *127*, 15−21. Available from https://doi.org/10.1016/j.marpolbul.2017.11.036, http://www.elsevier.com/locate/marpolbul.

Azeko, S. T., Etuk-Udo, G. A., Odusanya, O. S., Malatesta, K., Anuku, N., & Soboyejo, W. O. (2015). Biodegradation of linear low density polyethylene by Serratia marcescens subsp. marcescens and its cell free extracts. *Waste and Biomass Valorization*, *6* (6), 1047−1057. Available from https://doi.org/10.1007/s12649-015-9421-0, http://www.springer.com/engineering/journal/12649.

Bombelli, P., Howe, C. J., & Bertocchini, F. (2017). Polyethylene bio-degradation by caterpillars of the wax moth Galleria mellonella. *Current Biology*, *27*(8), R292−R293. Available from https://doi.org/10.1016/j.cub.2017.02.060, http://www.elsevier.com/journals/current-biology/0960-9822.

Bonhomme, S., Cuer, A., Delort, A.-M., Lemaire, J., Sancelme, M., & Scott, G. (2003). Environmental biodegradation of polyethylene. *Polymer Degradation and Stability*, *81* (3), 441−452. Available from https://doi.org/10.1016/s0141-3910(03)00129-0.

Bulak, P., Proc, K., Pytlak, A., Puszka, A., Gawdzik, B., & Bieganowski, A. (2021). Biodegradation of different types of plastics by tenebrio molitor insect. *Polymers*, *13* (20). Available from https://doi.org/10.3390/polym13203508, https://www.mdpi.com/2073-4360/13/20/3508/pdf.

Chow, J., Perez-Garcia, P., Dierkes, R., & Streit, W. R. (2023). Microbial enzymes will offer limited solutions to the global plastic pollution crisis. *Microbial Biotechnology*, *16*(2), 195−217. Available from http://onlinelibrary.wiley.com/journal/10.1111/(ISSN)1751-7915.

Di Rocco, G., Taunt, H. N., Berto, M., Jackson, H. O., Piccinini, D., Carletti, A., Scurani, G., Braidi, N., & Purton, S. (2023). A PETase enzyme synthesised in the chloroplast of the microalga Chlamydomonas reinhardtii is active against post-consumer plastics. *Scientific Reports*, *13*(1). Available from https://doi.org/10.1038/s41598-023-37227-5, https://www.nature.com/srep/.

do Canto, V. P., Thompson, C. E., & Netz, P. A. (2019). Polyurethanases: Three-dimensional structures and molecular dynamics simulations of enzymes that degrade polyurethane. *Journal of Molecular Graphics and Modelling*, *89*, 82−95. Available from https://doi.org/10.1016/j.jmgm.2019.03.001, http://www.elsevier.com/inca/publications/store/5/2/5/0/1/2/index.htt.

Ekanayaka, A. H., Tibpromma, S., Dai, D., Xu, R., Suwannarach, N., Stephenson, S. L., Dao, C., & Karunarathna, S. C. (2022). A review of the fungi that degrade plastic. *Journal of Fungi, 8*(8). Available from https://doi.org/10.3390/jof8080772, http://www.mdpi.com/journal/jof.

Falah, W., Chen, F. J., Zeb, B. S., Hayat, M. T., Mahmood, Q., Ebadi, A., Toughani, M., & Li, E. Z. (2020). Polyethylene terephthalate degradation by Microalga Chlorella vulgaris along with pretreatment. *Materiale Plastice, 57*(3), 260−270. Available from https://doi.org/10.37358/MP.20.3.5398, https://revmaterialeplastice.ro/pdf/24%20WAJEEHA%203%2020.pdf.

Gewert, B., Plassmann, M. M., & Macleod, M. (2015). Pathways for degradation of plastic polymers floating in the marine environment. *Environmental Sciences: Processes and Impacts, 17*(9), 1513−1521. Available from https://doi.org/10.1039/c5em00207a, http://www.rsc.org/publishing/journals/em/about.asp.

Hajighasemi, M., Nocek, B. P., Tchigvintsev, A., Brown, G., Flick, R., Xu, X., Cui, H., Hai, T., Joachimiak, A., Golyshin, P. N., Savchenko, A., Edwards, E. A., & Yakunin, A. F. (2016). Biochemical and structural insights into enzymatic depolymerization of polylactic acid and other polyesters by microbial carboxylesterases. *Biomacromolecules, 17*(6), 2027−2039. Available from https://doi.org/10.1021/acs.biomac.6b00223, http://pubs.acs.org/journal/bomaf6.

Hassan, M. W., Qasim, M. U., Iqbal, J., & Jamil, M. (2014). Study of penetration ability by Tribolium castaneum (Herbst.)(Coleoptera: Tenebrionidae) through different loose plastic packaging. *Journal of Pure and Applied Sciences, 24*(33), 1−2.

Hou, L., & Majumder, E. L. W. (2021). Potential for and distribution of enzymatic biodegradation of polystyrene by environmental microorganisms. *Materials, 14*(3), 1−20. Available from https://doi.org/10.3390/ma14030503, https://www.mdpi.com/1996-1944/14/3/503/pdf.

Howard, G. T., Ruiz, C., & Hilliard, N. P. (1999). Growth of Pseudomonas chlororaphis on a polyester-polyurethane and the purification and characterization of a polyurethanase-esterase enzyme. *International Biodeterioration and Biodegradation, 43*(1-2), 7−12. Available from https://doi.org/10.1016/S0964-8305(98)00057-2.

Iwamoto, T., & Nasu, M. (2001). Current bioremediation practice and perspective. *Journal of Bioscience and Bioengineering, 92*(1), 1−8. Available from https://doi.org/10.1016/S1389-1723(01)80190-0, https://www.sciencedirect.com/science/article/pii/S1389172301801900.

Juwarkar, A. A., Singh, S. K., & Mudhoo, A. (2010). A comprehensive overview of elements in bioremediation. *Reviews in Environmental Science and Biotechnology, 9*(3), 215−288. Available from https://doi.org/10.1007/s11157-010-9215-6.

Kjeldsen, F., & Zubarev, R. A. (2011). Effects of peptide backbone amide-to-ester bond substitution on the cleavage frequency in electron capture dissociation and collision-activated dissociation. *Journal of the American Society for Mass Spectrometry, 22*(8), 1441−1452. Available from https://doi.org/10.1007/s13361-011-0151-7.

Kundungal, H., Gangarapu, M., Sarangapani, S., Patchaiyappan, A., & Devipriya, S. P. (2021). Role of pretreatment and evidence for the enhanced biodegradation and mineralization of low-density polyethylene films by greater waxworm. *Environmental Technology (United Kingdom), 42*(5), 717−730. Available from https://doi.org/10.1080/09593330.2019.1643925, http://www.tandf.co.uk/journals/titles/09593330.asp.

Lou, H., Fu, R., Long, T., Fan, B., Guo, C., Li, L., Zhang, J., & Zhang, G. (2022). Biodegradation of polyethylene by Meyerozyma guilliermondii and Serratia marcescens isolated from the gut of waxworms (larvae of Plodia interpunctella). *Science of the Total Environment*, *853*. Available from https://doi.org/10.1016/j.scitotenv.2022.158604, http://www.elsevier.com/locate/scitotenv.

Luo, L., Wang, Y., Guo, H., Yang, Y., Qi, N., Zhao, X., Gao, S., & Zhou, A. (2021). Biodegradation of foam plastics by Zophobas atratus larvae (Coleoptera: Tenebrionidae) associated with changes of gut digestive enzymes activities and microbiome. *Chemosphere*, *282*. Available from https://doi.org/10.1016/j.chemosphere.2021.131006, http://www.elsevier.com/locate/chemosphere.

Luz, J. M. R., da Silva, M. d C. S., dos Santos, L. F., & Kasuya, M. C. M. (2019). *Plastics polymers degradation by fungi. Microorganisms*. IntechOpen.

Lu, H., Diaz, D.J., Czarnecki, N.J., Zhu, C., Kim, W., Shroff, R., Acosta, D.J., Alexander, B., Cole, H., Zhang, Y.J., Lynd, N., Ellington, A.D., & Alper, H.S. (2021). *Deep learning redesign of PETase for practical PET degrading applications*. bioRxiv, https://www.biorxiv.org. https://doi.org/10.1101/2021.10.10.463845.

Maurya, A., Bhattacharya, A., & Khare, S. K. (2020). Enzymatic remediation of polyethylene terephthalate (PET)-based polymers for effective management of plastic wastes: An overview. *Frontiers in Bioengineering and Biotechnology*, *8*. Available from https://doi.org/10.3389/fbioe.2020.602325, http://journal.frontiersin.org/journal/bioengineering-and-biotechnology#archive.

Mealworms Digest Plastic. (2023). *Mealworms digest plastic*. http://news.sina.com.cn/c/2003-12-30/16251466914s.shtml.

Megharaj, M., Venkateswarlu, K., & Naidu, R. (2014). *Bioremediation encyclopedia of toxicology* (Third edition, pp. 485−489). Australia: Elsevier. Available from http://www.sciencedirect.com/science/book/9780123864550, https://doi.org/10.1016/B978-0-12-386454-3.01001-0.

Min, K., Cuiffi, J. D., & Mathers, R. T. (2020). Ranking environmental degradation trends of plastic marine debris based on physical properties and molecular structure. *Nature Communications*, *11*(1). Available from https://doi.org/10.1038/s41467-020-14538-z, http://www.nature.com/ncomms/index.html.

Mohanan, N., Montazer, Z., Sharma, P. K., & Levin, D. B. (2020). Microbial and enzymatic degradation of synthetic plastics. *Frontiers in Microbiology*, *11*. Available from https://doi.org/10.3389/fmicb.2020.580709, https://www.frontiersin.org/journals/microbiology#.

Przemieniecki, S. W., Kosewska, A., Ciesielski, S., & Kosewska, O. (2020). Changes in the gut microbiome and enzymatic profile of Tenebrio molitor larvae biodegrading cellulose, polyethylene and polystyrene waste. *Environmental Pollution*, *256*113265. Available from https://doi.org/10.1016/j.envpol.2019.113265.

Ribitsch, D., Acero, E. H., Greimel, K., Eiteljoerg, I., Trotscha, E., Freddi, G., Schwab, H., & Guebitz, G. M. (2012). Characterization of a new cutinase from Thermobifida alba for PET-surface hydrolysis. *Biocatalysis and Biotransformation*, *30*(1), 2−9. Available from https://doi.org/10.3109/10242422.2012.644435, 10292446 Austria.

Russell, J. R., Huang, J., Anand, P., Kucera, K., Sandoval, A. G., Dantzler, K. W., Hickman, D., Jee, J., Kimovec, F. M., Koppstein, D., Marks, D. H., Mittermiller, P. A., Nunez, S. J., Santiago, M., Townes, M. A., Vishnevetsky, M., Williams, N. E., Vargas, M. P., Boulanger, L. A., … Devipriya, S. P. (2019). Efficient biodegradation of polyethylene (HDPE) waste by the plastic-eating lesser waxworm (Achroia grisella).

Environmental Science and Pollution Research, *26*(18), 18509−18519. Available from https://doi.org/10.1007/s11356-019-05038-9, http://www.springerlink.com/content/0944-1344.

Russell, J. R., Huang, J., Anand, P., Kucera, K., Sandoval, A. G., Dantzler, K. W., Hickman, D. S., Jee, J., Kimovec, F. M., Koppstein, D., Marks, D. H., Mittermiller, P. A., Núñez, S. J., Santiago, M., Townes, M. A., Vishnevetsky, M., Williams, N. E., Vargas, M. P. N., Boulanger, L. A., . . . Strobel, S. A. (2011). Biodegradation of polyester polyurethane by endophytic fungi. *Applied and Environmental Microbiology*, *77* (17), 6076−6084. Available from https://doi.org/10.1128/AEM.00521-11, http://aem. asm.org/cgi/reprint/77/17/6076.pdf, United States.

Ru, J., Huo, Y., & Yang, Y. (2020). Microbial degradation and valorization of plastic wastes. *Frontiers in Microbiology*, *11*. Available from https://doi.org/10.3389/ fmicb.2020.00442, https://www.frontiersin.org/journals/microbiology#.

Sangeetha Devi, R., Rajesh Kannan, V., Nivas, D., Kannan, K., Chandru, S., & Robert Antony, A. (2015). Biodegradation of HDPE by Aspergillus spp. from marine ecosystem of Gulf of Mannar, India. *Marine Pollution Bulletin*, *96*(1-2), 32−40. Available from https://doi.org/10.1016/j.marpolbul.2015.05.050, http://www.elsevier.com/locate/ marpolbul.

Sarmah., & Rout, J. (2018). Algal colonization on polythene carry bags in a domestic solid waste dumping site of Silchar town in Assam. *Phykos.*, *48*(67).

Singh, S., Kang, S. H., Mulchandani, A., & Chen, W. (2008). Bioremediation: environmental clean-up through pathway engineering. *Current Opinion in Biotechnology*, *19*(5), 437−444. Available from https://doi.org/10.1016/j.copbio.2008.07.012.

Vague, M., Chan, G., Roberts, C., Swartz, N.A., & Mellies, J.L. (2019). *Pseudomonas isolates degrade and form biofilms on polyethylene terephthalate (PET) plastic.* bioRxiv, https://www.biorxiv.org, https://doi.org/10.1101/647321.

Verschoor, J. A., Kusumawardhani, H., Ram, A. F. J., & de Winde, J. H. (2022). Toward microbial recycling and upcycling of plastics: prospects and challenges. *Frontiers in Microbiology*, *13*. Available from https://doi.org/10.3389/fmicb.2022.821629, https:// www.frontiersin.org/journals/microbiology#.

Vidali, M. (2001). Bioremediation. An overview. *Pure and Applied Chemistry*, *73*((7)), 1163−1172. Available from https://doi.org/10.1351/pac200173071163, 00334545, http://www.degruyter.com/view/j/pac.

Volke-Sepúlveda, T., Saucedo-Castañeda, G., Gutiérrez-Rojas, M., Manzur, A., & Favela-Torres, E. (2002). Thermally treated low density polyethylene biodegradation by Penicillium pinophilum and Aspergillus niger. *Journal of Applied Polymer Science*, *83* (2), 305−314. Available from https://doi.org/10.1002/app.2245.

Wallace, N. E., Adams, M. C., Chafin, A. C., Jones, D. D., Tsui, C. L., & Gruber, T. D. (2020). The highly crystalline PET found in plastic water bottles does not support the growth of the PETase-producing bacterium Ideonella sakaiensis. *Environmental Microbiology Reports*, *12*(5), 578−582. Available from https://doi.org/10.1111/1758-2229.12878, http://onlinelibrary.wiley.com/journal/10.1111/(ISSN)1758-2229.

Wu, Q., Tao, H., & Wong, M. H. (2019). Feeding and metabolism effects of three common microplastics on Tenebrio molitor L. *Environmental Geochemistry and Health*, *41*(1), 17−26. Available from https://doi.org/10.1007/s10653-018-0161-5, http://www.wkap. nl/journalhome.htm/0269-4042.

Yamada-Onodera, K., Mukumoto, H., Katsuyaya, Y., Saiganji, A., & Tani, Y. (2001). Degradation of polyethylene by a fungus, Penicillium simplicissimum YK. *Polymer*

Degradation and Stability, *72*(2), 323−327. Available from https://doi.org/10.1016/S0141-3910(01)00027-1.

Yang, S. S., Brandon, A. M., Andrew Flanagan, J. C., Yang, J., Ning, D., Cai, S. Y., Fan, H. Q., Wang, Z. Y., Ren, J., Benbow, E., Ren, N. Q., Waymouth, R. M., Zhou, J., Criddle, C. S., & Wu, W. M. (2018). Biodegradation of polystyrene wastes in yellow mealworms (larvae of Tenebrio molitor Linnaeus): Factors affecting biodegradation rates and the ability of polystyrene-fed larvae to complete their life cycle. *Chemosphere*, *191*, 979−989. Available from https://doi.org/10.1016/j.chemosphere.2017.10.117, http://www.elsevier.com/locate/chemosphere.

Yang, Y., Wang, J., & Xia, M. (2020). Biodegradation and mineralization of polystyrene by plastic-eating superworms Zophobas atratus. *Science of The Total Environment*, *708*135233. Available from https://doi.org/10.1016/j.scitotenv.2019.135233.

Yang, J., Yang, Y., Wu, W. M., Zhao, J., & Jiang, L. (2014). Evidence of polyethylene biodegradation by bacterial strains from the guts of plastic-eating waxworms. *Environmental Science and Technology*, *48*(23), 13776−13784. Available from https://doi.org/10.1021/es504038a, http://pubs.acs.org/journal/esthag.

Yang, Y., Yang, J., Wu, W. M., Zhao, J., Song, Y., Gao, L., Yang, R., & Jiang, L. (2015). Biodegradation and mineralization of polystyrene by plastic-eating mealworms: Part 1. Chemical and physical characterization and isotopic tests. *Environmental Science and Technology*, *49*(20), 12080−12086. Available from https://doi.org/10.1021/acs.est.5b02661, http://pubs.acs.org/journal/esthag.

Yeom, S. J., Le, T. K., & Yun, C. H. (2022). P450-driven plastic-degrading synthetic bacteria. *Trends in Biotechnology*, *40*(2), 166−179. Available from https://doi.org/10.1016/j.tibtech.2021.06.003, http://www.elsevier.com/locate/tibtech.

Yoshida, S., Hiraga, K., Takehana, T., Taniguchi, I., Yamaji, H., Maeda, Y., Toyohara, K., Miyamoto, K., Kimura, Y., & Oda, K. (2016). A bacterium that degrades and assimilates poly(ethylene terephthalate). *Science (New York, N.Y.)*, *351*(6278), 1196−1199. Available from https://doi.org/10.1126/science.aad6359, http://science.sciencemag.org/content/sci/351/6278/1196.full.pdf.

Zheng, Y., Li, J., Li, G., & Shi, H. (2023). Burrowing invertebrates induce fragmentation of mariculture Styrofoam floats and formation of microplastics. *Journal of Hazardous Materials*, *447*, 130764. Available from https://doi.org/10.1016/j.jhazmat.2023.130764.

Policy priorities: emerging trends in a global response 10

10.1 Introduction: plastics and the UN Sustainable Development Goals

In earlier chapters, we have examined the impacts of microplastics on species and habitats and considered the secondary effects that this may have on ecosystem services and wider human wellbeing. Taking into account these many primary and secondary effects, we may also explore the problem of the unsustainable use of plastic through the lens of the UN Sustainable Development Goals (SDGs). The SDGs, as initially set out in the United Nations General Assembly's Post-2015 Development Agenda, represent far-reaching, social, economic and environmental challenges. Alongside the identified issues are a set of targets to be achieved by 2030, as well as measurable indicators of either progress or success.

As many plastic products represent an essential component of both infrastructure and services worldwide, our efforts to reduce or manage plastics may subsequently be linked to many of the SDGs. Thus, the changes in policy, practice and behaviour required to reduce the production and mishandling of plastic wastes may advance or retard progress toward associated global targets. Obviously, a number of these goals, for example Goal 1: *No Poverty*, 2: *Zero Hunger*, 10: *Reduced inequalities* and 16: *Peace, Justice, and Strong Institutions*, are only tangentially linked to the issue of plastics (with either minor reliance on plastic products or limited scope for negative impacts as a result of plastic wastes). However, other goals are more closely related, with measured targets that directly pertain to the management methods discussed in previous chapters. Links between environmental concerns and the SDGs have previously considered the issue of energy production (represented under SDG 7 and SDGs 4 and 5) (McCollum et al., 2018). In the following sections, the various SDGs are grouped as having 'primary', 'secondary' and 'tertiary' relevance; with all of these underpinned by SDG 17: *Partnerships for the Goals*, which seeks to ensure adequate access to financial resources, investment and environmentally sound technologies to facilitate the attainment of wider targets.

10.1.1 Plastics and the SDGs – tertiary links

SDGs which have are only tangentially linked to the issue of plastic and microplastic pollution include Goal 5: *Gender Equality*. Numerous aspects of the plastic production

Microplastics. DOI: https://doi.org/10.1016/B978-0-443-13324-4.00010-8

chain may have disproportionate impacts on either men and women, for example, the textiles industry. Various processes in the production of garments, such as spinning and weaving and cutting and sewing, are typically undertaken by a higher proportion of either men or women, resulting in differing levels of microplastic fibre exposure in each group. Similarly, secondary contaminants released during the heating or burning of plastics may have greater impacts on those primarily involved in cooking (predominantly women), or burning for waste control (predominantly men). Additional tertiary links may be made with the targets underpinning Goal 4: *Quality Education*, is 4.7, to 'ensure that all learners acquire the knowledge and skills needed to promote sustainable development, including, among others, through education for sustainable development and sustainable lifestyles... ...[and a] culture's contribution to sustainable development'. The problem of micro- and nanoplastics has led to widespread information campaigns and has been adopted as a case study in many curricula, particularly in relation to the principles of Learning for Sustainability, and the use of plastic case studies may support the aquisition of key skills in maths, science and technology.

10.1.2 Plastics and the SDGs — secondary links

More closely linked goals and targets include SDGs 3: Good health and wellbeing, 6: Clean water and sanitation, 7: Energy, 8: Economy, 9: Infrastructure and 11: Cities and communities.

Under SDG 3: *Good Health and Well-being*, relevant targets include '3.9: substantially reduce the number of deaths and illnesses from hazardous chemicals and air, water and soil pollution and contamination'. As we highlighted in Chapter 4, there are a number of ways in which microplastics may affect wellbeing. Ingestion, inhalation and the effects of secondary contamination may all represent routes of uptake through which human health may be influenced. However, plastics also have a significant role in the protection of human health. For example, medical equipment has a substantial single-use plastic footprint, with significant proportions of this waste only suitable for incineration. Additionally, single-use plastics may be used to distribute vital goods, such as clean water or food aid, and are essential in both conflict zones and disaster response.

Inherently linked to health and wellbeing is Goal 6: *Clean Water and Sanitation*, which seeks to enable equitable and efficient access to clean water and sanitation by 2030, improve water management and reduce pollution, as well as promoting the preservation of key related ecosystems which underpin water availability and quality. We have already discussed the role of microplastics as an aquatic pollutant, as well as its occurrence in samples of tap and bottled water. In regions where water is collected directly from surface waterbodies or wells, the number of plastic sources within the catchment and the mode of collection may have significant impacts on both micro- and nanoplastic loads. In addition to these observations of microplastic in drinking water, where water is collected or obtained from local suppliers, second-hand or repurposed vessels are frequently used. These vessels are often plastic and may represent a source of microplastics in their own right.

Many of the targets pertaining to Goal 7: *Affordable and Clean Energy* also have implications for the use of energy from waste (EfW) programmes. For example, target 7.a sets out the requirement to 'enhance international cooperation to facilitate access to clean energy research and technology, including renewable energy, energy efficiency and advanced and cleaner fossil-fuel technology, and promote investment in energy infrastructure and clean energy technology'. Management via EfW is the eventual fate of a significant proportion of challenging plastic wastes, however, as previously noted, these processes are also associated with the release of potentially damaging contaminants (Hahladakis et al., 2018). Policy decisions associated with more sustainable energy production may necessitate a reduction in EfW systems, however, suitable routes must be put in place to manage the material which would previously have been disposed of in this manner. It may be that such policy changes may help in the promotion of circular technologies for waste prevention, however, these must be put into the global context of waste exports, technological advancement and consumer demand.

Of course, the disruption of any industry, and the development of novel products and associated services may affect patterns of employment and the attainment of Goal 8: *Decent Work and Economic Growth*. In Chapter 7 we discussed the potential impacts of the international transport of plastic wastes on the receiving country and local people, as well as the response of previous exporters of plastic wastes, such as China, in terms of policy development. If we again consider a movement away from EfW in addition to increased resource circularity, the adoption of new technologies and services for the management and handling of plastic waste may represent an additional route towards Target 8.2 which highlights the need to 'achieve higher levels of economic productivity through diversification, technological upgrading and innovation, including through a focus on high-value added and labour-intensive sectors'. Similarly, in relation to Target 8.4, such a move toward a more circular economy and more sustainable manufacturing will benefit 'global resource efficiency in consumption and production and [...] decouple economic growth from environmental degradation'.

Related to the above issues, but also to wider society is the role of plastics in the development and maintenance of global infrastructure. Our ability to reduce our reliance on plastics (particularly single-use plastics) as well as the sustainable management of plastic waste is dependent on many of the targets associated with Goal 9: *Industry, Innovation and Infrastructure*. Under this goal, Target 9.1, encourages us to 'develop quality, reliable, sustainable and resilient infrastructure, including regional and transborder infrastructure, to support economic development and human well-being...'. Here, the implementation of sustainable infrastructure may relate to provisions for low-waste lifestyle changes, such as drinking water stations in cities, or the development of improved systems of waste infrastructure. In addition, Targets 9.2 and 9.4 look towards industry, seeking to 'Promote inclusive and sustainable industrialisation[...]' and highlight the importance of 'upgrad[ing] infrastructure and retrofit[ing] industries to make them sustainable, with increased resource-use efficiency and greater adoption of clean and environmentally sound technologies and industrial processes[...]'.

Alongside, the development of more sustainable industries described above, Goal 11 focuses on the development of *Sustainable Cities and Communities*. Plastics-associated targets include 11.6, 'to reduce the adverse per capita environmental impact of cities, including by paying special attention to air quality and municipal and other waste management'. Again, movement toward a more sustainable use of resources, the use of alternative products and methods for more effective resource recovery set out in Chapter 7 are key in this regard.

Finally, the push for a change in approach to the use of plastics may also impact our ability to achieve Goal 13: *Climate Action*, whereby target 13.2 seeks to 'integrate climate change measures into national policies, strategies and planning'. We have already indicated that movement away from plastic to alternative products may have significant effects on land and water footprints as well as the creation of secondary pollutants, however, there may also be a change in the carbon footprint of products as highlighted in other chapters. Here, alteration in product weight has implications for transport costs, as does the implementation of deposit returns schemes.

10.1.3 Plastics and the SDGs — primary links

Perhaps most obvious in their links to the impacts of both plastics and microplastics are Goals 14 and 15, *Life Below Water* and *Life on Land* which seek to prevent or reverse environmental damage. In Chapters 4–6, we highlighted the many and varied observed effects of plastics on aquatic organisms and communities. Understanding of these impacts has led to significant pressure to limit the sources and scale of plastic, microplastic and nanoplastic debris, although, the collection and removal of plastics may have a significant effect on these same communities. Relevant targets relating to these goals include 14.1 and 15.1, which seek to 'reduce marine pollution' and 'conserve and restore terrestrial and freshwater ecosystems' respectively. However, success in these areas is intrinsically linked to our ability to address many of the targets identified above in addition to our next closely linked SDG.

Perhaps most directly linked in the relationship between plastics and sustainable development is SDG 12: *Responsible Consumption and Production*. Targets under this goal pertain to the sound management of wastes at all stages of the product lifecycle, reduced waste generation (focussing on the upper end of the waste hierarchy), improved industry practices, sustainable procurement in the public sector and increased public awareness. The indicators by which progress towards these targets is measured are set out in Table 10.1. The inclusion of microplastics in international agreements, and the associated number of signatories, demonstrates ongoing progress towards the first of these targets (12.4.1). National-level commitments, such as the apparent change in the proportion of plastic wastes sent to landfill or EfW schemes, demonstrate progress towards the second (12.4.2) and, by association, the third target (12.5.1) may be evidenced by the increasing proportion of plastics wastes recycled since the introduction of

Table 10.1 Targets and indicators associated with SDG 12: responsible consumption and production.

Target	Indicator	Plastic relevance
12.4: By 2020 achieve the environmentally sound management of chemicals and all wastes throughout their lifecycle, in accordance with agreed international frameworks, and significantly reduce their release to air, water and soil in order to minimise their adverse impacts on human health and the environment	12.4.1: Number of parties to international multilateral environmental agreements on hazardous waste, and other chemicals that meet their commitments and obligations in transmitting information as required by each relevant agreement 12.4.2: (a) Hazardous waste generated per capita; and (b) proportion of hazardous waste treated, by type of treatment	Commitment to international plastics targets Mass of plastics waste produced per capita; and proportion of plastic waste directed to each waste treatment type
12.5: By 2030 substantially reduce waste generation through prevention, reduction, recycling and reuse	12.5.1: National recycling rate, tons of material recycled	Plastic recycling rates in tons
12.6: Encourage companies, especially large and transnational companies, to adopt sustainable practices and to integrate sustainability information into their reporting cycle	12.6.1: Number of companies publishing sustainability reports	Companies implementing plastic pledges and reporting plastic waste Extended producer responsibility schemes
12.7: Promote public procurement practices that are sustainable, in accordance with national policies and priorities	12.7.1: Number of countries implementing sustainable public procurement policies and action plans	Public bodies implementing plastic pledges and reporting plastic waste
12.8: By 2030 ensure that people everywhere have the relevant information and awareness for sustainable development and lifestyles in harmony with nature	12.8.1: Extent to which (i) global citizenship education and (ii) education for sustainable development are mainstreamed in (a) national education policies; (b) curricula; (c) teacher education; and (d) student assessment	Inclusion of plastic and microplastic science into the curriculum (potentially through STEM and Learning for Sustainability initiatives)

these targets. However, while reporting has increased on corporate sustainability and on public procurement policies, it has fallen when it comes to sustainable consumption and monitoring sustainable tourism (U.N., 2023).

10.2 International treaties and the plastics issue

Progress towards SDG 12 and the wider goals is, in part, supported by the UN Environment Programme's guidance for legislators and policymakers, as well as a substantial body of research and wider initiatives to limit plastic waste and encourage its sustainable management. The scale of international concern and associated willingness to adopt sustainable plastic management measures is evidenced by the Group of Seven (G7) and Group of Twenty (G20), which have both highlighted plastic pollution as a pressing global issue and have made commitments to address it through various initiatives and policies.

Indeed, a variety of existing international treaties, agreements and conventions address either the management of waste plastics or the protection of environments into which plastics may be discharged. Most of these agreements have a broad environmental or socio-economic focus but include plastics as a named category of concern. In addition, 2022 saw the adoption of UNEA Resolution 5/14 — End plastic pollution: Towards an international legally binding instrument, which stresses the 'urgent need to strengthen the science-policy interface at all levels, improve understanding of the global impact of plastic pollution on the environment, and promote effective and progressive action at the local, national, and global levels, recognising the important role played by plastics in society'. Under this resolution, the International Negotiating Committee will develop an international instrument, which could include both binding and voluntary approaches to influence the full lifecycle of plastic. At time of writing talks are still ongoing, however, at the most recent talks in Nairobi, an initial draft of a legally binding instrument (UNEP/PP/INC.3/4) was introduced. This draft included many elements pertaining directly to the issue microplastics, such as the management of plastic feedstocks, the control of products containing intentionally added microplastics, as well as broader provisions for the sustainable use and handling of plastics.

Also at an international scale, in the European Union, the Single-Use Plastics Directive (EU) 72/2019/904 builds on existing waste regulations and sets out the requirement for (among other things) members to reduce the consumption of single-use cups for beverages and food containers and to restrict the sale of cotton buds, cutlery, expanded polystyrene (EPS) food packaging and similar products, as well as outlining requirements for plastics as a component of other products. Another example of international cooperation for the sustainable plastic use is that of the Association of Southeast Asian Nations' Regional Action Plan for Combating Marine Debris in the ASEAN Member States (2021–2025). This policy seeks to reduce plastic leakage and increase waste recovery by increasing the value of plastic wastes, limiting reliance on single use plastics and harmonising recycling standards across the region (World Bank, 2021). However, the effectiveness of these policies and their subsequent impact on environmental concerns is often dependent on their extent and level of enforcement, indeed, the previous EU Directive sets out restrictions on a number of products comprised of EPS, but not their extruded polystyrene (XPS) counterparts. Increased use of XPS single-use products may undermine the stated goals of increasing resource circularity and reducing waste generation (Troya et al., 2022).

10.2.1 Limiting the international movement of plastic wastes

Historically, many forms of wastes containing potentially valuable secondary materials have been traded and transported internationally. For example, in 2016 approximately 74% of the world's imports of waste plastics were sent to Asian countries (Liang et al., 2021). In addition to the energy cost of moving bulk wastes in this manner, a common criticism of such exports is that they move the health and environmental impacts of waste generation from the producing country (often from OECD countries) to the receiving country (frequently to non-OECD countries), relocating wastes to locations with weaker waste infrastructure. An example of international efforts to address these problems albeit at a much larger scale than that of plastics alone, is the Basel Convention, adopted in 1989 and coming into force in 1992, which governs the 'Control of Transboundary Movements of Hazardous Wastes and Their Disposal'. Under this convention, the rules and reporting procedures for the movement of such material are set out (e.g. banning certain exports following 1995), as are prior informed consent procedures (PIC) to ensure that those countries receiving the waste are aware of its content and able to safely handle and treat it. Notably, plastic wastes were added to the list of materials considered under the convention in 2019.

In Europe, the export of certain wastes identified under Annex III or IIIA to Regulation (EC) No 1013/2006, which includes substances identified under the Basel Convention, is managed under Regulation (EC) 1418/2007. Among other challenging waste materials, this European regulation seeks to control the movement of solid plastic wastes and mixtures of wastes from member states to non-OECD countries, preventing worker exposure and environmental contamination at the receiving locations. However, there remain concerns that waste traded with countries outside the EU that are ostensibly able to safely and sustainably manage such imports may be shipped to less suitable or more vulnerable secondary locations at a later date. Subsequently, some organisations have called for outright bans on the trade in such material.

Under the above agreements, plastics intended for export must be assessed to determine the level of contamination. While important in preventing the export of challenging or harmful materials that may cause environmental damage in the receiving country, it also has secondary implications regarding the ability of some producers and industries to find recyclers to accept some wastes. For example, agricultural plastics such as mulch films are frequently contaminated with earth and chemical residues (such as pesticides and fertilisers). As a result, farmers subject to restrictions on the on-farm management of wastes may struggle to access 'sustainable' management options (Chen, 2022).

At the other end of this issue, some countries have aimed to limit the impact of traded wastes by banning imports. Perhaps the most well-known example of this approach is that taken by China in 2018, ostensibly as a response to concerns over the potential negative effects on the country's environmental, economic and public health. This move reflected China's desire to reduce the proportion of low-value wastes in the system, minimise energy-intensive recycling processes, improve domestic waste management and sustainability and move towards a more circular model. The policy, termed 'National Sword', has had far-reaching effects on global waste streams. Prior to its implementation, in 2017, China was estimated to take in approximately 5.8 to 8.3 Mt plastic waste and other recyclables each year, which subsequently fell to just 52 kt in 2018 (Liang et al., 2021). Subsequently, countries that commonly exported to China have been forced to find alternative solutions to the management of recyclables and other wastes. Indeed, exports from the largest exporting countries or regions such as Hong Kong (China), the United States, Japan and Germany decreased. However, the remaining waste has to go somewhere and increased imports have subsequently been observed in Vietnam, and Malaysia, among others (Liang et al., 2021).

Of course, the import of other types of waste may also result in the spread of micro- and nanoplastic pollution. For example, the movement of sewage sludges between countries may act as a direct route for the introduction of plastic waste onto agricultural land and recent reporting in the United Kingdom highlighting the import of 27,000 tonnes of sludge from the Netherlands has provoked widespread condemnation. Studies have previously noted microplastic abundance in sewage sludge of up to 301,400 MP/kg (Nguyen et al., 2022), however, microplastics are not included in the potentially toxic element tests that must be carried out on both land and soil to ensure that damaging levels are not reached.

10.2.2 Protecting environments

A variety of directives and conventions seek to prevent and address the pollution of specific ecosystems, for example, the pollution of marine environments is addressed under the London Convention and Protocol. The London Convention, adopted in 1972, and its 1996 Protocol regulate the dumping of wastes and other matter at sea. Although the focus is not exclusively on plastic waste, this agreement plays a role in preventing marine pollution, including plastic debris, from being disposed of by shipping. More locally, the OSPAR convention includes guidance for the onshore and offshore pollution sources previously included in the 1972 Oslo and 1974 Paris Conventions on the protection of the North-East Atlantic. Similarly, the EU's Marine Strategy Framework Directive (MSFD) 2008/56/EC aims to achieve Good Environmental Status in its marine waters, setting out targets and indicators by which progress may be monitored. These include Descriptor 10, which focuses on marine litter, the targets of which include reductions in both environmental levels of macro- and micro-litter, and the associated occurrence of entanglement and ingestion, to below damaging levels (Table 10.2). By such measures, direct inputs of waste and their drivers may be managed, and requirements for the routine monitoring of environmental contamination set out.

10.3 **National policymaking for plastic management**

The international focus on the negative impacts of plastic wastes has also resulted in a variety of related policies at national scales. The impetus behind these policies has frequently been driven by public interest, increased by media coverage of both macro- and microplastic impacts (Dunn & Welden, 2023). as well as by the aforementioned international agreements.

Policy-making decisions are frequently split into 'upstream' options, which address the top of the waste hierarchy and seek to encourage both sustainable design practices or influence consumer purchasing behaviour (limiting resource use and

Table 10.2 Relevant descriptors from the Marine Strategy Framework Directive.

Descriptor	Hierarchy	Detail
D10C1	Primary	The composition, amount and spatial distribution of litter on the coastline, in the surface layer of the water column, and on the seabed, are at levels that do not cause harm to the coastal and marine environment. Member States shall establish threshold values for these levels through cooperation at Union level, taking into account regional or subregional specificities
D10C2	Primary	The composition, amount and spatial distribution of micro-litter on the coastline, in the surface layer of the water column, and in seabed sediment, are at levels that do not cause harm to the coastal and marine environment Member States shall establish threshold values for these levels through cooperation at Union level, taking into account regional or subregional specificities
D10C3	Secondary	The amount of litter and micro-litter ingested by marine animals is at a level that does not adversely affect the health of the species concerned. Member States shall establish threshold values for these levels through regional or subregional cooperation.
D10C4	Secondary	The number of individuals of each species which are adversely affected due to litter, such as by entanglement, other types of injury or mortality or health effects. Member States shall establish threshold values for the adverse effects of litter, through

waste production). Conversely, 'downstream' options, may affect municipal waste treatment or drive public behaviour at point of disposal, This latter approach seeks to encourage the use of options higher up Lasink's Ladder, limit landfilling and EfW, as well as reducing the mishandling of wastes. The priorities and principles set out in these policies provide a framework for the introduction of specific regulations and targets which may be implemented by relevant agencies or departments, often with associated penalties for noncompliance, such as fines or sanctions.

10.3.1 Upstream: reducing waste production

Reduction in the mass of plastics produced annually may be brought about by a variety of legislantive measures, including incentives for the production or purchase of sustainable, disincentives for the production and purchase of less sustainable products, or the outright ban of certain products or product categories. These approaches have previously been applied with apparent success (and subsequent positive environmental and human health outcomes) in relation to a range of harmful pollutants, for example, various sources of air pollution as set out in Box 10.1. Numerous countries have introduced measures designed to reduce both plastic production and plastic waste by restricting the sale of many products. Early

---

Box 10.1 Case study: learning from past pollution policy

In the United Kingdom, and further afield, the issues of widespread pollution and environmental change have their roots in the Industrial Revolution (1760—1840). During the period developments in technology, living conditions and population growth drove additional demands on environmental resources, as well as increases in both solid waste and air pollution. However, while the scale and impact of solid waste pollution initially received little attention, recognition of rising air pollution resulted in some of our earliest global environmental policies. Below we will consider the development of air pollution policy in the United Kingdom, including the drivers and development of national policy and its interaction with international priorities, identifying common themes of relevance to our attempts to minimise microplastic pollution in the modern era.

The following will consider UK air pollution from the 13th century onward, exploring emerging polices and classifying them according to their apparent targets. Using this information, we will look for areas for development in the existing response to plastic pollution.

Policy for the prevention of air pollution

The United Kingdom's first actions against air pollution were in 1273 and 1306, in relation to the burning of bituminous sea coal (recognised for its particularly noxious smoke), the use of which would result in, initially, fines or the demolition of the furnace in which it was used (Mister, 1970). These ordinances were taken seriously, resulting in torture and at least one execution, although these penalties were later replaced with taxation and control. The country's next forays into air pollution control would not be for another 500 years, pertaining to a widespread technological advancement, the steam train. In 1845 the Railway Clauses Consolidated Act required that steam trains consume their own smoke, rather than it being released to the surrounding environs. This was followed shortly by the 1847 Improvement Clauses

(Continued)

Box 10.1 Case study: learning from past pollution policy (Continued)

Act which, among other things, sought to reduce the impact of factory smoke. Clearly getting into their stride, parliament then passed the 1863 Alkali Works Regulation Act, which set 95% reduction targets on these industries, and the 1866 Sanitary Act, which gave authorities legal grounds to intervene in the event of undue smoke pollution. The introduction of relevant air pollution measures in the 19th century was rounded off in 1875, when the Public Health Act was passed, considered among its nuisances 'any fireplace or furnace which does not as far as practicable consume the smoke arising from the combustible used therein [...]' and '[...] any chimney (not being the chimney of a private dwelling-house) sending forth black smoke in such quantity as to be a nuisance'.

The 20th century built on many of these initial policies and, in its latter half, was partially driven by the international targets set out by the European Union. In 1906, the Alkali Act was expanded to enable the prevention of discharge of noxious or offensive gases by way of novel technologies and improved behaviours, followed by extensions and amendments to the Public Health Act in 1926 and the Clean Air Act in 1956 and 1968. At the local level, 1946 saw the introduction of the Manchester Corporation Act specifying conditions for the 'prohibition of smoke in certain city areas', and the creation of the United Kingdom's first smokeless zone.

Efforts to minmise air pollution then focus on vehicle outputs, with the introduction of European (EC) Directives 70/220/EEC and 72/306/EEC, which limited carbon monoxide, hydrocarbons and particulates from petrol and diesel engines. These were followed by the United Kingdom's Motor Vehicles (Construction and Use) Regulations 1973 and the Control of Pollution in 1974.

Subsequently, in 1975 EC Directive 75/441/EEC set out guidelines for air quality data standards to enable information sharing throughout the region, prior to the introduction of measures to limit the level of harmful chemicals such as sulphur and lead in fuels in Directives 75/716/EEC and 78/611/EEC. These directives were later translated into local policy with the 1981 Motor Fuel (Lead Content of Petrol) Regulation, which set maximum levels of lead in fuels to 0.4 g/l.

Due to increasing awareness of the transboundary effects of many airborne pollutants and the associated International Convention on Long Range Transboundary Pollution, further EC directives sought to set limits on the values of sulphur dioxide, suspended particulates (80/779/EEC), lead (82/884/EEC), as well as harmonised methods for the measurement of the above (89/427/EEC). These were transcribed into UK law in 1989 in the Air Quality Standards Regulations, and also relate to other sources of air pollution including industrial facilities, combustion plants, incinerators and ozone sources (84/360/EEC, 88/609/EEC, 89/429/EEC and 92/72/EEC).

Subsequently, in the United Kingdom, small air pollution sources were drawn together under the Environmental Protection Act in 1990, and further management of vehicle emissions was legislated for within the Road Vehicles Regulations of 1991 which introduced emission standards into routine vehicle roadworthiness tests. More broadly, the Environment Act (1995) set down a requirement for air quality standards and targets as part of a national strategy, itself published in 1997. This was updated 3 years later with the Air Quality Strategy for England, Scotland, Wales and Northern Ireland and its objectives for local authorities.

Throughout the above policy, we see several key approaches to the management of air pollution:

- Product bans
- Alternative materials/changing product composition
- Setting emissions targets
- Enabling routine monitoring and enforcement

(*Continued*)

Box 10.1 Case study: learning from past pollution policy (Continued)

Comparisons with plastic policy

As with the sources of air pollution highlighted above, microplastics are a mixture of chemical substances with differing origins and impacts and their regulation and management have many similarities. For example, we may also compare the banning of microbeads in personal care products or the replacement of plastics in products with renewable alternatives to the removal of lead from fuels, whereby there is an attemp to limit the use of the pollutant without affecting behaviours. Similarly, the development of regional smoke-free zones may be analogous to the early localised regulations around the use of particular plastic types (such as single-use takeaway plastics). Finally, there is the targeting of harmful activities or industries, for example in the use of packaging recovery notes to ensure that wastes are appropriately managed. However, from the above narrative, we can see that there are a number of areas in which our response to plastic pollution has yet to develop, namely, the setting of targets and enforcing regulations.

1. *SMART targets*

    Effective policy sets out the conditions for its own success, identifying the outcomes desired within a certain timescale. For preference these should be 'SMART': Specific, Measurable, Achievable, Relevant and Time-Bound. Implementation of the Clean Air Act and the achievement of its subsequent goals and targets therefore necessitated standardised monitoring in order to determine its effectiveness in controlling these negative effects. In 1961 the United Kingdom-wide monitoring network, the National Survey, was implemented in order to monitor levels of sulphur dioxide and fine particulates known as 'black smoke'. The creation of this network was able to identify the various successes of this and subsequent air pollution controls, with later iterations of this national monitoring system examining the pollution arising from a range of sources, including ozone and nitrogen dioxide. Latterly, in 1987 it evolved into automated systems (Automatic Urban and Rural Network) as a result of the introduction of related EC Directives concerning pollutant monitoring. While progess is being made to develop descriptors of environmental health in relation to the microplastic issue, at this time, we lack the enforceable targets and widely accepted standardised testing required for determining the effectiveness of emerging policy.

2. *Enforcing regulation*

    Globally, compliance with plastic related regulations and their resulting enforcement is mixed. Levies or bans on plastic products have been applied globally, however, these are unevenly successful. In the United Kingdom and other countries, regular reporting of production and sales enables compliance to be efficiently monitored but this is not the case internationally, and where noncompliance is detected, there is little capacity for punitive measures.

Learning and knowledge outcomes

The global policy response to the problem of plastics has occurred over a much shorter timeline than that relating to air pollution, however, there remain many areas for development. Our understanding of plastic and microplastic pollution and distribution requires broad and comparable monitoring, and this information should be used in both the setting of effective targets and the implementation of reliable enforcement regimes.

---

legislation regarding the use of plastic bags appeared in the early nineties, whereas bans on the use of microbeads in products were first introduced in 2014 (Xanthos & Walker, 2017). Subsequently, may authors have explored the focus, spread and success of national-level policies concerning the production, use, disposal and management/circularity of plastics. By comparing the focus of this literature (published between 2016 and 2021) it has been observed that bans of single-use plastic items were most broadly evaluated, with plastic bags being the most common target (Knoblauch & Mederake, 2021). Other common products affected by this legislation include disposable straws, stirrers and cotton buds (Schnurr et al., 2018). Examples illustrating the proliferation of regulatory approaches to plastic production and use at multiple scales are outlined in Table 10.3.

The introduction of these measures may be linked to existing public pressure, for example a photo campaign covering plastic bag waste and pollution in Kenya resulted in a positive response from the then Environment Secretary ahead of the introduction of their bag ban in 2017 (Schnurr et al., 2018). As of 2018, about 60% of countries (127 out of 192) had regulations in relation to the manufacture, importing, use and disposal of single use plastics, including a combination of both product bans and tax/levies, although these vary in scope and enforcement. Conversely, controls on the use of microbeads were in place in just 8 out of 192 countries, however, these controls do not account for all microbead containing products (Xanthos & Walker, 2017). Nevertheless, while such legislation has been seen to yield some reduction in the use of the plastic items targeted (Lam et al., 2018), the progress of such measures has not been without its challenges. For example, at time of writing, the Canadian Federal Court has ruled that the listing of all plastics (rather than those products known to be environmentally damaging) as toxic under national pollution legislation overstepped government authority. While this has resulted in the overturning of some of the measures listed in Table 10.3, it does not preclude the adaption or introduction of further legislation under a more narrow, category-by category approach.

Evidence of the positive outcomes of such legislation has been reported in relation to Ireland's 2002 levy on plastic bags, which was reported have to reduced plastic bag use by approximately 90% (Convery et al., 2007) and Botswana's bag charge which also saw usage reduction (Dikgang & Visser, 2012). Subsequently, in China, it has been observed that the 'tougher' the ban (in this case, implementing increased restrictions in the categories of bags available for sale and introducing significant fines for those that contravene these measures), the greater the reduction in the number of bags sold and the more frequent the reuse of bags (Wang et al., 2021). However, South Africa's bag charge has been proven to be less effective, despite a sharp initial decrease after the levy introduction, demand for bags resurged less than 2 years later as customers became accustomed to the extra charge, though still at lower levels than before the levy (Dikgang & Visser, 2012).

An alternative approach to managing the production of virgin plastics is the taxation of plastic products which do not contain a minimum proportion of recycled material. These approaches have the added benefit of increasing the value of recycled materials, incentivizing producers to invest in the recyclability of their products and increasing support for improved waste management, which

Table 10.3 Examples of emergent plastic policies at regional, national and international scales, indicating both increasing interest and targeted products[1] (single-use plastic plates, trays, bowls, cutlery, balloon sticks and certain types of polystyrene cups and food containers,[2] Ring carriers, [3] Checkout bags, cutlery, foodservice ware, stir sticks, straws).

Date (enforced from)	Legislation
2008	China: Ban plastic grocery bags
2011	Wales, UK: Levy on plastic bags
2013	Northern Ireland, UK: Levy on plastic bags
2014	Scotland, UK: Levy on plastic bags
	California, USA: Ban plastic bags
2015	England: Levy on plastic bags
	Hawaiian Islands: Ban plastic bags
2016	Illinois, USA: Levy on plastic bags
2018	Seattle, WA, USA: Ban plastic straws
	Montreal, Canada: Ban on plastic bags
2019	Victoria, BC, Canada; Ban single-use plastic bags
	Rhode Island, USA: Ban polystyrene 'clam-shell'-type packaging
	Maine, VA, USA: Ban single-use plastic bags
2020	United Kindgdom: Ban on plastic straws
	New York, USA: Ban single-use plastic bags
	Hoboken, NJ, USA: Ban single-use styrofoam and plastic bags
	Vancouver, BC, USA: Ban plastic straws and styrofoam cups/containers
	Berkley, CA, USA: Levy on take-away food packaging
	New Jersey, Oregon and Washington, USA: Ban single-use plastic bags
	South Australia: Single-use and Other Plastic Products (Waste Avoidance) Act 2020
2021	Colorado and Connecticut, USA: Ban single-use plastic bags
	Banff, AB, Canada: ban multiple single-use plastics
	New Jersey, USA: ban on styrofoam take-away containers
	South Australia: Ban on plastic straws
2022	Scotland, UK: Ban on single-use plastics
	South Australia: Ban on single-use take-away plastics and oxodegradables
2023	Europe: Commission Regulation (EU) 2023/2055 — Restriction of microplastics intentionally added to products
	England, UK: Ban on single-use plastics[1]
	Canada: Prohibition on Single-use Plastics[2] (*see text*)
	Canada: Prohibition on Single-use Plastics[3] (*see text*)
	South Australia: Ban on single-use plastic bowls, cotton bud sticks and pizza savers

we will discuss further below. Such legislative approaches to the management of plastics are frequently associated with awareness campaigns which seek to bring about a voluntary change in consumer behaviour. A comparative analysis of the success of awareness campaigns and alternative policies in Australia has indicated that council investment in publicising activity led to greater reductions of waste in the environment than did investment in policies, resulting in significantly reduced beach litter within the region (Willis et al., 2018). Nevertheless, a review of existing measures intended to encourage voluntary sustainable behaviours in relation to single-use plastic indicates that communication and informational campaigns were the most common methods employed, however, the approaches used limited links to existing literature in the development and were seen to limit individual autonomy (Mathew et al., 2023).

In Thailand, where efforts to meet SDG 12 have focused on the reduction of single-use plastics, an assessment of stakeholder perceptions regarding a proposed ban on single-use plastics revealed concerns in relation to the impact that reducing or replacing plastics may have on trust in food quality, cost and accessibility of goods and convenience (Sedtha et al., 2023). Consumer opinions and apparent compliance with legislative interventions have also been seen to vary in other locations. In Canada, analysis of consumer responses to single-use plastic bag bans revealed widespread support for the intentions of the policies (77%). However, the extent of this support was variable depending on participant demographics and region, and it has been suggested that the effectiveness of policies could be improved by the standardisation of restrictions supported by the implementation of consumer awareness campaigns and prior run-in periods (Molloy et al., 2022). Similarly, in relation to the aforementioned policies in the Caribbean region, the most successful employed a tiered approach to policymaking, utilising primary stakeholder engagement, a run-in period prior to implementation, and associated stakeholder information campaigns (Clayton et al., 2021). At other sites, the success of many of these measures is limited by state capacity to monitor and regulate outcomes (Muposhi et al., 2022). For example, of the eleven Caribbean countries that have introduced legislative policies, seven states also included fines and penalties for noncompliance (Clayton et al., 2021).

10.3.2 Downstream: improved waste management

As with preventing or limiting the production of plastic wastes, waste management may be addressed by way of a variety of small- and large-scale legislative interventions aimed at either single products or whole waste streams. Of course, in many countries, households are taxed in order to cover the cost of municipal waste facilities, however, accessiblity and type of treatement varies globally and more targeted approaches are required to enable the movement away from landfilling and EFW and towards recycling and other sustainable waste management measures.

As with bans and levies, frequent targets of these schemes are single-use plastics, such as food and beverage containers, as well as generic packaging. This is not surprising as, at the time of writing, approximately 40% of plastic waste arises from packaging, 12% from consumer goods and 11% from clothing and textiles (Global Plastics Outlook: Economic Drivers, Environmental Impacts and Policy Options, 2022). These measures may seek to influence either public behaviour at point of disposal or the scale and effectiveness of the waste management system. Under the first category are deposit return schemes (DRSs), which attempt to increase public interest in recycling by including a small cash deposit in the price of a product which is returned to the consumer when it is returned for recycling, either to a machine or nominated collection centre. Such schemes have historically been very widespread, for example the bottle schemes of soft drinks companies or doorstep milk deliveries. Contemporary examples include the bottle returns schemes in the Netherlands which placed a 25- and 15-cent deposit on large and small plastic bottles respectively, aiming for a 95% recycling rate. A similar scheme is planned for the recycling of beverage containers in the United Kingdom in 2025 (DEFRA, 2021); however, this and associated programs require the introduction of suitable collection, transport and cleaning infrastructure, as well as new labelling conventions, and an assessment of the impacts of changing product costs on consumers. In the Netherlands, mandatory collection sites include petrol stations and supermarkets with a footprint of over 200 m^2, and other locations may register as voluntary collection points. Additionally, the handling and administration costs which underpin DRSs may be prohibitive (Hogg et al., 2011).

Alternatively, related regulations pertaining to disposal may focus on particular products (e.g. packaging) or industry wastes (e.g. from construction). For example, as briefly mentioned above, regulations introduced have restricted the ability of farmers in the United Kingdom to bury or burn farm wastes, which will include a range of single-use agriplastics such as mulch films and feed containers. Here farmers must identify, classify and separate their waste, using a waste classification code, before sending it for recycling or disposal by a registered handler (retaining records for a minimum of 2 years). Food packaging-related initiatives in the same region include ongoing trials of the recycling of plastic bags and food wrappers (running between 2023 and 2025), which include low-value plastic films that have also been of limited interest to recyclers in the past.

10.3.2.1 Fibres

More specific to the issue of microplastics is the control of the release of fibres with wastewater and the prevention of their distribution onto agricultural lands and other receiving environments. One option for the capture of microplastics in effluent waters is the requirement for increased coverage and intensity of wastewater treatment, as well as for the secondary treatment of sewage sludges in order to minimise or prevent redistribution. Wastewater treatment systems range from simple 'Primary' filtration, to complex mechanical, physical, or chemical 'Tertiary' treatment (Fig. 10.1). A range of tertiary treatment measures have been seen to elevate the capture of fibres to over 98% (Iyare et al., 2020). The drawbacks of these measures include the large financial outlay required in order to introduce or refit suitable water treatment infrastructure for what may, in some

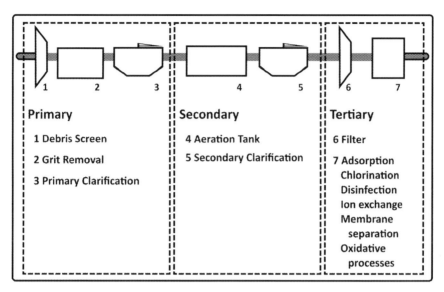

FIGURE 10.1

Wastewater treatment steps processes at the various stages of wastewater treatment.

locations, represent only a small percentage increase in efficiency. Additionally, as fibre capture efficiency in WWT increases, so will the level of fibre in sewage sludges, further necessitating improvements in sludge treatment to limit the impacts on the receiving environment.

An alternative regulatory approach which may be applied in the management of microplastic fibres in wastewater is the introduction of mandatory fibre filtration in washing machines. Intervention during domestic washing has been highlighted as a highly efficient way in which to control fibre loss via wastewater systems. Such fibre catchers have the potential to recover large percentages of microplastics and other material per wash, reducing the burden on WWT. Subsequently, a number of countries, such as France, have already announced or implemented such methods. As with the implementation of deposit return schemes, the introduction of mandatory filtration necessitates changes in both design standards and manufacture which have implications for production costs that may be passed on to the consumer. Additionally, there is currently no recommended disposal route for captured fibres, and legislators may have to define a recommended disposal pathway to divert this waste away from landfills.

10.3.3 **Holistic approaches**

Some policy interventions influence many parts of the product lifecycle, targeting single groups (such as producers) in a way that affects other stakeholders. For example, the creation of 'wasteful' products and the barriers to sustainable management of plastic wastes may be reduced by the implementation of Extended Producer Responsibility Schemes (EPRs), which disincentivise unsustainable industry practices while funding waste management. Typically introduced in relation to specific product categories which are known to result in significant environmental of health impacts, EPRs seek to hold industry accountable for the effects of their products and services from production to end-of-life, removing the apparent burden from the consumer or municipal authority. As such, the producer may be responsible for any costs (via taxes or other levies) or to provide suitable operations for the handling of wastes, which may be covered by the creation of specific funds or the introduction of abatement measures by the originator. Many EPRs exist for the handling of used tires, specifying requirements for collection, transport and recycling, and are also widespread in relation to electrical goods in the form of the EU Waste Electrical and Electronic Equipment (WEEE) Directive 2012/19/EU. Under this directive, producers provide financial support for the collection and handling of WEEE wastes at end-of-life. Specific recycling targets are set for various product groups within the WEEE category, and producers must provide information regarding both their sales and their contributions to waste management efforts. In addition, WEEE recycling schemes frequently include public education elements which seek to inform consumers to encourage and enable their participation.

The scope of an EPR is dependent on the characteristics of the product considered as well as the scale of preset targets designed to encourage either sustainable innovation or a reduction in harmful practices. Progress towards these goals may be measured via a governmental body (as under the WEEE Directive) or it may be self-reported. Failure to comply with an EPR may be associated with a range of penalties, from operating restrictions to fines. However, the scope, regulation and enforcement of EPRs vary greatly in relation to the priorities of the government and the legislative approach employed.

Similar schemes have been suggested in relation to plastic wastes that may affect the production and spread of macro-, micro- and nanoplastics. For example, in the United Kingdom, the Producer Responsibility Obligations (Packaging Waste) Regulations 2007[1] require producers or associated Producer Compliance Schemes (PCSs) to purchase either a Packaging Recycling Note (PRN) or a Packaging Export Recycling Note (PERN) to cover the cost of managing each tonne of material produced. PRNs and PERNs are purchased from an accredited reprocessor or exporter, who becomes responsible for ensuring that the packaging is handled in such a way as it meets the requirements for 'end-of-waste' status.

End-of-waste status is met when the material is recycled or recovered (although changes to the regulations now no longer include recovery in their targets) to eliminate any characteristics that would prevent its use in place of a

---

[1] Which implements the requirements of the Packaging and Packaging Waste European Directive (94/62/EC).

nonwaste comparator. For example, if a beverage container may be decontaminated and shredded and the resulting plastic is suitable for use as a feedstock in the manufacture of another bottle or similar product. Wider conditions for end-of-waste status are set out in Article 6 of Directive 2008/98/EC of the European Parliament and of the Council. For those organisations exporting waste, they also become responsible for adhering to 'local controls', country-specific rules pertaining to the import and export of wastes, including the acquisition of export licences, and associated monitoring and reporting. Across Europe, recycling rates of plastics remain comparatively low, partially the result of limited demand and value. It has previously been suggested that increased recycling rates may be incentivised by the further implementation of EPR schemes for a wider range of plastics (Leal Filho et al., 2019).

Indeed, in the United Kingdom, an updated EPR for packaging is planned for implementation in 2024, with producer data collection beginning in 2023. Under this scheme, the 'producers' include brands who package goods and sell them under their brand name, importers who trade in filled packaging, service providers who hire out or lend reusable packaging and distributors who manufacture or import and sell empty packaging to anyone who is not an obligated producer. Also included are online marketplaces which enable non-UK sellers to sell to UK consumers, as well as sellers who sell filled packaging to the end consumer.

Under the revised system, larger producers must, where relevant, record and report data about the empty packaging and packaged goods supplied or imported in addition to paying waste management fees, scheme administrator costs, and charges to the environmental regulator, and purchasing PRNs or PERNs to meet recycling obligations. Small producers must record data about all the empty packaging and packaged goods supplied or imported during the initial data collection period and, subsequently, pay a charge to the environmental regulator and report ongoing data on the empty packaging and packaged goods supplied or imported. The distinction between these groups is based on commercial income and the weight of packaging produced, as indicated in Table 10.4.

It is planned that these actions will result in a recycling rate of 51% in 2024 and 62% by 2030, with the additional aims that the scheme will encourage manufacturers to design packaging that is more easily recyclable, improve the quality of the recycled material and strengthen and standardise recycling services

Table 10.4 Classifying producers of packaging waste.

Organisation type	Income	Packaging import/supply
Exempt	< £1 million	<25 tonnes
	Charities	
Small	> £1 million	25–50 tonnes
	£1 million to £2 million	> 25 tonnes
Large	>£2 million	>50 tonnes

between regional authorities. Interestingly, as compostable and degradable plastics are typically nonrecyclable, these may result in a higher levy on producers. A similar scheme is already underway in Germany, with producers reporting packaging generated to the Central Packing Registry, and companies paying based on a 'Green Dot' system, whereby payment is determined by the type and volume of packaging produced. Similarly, EPR schemes may also be introduced at regional scales, for example various schemes have been implemented at the regional level in Canada, with some being managed by stewardship bodies. Although these schemes vary considerably by scale and waste target.

The use of EPRs as a management approach is an increasingly popular one, and recommendations for the effective development of EPRs are outlined by the WWF (WWF-Malaysia, 2020). These include the need to:

1. Establish a legal framework of a mandatory EPR system and strengthening an institutional framework. To make the law practicable and effective, agreements and discussions between competent authorities and the private industry are required.
2. Establish a voluntary, pre-Producer Responsibility Organisation (PRO) basis facilitating the development of a mandatory EPR. It is recommended that a voluntary PRO is set up on an interim basis. Through such actions, voluntary companies and organisations can cooperate and negotiate with the policymakers about the setup of the mandatory system regarding organisational and regulatory foundation as well as control mechanisms.
3. Define mechanism for continuous improvement and optimisation after the mandatory EPR scheme is launched. After the mandatory EPR is in place, steps must be taken that ensure the EPR system and PRO are continuously being optimised and evolved.

As of 2018, EPR shemes pertaining to single-use plastics had been passed into law by 62 countries, a figure that includes both recycling targets, deposit refunds and product take-back (Xanthos & Walker, 2017).

10.4 **Monitoring**

As with all approaches to regulation, strategies for the reduction of plastic and microplastic waste and environmental contamination (such as fibre capture) should be linked to suitable targets and monitoring schemes, an essential step in determining the efficacy of these measures, enabling iterative policy development (Box 10.1) and encouraging further participation. Indeed, in their recent report, the University of Portsmouth's Global Plastic Policy Centre highlights a lack of evidence by which the effectiveness of plastic policies may be determined, with many of the less well-documented policies being among the longest established (Global Plastics Policy Centre, 2022).

However, as with environmental contamination, monitoring is decidedly challenging and the subsequent imapct of regulation on the levels of plastic contamination are often unclear. Nevertheless, there have been some efforts to implement long-term monitoring schemes to determine the levels of plastics in key environmental compartments, and to identify the potential impact of these plastics on the wider ecosystem. For example, long-term monitoring as part of the OSPAR assessment of the number of plastic particles in the stomachs of fulmars, *F. glacialis*. In this case, the current target for the number of stranded fulmars with over 0.1 g of plastic in their stomach is 10%, however, the current level is approximately 58% (Van Franeker et al., 2011), highlighting the need for increased intervention.

Fortunately, despite the limited number of widespread, consistent plastic monitoring routines, the apparent efficacy of some regulatory measures has been confirmed by citizen scientists. For example, the apparent effectiveness of the introduction of plastic bag charges in the United Kingdom's was recorded by volunteers participating in the Marine Conservation Society's Big Beach Clean, which reveals a large drop in the number of bags recovered in standardised beach cleans performed around the UK coast. Clearly, the quantification of policy impacts is not wholly out of reach.

10.5 Conclusions

The suite of available policy options for the management of microplastic wastes and those of their parent materials are substantial, and their use is increasing globally in line with local, national and international priorities. Of course, in addition to legally binding frameworks for the management of plastic wastes, many governments, organisations, interest groups and consortia have committed to abide by voluntary agreements concerning the sustainable use of plastics and plastic products. For example, the UK Plastics Pact, a collaborative initiative involving businesses, governments and NGOs which aim to bring about a circular economy for plastics by setting targets to reduce plastic waste and increase recycling rates.

Nevertheless, regulation of this kind may be by no means simple. Firstly, as indicated in Chapters 7 to 10, suitable targets for policy intervention must be selected based on existing research and best available data. Subsequently, associated industries, fearing external controls, may seek to negatively influence the development of suitable policy and regulations. For example, delays to the implementation of EPRs have been achieved by counter-lobbying for reduced limits or targets, extended trials and run-in periods or obfuscated monitoring, the implementation of (often lenient and unpoliced) voluntary targets, the obscuring of data on plastic and waste production and recycling, or advocacy for less-impactful alternative management measures or mitigation methods, or legal challenges of their own (Tangpuori et al., 2020). Finally, public compliance may vary in relation to a range of factors, some of which will be discussed in the following chapter. Conversely, improvements in policy may be achieved by consideration of location-specific factors, standardisation across regions and integration with existing policies, improved public engagement, routine monitoring and regular evaluation.

References

Chen, S. (2022). British dairy farmers' management attitudes towards agricultural plastic waste: reduce, reuse, recycle. *Polymer International*, *71*(12), 1418−1424.

Clayton, C. A., Walker, T. R., Bezerra, J. C., & Adam, I. (2021). Policy responses to reduce single-use plastic marine pollution in the Caribbean. *Marine Pollution Bulletin*, *162*. Available from https://doi.org/10.1016/j.marpolbul.2020.111833, http://www.elsevier.com/locate/marpolbul.

Convery, F., McDonnell, S., & Ferreira, S. (2007). The most popular tax in Europe? Lessons from the Irish plastic bags levy. *Environmental and Resource Economics*, *38* (1), 1−11. Available from https://doi.org/10.1007/s10640-006-9059-2.

DEFRA (2021). 2023 Consultation outcome: Introduction of a deposit return scheme in England, Wales and Northern Ireland https://www.gov.uk/government/consultations/introduction-of-a-deposit-return-scheme-in-england-wales-and-northern-ireland.

Dikgang, J., & Visser, M. (2012). Behavioural response to plastic bag legislation in Botswana. *South African Journal of Economics*, *80*(1), 123−133. Available from https://doi.org/10.1111/j.1813-6982.2011.01289.x.

Dunn, R. A., & Welden, N. A. (2023). Management of environmental plastic pollution: A comparison of existing strategies and emerging solutions from nature. *Water, Air, and Soil Pollution*, *234*(3). Available from https://doi.org/10.1007/s11270-023-06190-2, https://www.springer.com/journal/11270.

Global plastics outlook: Economic drivers, environmental impacts and policy options. OECD, 2022. doi: 10.1787/de747aef-en.

Global Plastics Policy Centre. (2022). A global review of plastics policies to support improved decision making and public accountability.

Hahladakis, J. N., Velis, C. A., Weber, R., Iacovidou, E., & Purnell, P. (2018). An overview of chemical additives present in plastics: Migration, release, fate and environmental impact during their use, disposal and recycling. *Journal of hazardous materials*, *344*, 179−199.

Hogg, D., Sherrington, C., & Vergunst, T. (2011). A comparative study on economic instruments promoting waste prevention. *Final Report to Bruxelles Environment*.

Iyare, P. U., Ouki, S. K., & Bond, T. (2020). Microplastics removal in wastewater treatment plants: a critical review. *Environmental Science: Water Research & Technology*, *6*(10), 2664−2675.

Knoblauch, D., & Mederake, L. (2021). Government policies combatting plastic pollution. *Current Opinion in Toxicology*, *28*, 87−96. Available from https://doi.org/10.1016/j.cotox.2021.10.003, https://www.journals.elsevier.com/current-opinion-in-toxicology/.

Lam, C. S., Ramanathan, S., Carbery, M., Gray, K., Vanka, K. S., Maurin, C., Bush, R., & Palanisami, T. (2018). A comprehensive analysis of plastics and microplastic legislation worldwide. *Water, Air, and Soil Pollution*, *229*(11). Available from https://doi.org/10.1007/s11270-018-4002-z, http://www.kluweronline.com/issn/0049-6979/.

Leal Filho, W., Saari, U., Fedoruk, M., Iital, A., Moora, H., Klöga, M., & Voronova, V. (2019). An overview of the problems posed by plastic products and the role of extended producer responsibility in Europe. *Journal of Cleaner Production, 214*, 550−558. Available from https://doi.org/10.1016/j.jclepro.2018.12.256, https://www.journals.elsevier.com/journal-of-cleaner-production.

Liang, Y., Tan, Q., Song, Q., & Li, J. (2021). An analysis of the plastic waste trade and management in Asia. *Waste Management, 119*, 242−253. Available from https://doi.org/10.1016/j.wasman.2020.09.049, http://www.elsevier.com/locate/wasman.

Mathew, A., Isbanner, S., Xi, Y., Rundle-Thiele, S., David, P., Li, G., & Lee, D. (2023). A systematic literature review of voluntary behaviour change approaches in single use plastic reduction. *Journal of Environmental Management, 336*. Available from https://doi.org/10.1016/j.jenvman.2023.117582, https://www.sciencedirect.com/journal/journal-of-environmental-management.

McCollum, D. L., Echeverri, L. G., Busch, S., Pachauri, S., Parkinson, S., Rogelj, J., Krey, V., Minx, J. C., Nilsson, M., Stevance, A. S., & Riahi, K. (2018). Connecting the sustainable development goals by their energy inter-linkages. *Environmental Research Letters, 13*(3). Available from https://doi.org/10.1088/1748-9326/aaafe3, http://iopscience.iop.org/journal/1748-9326.

Mister, A. A. (1970). Britain's Clean Air Acts. *The University of Toronto Law Journal, 20*(2), 268. Available from https://doi.org/10.2307/824870.

Molloy, S., Medeiros, A. S., Walker, T. R., & Saunders, S. J. (2022). Public perceptions of legislative action to reduce plastic pollution: A case study of Atlantic Canada. *Sustainability (Switzerland), 14*(3). Available from https://doi.org/10.3390/su14031852, https://www.mdpi.com/2071-1050/14/3/1852/pdf.

Muposhi, A., Mpinganjira, M., & Wait, M. (2022). Considerations, benefits and unintended consequences of banning plastic shopping bags for environmental sustainability: A systematic literature review. *Waste Management and Research, 40*(3), 248−261. Available from https://doi.org/10.1177/0734242X211003965, https://journals.sagepub.com/home/WMR.

Nguyen, M. K., Hadi, M., Lin, C., Nguyen, H. L., Thai, V. B., Hoang, H. G., Vo, D. V. N., & Tran, H. T. (2022). Microplastics in sewage sludge: Distribution, toxicity, identification methods, and engineered technologies. *Chemosphere, 308*. Available from https://doi.org/10.1016/j.chemosphere.2022.136455, http://www.elsevier.com/locate/chemosphere.

Van Franeker, J. A., Blaize, C., Danielsen, J., Fairclough, K., Gollan, J., Guse, N., Hansen, P. L., Heubeck, M., Jensen, J. K., Le Guillou, G., Olsen, B., Olsen, K. O., Pedersen, J., Stienen, E. W. M., & Turner, D. M. (2011). Monitoring plastic ingestion by the northern fulmar Fulmarus glacialis in the North Sea. *Environmental Pollution, 159*(10), 2609−2615. Available from https://doi.org/10.1016/j.envpol.2011.06.008, https://www.journals.elsevier.com/environmental-pollution.

Schnurr, R. E. J., Alboiu, V., Chaudhary, M., Corbett, R. A., Quanz, M. E., Sankar, K., Srain, H. S., Thavarajah, V., Xanthos, D., & Walker, T. R. (2018). Reducing marine pollution from single-use plastics (SUPs): A review. *Marine Pollution Bulletin, 137*, 157−171. Available from https://doi.org/10.1016/j.marpolbul.2018.10.001, http://www.elsevier.com/locate/marpolbul.

Sedtha, S., Nitivattananon, V., Ahmad, M. M., & Cruz, S. G. (2023). The first step of single-use plastics reduction in Thailand. *Sustainability (Switzerland), 15*(1). Available from https://doi.org/10.3390/su15010045, http://www.mdpi.com/journal/sustainability/.

Tangpuori, A., Delemare, H.R., George, U., Nusa, B., & Zallio, X.P. (2020). Changing Markets Foundation Talking Trash: The corporate playbook of false solutions to the plastic crisis. https://talking-trash.com/wp-content/uploads/2020/09/TalkingTrash_FullReport.pdf.

Troya, M. D. C., Power, O. P., & Kopke, K. (2022). Is it all about the data? How extruded polystyrene escaped single-use plastic directive market restrictions. *Frontiers in Marine Science*, 8. Available from https://doi.org/10.3389/fmars.2021.817707, https://www.frontiersin.org/journals/marine-science#.

U.N. (2023). The Sustainable Development Goals Report 2023: Special edition towards a rescue plan for people and planet. https://unstats.un.org/sdgs/report/2023/The-Sustainable-Development-Goals-Report-2023.pdf.

Wang, B., Zhao, Y., & Li, Y. (2021). How do tougher plastics ban policies modify people's usage of plastic bags? A case study in China. *International Journal of Environmental Research and Public Health*, 18(20). Available from https://doi.org/10.3390/ijerph182010718, https://www.mdpi.com/1660-4601/18/20/10718/pdf.

Willis, K., Maureaud, C., Wilcox, C., & Hardesty, B. D. (2018). How successful are waste abatement campaigns and government policies at reducing plastic waste into the marine environment? *Marine Policy*, 96, 243−249. Available from https://doi.org/10.1016/j.marpol.2017.11.037, http://www.elsevier.com/inca/publications/store/3/0/4/5/3/.

World Bank. (2021). 2023 ASEAN member states adopt regional action plan to tackle plastic pollution. https://www.worldbank.org/en/news/press-release/2021/05/28/asean-member-states-adopt-regional-action-plan-to-tackle-plastic-pollution.

WWF-Malaysia. (2020). WWF Study on EPR scheme assessment for packaging waste in Malaysia. https://wwfmy.awsassets.panda.org/downloads/study_on_epr_scheme_for_-packaging_waste_in_malaysia_wwfmy2020.pdf.

Xanthos, D., & Walker, T. R. (2017). International policies to reduce plastic marine pollution from single-use plastics (plastic bags and microbeads): A review. *Marine Pollution Bulletin*, 118(1−2), 17−26. Available from https://doi.org/10.1016/j.marpolbul.2017.02.048, http://www.elsevier.com/locate/marpolbul.

'Solutions' versus sustainability

11

'And I am a weapon of massive consumption, And it's not my fault, it's how I'm programmed to function'
Lily Allen

11.1 Introduction

The previous chapters have introduced a range of interventions intended to reduce the production of plastic wastes, encourage the reuse and reclamation of resources and minimise the requirements for EfW and landfilling. We also highlight emerging technologies to enable their recovery and removal from the environment in the event of loss from these systems. By reducing production, improving handling, and enabling the extraction of macroplastics and primary microplastics, we limit environmental microplastic pollution. However, it has also been highlighted that these measures are not without their drawbacks, and there are numerous apparent trade-offs between pre-existing environmental concerns (water availability, land-use and energy use) and the solutions to the plastic problem.

11.2 Technological barriers to successful implementation of existing and emerging solutions

Despite the introduction of numerous replacements for traditional polymers, the transition away from plastics has thus far been limited. A key barrier to the uptake of novel alternative materials, such as bioplastics, is their functional properties. As we indicated in the first chapter, the widespread adoption of plastics was driven by their durability, flexibility, low weight and cost. Many plastic alternatives, such as natural fibres and bio-based polymers, have lower durability and shorter useful lives compared to their fossil fuel-based counterparts. This may be desirable in the case of single-use items sent to landfill; however, in other products, this could result in more frequent replacement, leading to alternative waste issues. Similarly, the need for key functional properties, such as flexibility and water resistance can be challenging to accommodate with alternative materials.

In addition to the energy costs associated with the production of the alternative products, the increase in product weight associated with alternative materials such as glass and metal may result in a commensurate increase in energy required to transport the

Microplastics. DOI: https://doi.org/10.1016/B978-0-443-13324-4.00011-X

resulting products, for example, when comparing the use phase of wooden or plastic pallets (Khan et al., 2021). Thus the carbon footprint generated by the sourcing of alternative materials, the processes involved in their production and reuse or management at end of life must be assessed in order to ensure that the proposed solutions are as sustainable as they are made out to be. In addition, other environmental footprints may also be affected by changing material choices, for example bio-based plastics may also lead to competition with agriculture, changes in land use and secondary environmental impacts. As a result, movement from traditional polymers to sustainable alternatives may affect the performance of products in certain applications or conditions.

Finally, we must consider the impacts of these new products upon disposal. Adoption of either novel or alternative materials has implications for the suitability of existing waste streams, resulting in challenges in, and potential changes to, waste management infrastructure which allow us to handle these new materials (Paul et al., 2023). If existing facilities to properly dispose of the resulting wastes do not exist, then the new materials could end up in landfills, negating environmental benefits. The development of new facilities may also result in a cost to consumer via local taxes to pay for municipal services.

11.3 Secondary environmental impacts of plastic management

Resource intensity concerns the assets and materials associated with the lifespan of a product, the completion of a process, or provision of a service, and is a key issue in the use of apparently sustainable alternative materials. Although plastic production is associated with the issue of fossil fuel extraction and processing, alternative materials may require more resources to produce. As a result, transition away from the use of plastic may drive increased demand for raw materials such as wood, metals or agricultural crops, as well as additional water footprint, chemical loads and energy. These requirements may also result in secondary negative environmental impacts not considered when looking at the issue of waste production alone.

In order to capture these unintended secondary impacts, comparison between plastic products and their alternatives may be achieved by way of life cycle assessment (LCA). LCAs consider a product from a cradle-to-grave (or cradle-to-cradle) perspective, assessing impacts during production, transportation, use and at end-of-life. The results of LCAs may provide an increased understanding of the resource-intensive parts of a product's life cycle or a holistic overview of the relative sustainability of various products, enabling individuals and industries to accurately compare the available options.

Life cycle assessments are achieved in a number of stages. First, determining the scope of the assessment, the number of products or processes to be compared and what impacts or endpoints are of interest. For each product or process, an inventory is taken of the relevant stages of the product life cycle. This inventory considers aspects such as raw materials, energy costs, water and land use and transportation. For each stage in the product life cycle, an impact assessment is undertaken to determine the environmental and resource costs (Fig. 11.1). In reviewing the outcomes of these assessments, industries, waste managers and regulators seek to make informed choices, avoiding burden-shifting or the creation of a new negative impact as a result of the elimination of an existing one.

Of course, there are drawbacks to these approaches, for example the availability and extent of data regarding the resources and processes considered (Hetherington et al., 2014). Additionally, when an LCA is undertaken on a single product in isolation, the outcomes of the analysis may result in just a small change to the product where a shift to a completely different product may yield better results. Also, if comparing the impact of several product categories, averaged data is often applied within models. For example, in understanding the benefits of moving away from single-use carrier bags, we may compare average data from multiple single-use carriers, multiple reusable plastic carriers and multiple natural fibre carriers. In reality, there will be a high degree of variation between products in one category and the beneficial choice may be obscured.

Currently, LCAs regarding options for the replacement of plastic products are limited but increasing. However, some comparisons may still be drawn, for example, in the CO_2 outputs of reusable plastic (LDPE and PP), single-use plastic (HDPE), biodegradable plastic and paper bags[1]. This analysis revealed that the

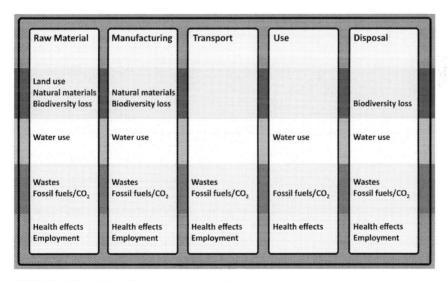

FIGURE 11.1 Structuring life cycle assessment.

Examples of domains and indicators that may be included in a product LCA.

---

[1] Based on average use of a bag over a year.

paper bags had the lowest CO_2 output, reusable PP bags (20 re-uses) had the second lowest impact, biodegradable bags had the third lowest impact, single-use PP bags had the second highest impact and reusable LDPE page had the highest impact even after 10 uses (Civancik-Uslu et al., 2019). A similar analysis of the global warming impacts of reusable food containers revealed a different picture, here all of the reusable options outperformed single-use options, breaking even in as little as four uses. These results were similar for energy, water consumption and cost, with the scale of the benefits dependent on the type of single-use container replaced (Hitt et al., 2023).

Frequently, the introduction of re-useable products also results in a need for cleaning or other secondary processing. A cradle-to-grave analysis of drinking straws made of plastics, stainless steel, paper bamboo and jute compared the effects of reuse, cleaning, accessories (such as cleaning kits) and disposal. This analysis revealed that single-use plastic straws were favourable. This was primarily the result of the accessories provided with reusable straws in addition to washing requirements. However, when accounting for handling (or mishandling) at end-of-life, the impacts of reusable alternatives appear to be similar to those of single use straws. In such cases, reducing straw use rather than replacing the product is undoubtedly the best option, followed by extended use of reusable straws with efficient/low impact washing methods (Zanghelini et al., 2020).

When comparing the production of greenhouse gasses, acidification factors and other chemicals[2] produced during the manufacture of glass and PET, the gram for gram comparison between the two favours the use of glass. However, there is considerably more glass used in the production of a glass bottle than plastic in a plastic bottle, as a result the per product costs favour the use of plastic in bottles. Fortunately, on reuse of the bottles, the impacts of the production phase as reduced, with glass displaying greater parity with the plastic bottle in the production of polluting factors (Vellini & Savioli, 2009).

Finally, LCA methods may also highlight options which may reduce the incidence of illness and disease. In exploring the effect of negative environmental changes on human health, a meta-analysis of LCA data concerning the recycling and reuse of food-contact plastics indicated that, after 30 uses, the reduction in damage to the environment bought about by increased plastic managements results in potential increases in wellbeing, and that health risks could be improved by further increasing recycling (Deeney et al., 2023).

Nevertheless, while LCAs can help in identifying harmful stages of the product life cycle, assist in choosing between products and highlight long-term effects, they seldom include the impact of these changes on social systems, such as employment and lifestyle. While technological fixes and regulatory measures have the benefits of eliminating specific plastic products or preventing their production and sale, these approaches do not address the behaviours that underpin our unsustainable use and disposal of plastics.

---

[2] CO, CO_2, CH_4 and N_2O, NO_x, SO_x, SO_2, VOC.

11.4 Social barriers to successful implementation of existing and emerging solutions

Of course, in addition to the apparent environmental impacts of plastics management and the production of alternative materials, various social factors may also influence the sustainable use of plastics. Not least amongst these is the availability of technologies, products and services to individuals and communities; however, where sustainable alternatives to plastic products are available, their uptake is not assured. Voluntarily change in consumer behaviour is dependent on a variety of factors, and the desire to act in a sustainable manner interacts with competing motivations. Below we will explore these factors and the manner in which they affect plastic use and sustainable choices.

11.4.1 Cost and convenience

The replacement of traditional plastics in products may have implications for cost, for example alternative materials may be more expensive to produce or process. Where passed on to the consumer, such costs may result in a decreased likelihood of individuals adopting the related product or behavioural change. In exploring consumer willingness to pay for biodegradable food containers rather than expanded PS ones, it was found that just 81.0% of respondents would support a ban on EPS containers, with participants prioritising water resistance, microwave suitability and local sources (Barnes et al., 2011). The nature of any trade-off is often related to the type and importance of the product, for example, while many parents expressed interest in toys made of bio-based plastics, this was limited by price, and the overall degree of acceptance reflected individuals' existing attitudes to environmental concerns. Where parents reported existing awareness of ecological issues, they also demanded that toys meet minimum ecological standards (Scherer et al., 2017).

Similarly, regulations regarding product specification may also lead to an additional financial burden. For example, the introduction of legislation requiring the inclusion of fibre filters into washing machines may result in increased materials and manufacturing costs.

Where regulation results either in additional cost or negatively influences productivity, and the company is unable to pass on costs to the consumer (as a result of market competitiveness or existing contracts), these impacts must be absorbed within the production chain. For example, we have previously noted that farms in the UK are required to separate, record and dispose their wastes via a registered handler. However, the impact of these requirements varies between location and farm size. In this case, the first impact is the time cost associated with the sorting and recording of wastes. The second related to the facilities for their collection. The distribution of which is geographically uneven and − due to the quality of the plastics as well as their secondary contamination, by soil, pesticides, antibiotics and other chemicals — the wastes are a less desirable to recyclers. This is not the only group to struggle with accessing suitable waste management routes; for example, analysis of the plastic waste trade in Asia indicates that the collection rates in most Asian countries are lower than 50% and, in some countries, there is a lack of technology and ability for plastic waste treatment even for the internally produced waste (Liang et al., 2021). Under certain circumstances, combinations of regulations around plastics recycling may prove prohibitive, acting as a barrier to effective recycling. This has previously been reported by recyclers in a study of the interactions between legislation on the Registration, Evaluation, Authorisation and Restrictions of Chemicals (REACH), the Water Framework Directive and food contact materials on the recycling on non-PET food contact plastics (De Tandt et al., 2021).

Returning to farming, one agricultural waste that represents a target for ongoing legislation is that of mulch films. In agriculture, PE mulches have typically been managed by removal and landfilling, the remnants of which are believed to lead to the build-up of plastics and microplastics in the soil. In order to identify suitable alternative management strategies, surveys of growers were undertaken. This revealed that 40% of respondents were unlikely to consider degradable alternatives, 35% were slightly likely, 15% moderately likely and just 10% very likely. While 97% of respondents wished for better ways to manage their waste, 77% were put off by the lack of a proven track record for the technology (DeVetter et al., 2021). However, farmers' acceptance of the use of biodegradable plastic mulches has been seen to be greater in the event that they may charge more for the resulting product, as well as in the result of confirmed benefits to soil health, reduction in apparent soil plastic pollution (Chen et al., 2020).

11.4.2 Consumer behaviour

The reduction of plastic waste generation and subsequent pollution is intrinsically linked with consumer behaviour. While some behaviours may be managed by technological fixes and legislative change, others rely on the values and interests of the public at large. Of course, not all consumers are inclined towards the voluntary uptake of sustainable approaches to the use of plastics. For example, following an apparently failed nationwide ban on plastic bags in India, questionnaire surveys of the opinions of residents of Kancheepuram district, Tamil Nadu, sought to gauge acceptance of a state-wide alternative. This revealed that, although 76% of respondent had a positive response to the ban, 52% continued to use plastics despite the 80% being aware of plastic's negative impacts (Sujitha et al., 2019). This is not surprising; similar differences have previously been observed between UK consumers and their engagement with the climate change issue. Here, the public was grouped into seven segments, ranging from positive greens (who do as much as they can and experience guilt for what they cannot) which made up approximately 18% of the respondents, to the honestly disengaged (who do not intend to make any changes to their lifestyle, even in the event of apparent risk) which also represented approximately 18% of the respondents (DEFRA, 2008).

There are, of course, a number of factors which may influence an individual's engagement with environmental issues, many of which we might consider in relation to Maslow's Hierarchy of Needs. This hierarchy visualises a prioritised list of an individual's essential requirements. At its most basic level, our physiological needs, such as food and water, must be met. These are followed by the needs for safety, emotional connection, esteem and self-actualisation (Fig. 11.2). For those individuals that are engaged with the issue of plastic pollution, at least the first three levels of the hierarchy must be met. However, for people who have a limited or no interest in plastic pollution, all five levels must be met before behavious of this kind may be reliably manifested. Challenges to the lower level of the hierarchy may be the result of geographic inequalities or periods of societal or economic unrest, for example the impact of the COVID-19 pandemic, conflict or national disaster. Indeed, in these latter examples, the use of some plastics may be expected to rise, such as the use of face masks or other sterile medical supplies.

Even where participants are willing, it can take a substantial and sustained effort to successfully adopt a new behaviour. Dual process theory explains why people may not take action to minimise plastic use, for example when selecting a product at the shops is based on habit or alternative preference rather than the conscious weight of the product's environmental footprints.

Human behaviours may be thought of as either automatic/implicit or system 1 thinking, or conscious/explicit or system 2 thinking (Evans, 2011). System 1 thinking enables us to reduce the time needed to act or make decisions quickly, such as buttoning a shirt or (perhaps more worryingly) driving to work and then not remembering the bulk of the journey. This type of thinking is based on the use of internal biases and heuristics, or simple rules by which enable swift decision making. Adopting new behaviours requires an increase in our system 2 thinking (although our internal heuristics may still play a role), such as remembering to bring reusable cups and bags, or altering our pruchase preferences while shopping (Evans & Stanovich, 2013).

FIGURE 11.2 Maslow's hierarchy of needs.

The needs in the tiers below must be fulfilled before those above can be met.

Nevertheless, many of our system 1 behaviours were once system 2; it is the repetition that makes them unconscious and instinctual.

In communicating the issue of sustainable plastic management, we may appeal to both types of thinking. Media coverage typically associates the issue of microplastics with emotionally charged messages about animal health and food security which many of us instinctually percieve as important. Hightlighting the issue in this manner can lead to the dominance of the intended narrative (that the production of microplastics and their loss to the environment should be avoided).

The way in which an issue is depicted in advertising and other media is known as framing. Framing is the context into which information is placed so that it appeals to a particular audience (Rakib et al., 2022). This may be by way of *metaphor*, comparing the build-up of plastics in the environment to other environmentally damaging or reprehensible acts, or through the use of simple *slogans and catchphrases*, such as Jack Johnson's musical recounting of 'reduce, reuse, recycle', increasing the memorability of the information.

Similarly, we might use *storytelling* methods such as the emotional accounts of whales and turtles that have suffered following the ingestion of plastic bags and other marine debris, or links may be made to *cultures and traditions* (or the negative effects that plastics may have on said traditions) and *artefacts*, objects (or potentially locations) with value to the target audience. For example, microplastics in what people traditionally view as either pristine or vulnerable habitats like the Arctic. Alternatively, we might seek to invoke *contrast*, comparing the problem of microplastic pollution to a state in which microplastics do not exist: at its simplest, 'clean' versus 'dirty' or 'healthy' versus 'unhealthy'. Finally, we may attempt to *spin* the issue. In this latter case, using descriptions which convey an inherent judgement and invoke or create bias in the intended audience (Fairhurst & Sarr, 1996).

One factor relevant to both System 1 and System 2 thinking is is confirmation bias. This phenomenon is described as the process of selective acceptance of information which supports our existing worldview. For example, environmentally-minded individuals may be predisposed to believe that 'natural' materials have a smaller environmental footprint and are more sustainable. Indeed, many companies who produce or sell goods utilising bio-based plastics or renewable materials rely heavily on these assumptions, emphasising them in their marketing campaigns by way of suitably aligned framing. This may be as simple as filling adverts with 'natural' and 'green' imagery, or by storytelling and comparison with the environmental issues that arise as a result of widespread plastic pollution. Take, for example, 'eco-friendly' plant-based floss picks (a strand of dental floss, held in a y-shaped pick). The plant-based alternative may have a shorter time until decomposition than plastic picks, but both are single-use products which may make a significant contribution to landfilling if they were to be used widely. If routinely used, a much lower waste footprint may made by floss spools which contain many meters of floss within one packet. Similarly, companies utilising recycled ocean plastics may frame their products as a solution to the problem of marine pollution and the plight of marine life, however, without a proper disposal route for these products at end of life, there may still be a secondary contribution to environmental waste or landfilling.

As discussed in earlier chapters, the best way in which to manage our waste is to avoid the overconsumption of unnecessary products and to encourage their reuse and recycling in order to minimise loss to the environment. Much of the advertising for the above products will utilise positive environmental framing, despite a lack of information regarding the environmental trade-offs being made. At their best, these approaches prove to be effective marketing for a novel product, at their worst, these measures may be considered 'greenwashing'. As highlighted by Franks et al. (2017), to be effective in achieving its sustainability goals, framing must be backed up using '*deep, relevant, broad and sustained set of interventions*', a list to which I might add 'good science'.

Finally, While some members of the public will take an extreme approach to the management of their waste footprint, others will select only a few apparently sustainable options, seeing these acts as their 'doing their bit'. In these cases, the sense of having engaged with an issue (and the resulting reduction in environmental guilt or associated satisfaction) may reduce the likelihood of an individual making subsequent concessions to sustainability. This set of behaviours is widely recognised and is referred to as single-action bias. Manifestations of such 'single actions' may include the appearance of a reuseable travel cup sat beside a trio of individually wrapped biscuits in a staff meeting, the mounds of reusable bags that is added to each time their owner forgets one on their trip to the shops, and other examples of the author's own cognitive dissonance.

Of course, not all sustainable products are equally appealing to consumers. As has been touched upon in the previous section, we can explore the apparent openness of a community to adopt new measures for the sustainable use or management of plastics via a number of metrics, including willingness to pay and willingness to accept, as well as the factors that influence these metrics. For example, the effect of plastic 'warning labels' on food packaging has been explored in relation to consumers' willingness to pay for different egg box constructions. While all labels decreased the price that consumers were willing to pay for the plastic-packed option, the amount that this value was reduced was higher when human health concerns were included than when labelling highlighted environmental or safety concerns (Van Asselt et al., 2022), demonstrating the value of one frame over another. Suggested guidelines from produced by Polyportis et al. (2023) for increasing consumer acceptance of recycled materials include:

- Ensuring trust in environmental impact,
- Bringing environmental impact closer to the individual consumer (reduce abstraction),
- Informing consumers about recycling content history,
- Signalling a person's environmental identity—on the role of recognisability,
- Tackling the perceived quality and performance risks,

- Reducing perceived contamination risk, and
- Highlighting the innovativeness of using recycled content.

11.5 Compliance

We have previously highlighted that regulations regarding plastic use and disposal may have unintentional impacts on both plastic producers and consumers, however, the introduction of policies designed to bring about behavioural change do not necessarily result in the desired outcome when applied in isolation. Both individuals and industry may intentionally or unknowingly persist in environmentally damaging behaviours despite the introduction of apparently well designed regulatory controls. For example, in Nepal, analysis of the factors influencing retailer compliance with the country's plastic bag ban revealed that the type of business (what is being sold) influences compliance. This is believed to be, in part, the result of the suitability of the alternative packaging available. Also, importantly, while retailer perceptions of the plastic problem influenced compliance, so did the level of enforcement (Bharadwaj et al., 2023). Similarly, in India, compliance with EPR regulations, particularly the submission of data was improved between CPCB reports only in the event of legal action (Pani & Pathak, 2021), and in the UK, anecdotal reports of illegal on-farm burning are commonplace. As with so many environmental protections, legal statutes are only as good as the authorities and penalties that are associated with them.

11.6 Conclusions

From the exploration of the secondary environmental impacts of plastic management described above and in the previous chapters, it is clear that a holistic approach to resources use, manufacturing and waste generation is required. One way to achieve these aims is by way of comparative life cycle assessments that seek to identify and compare the scale of various environmental footprints across a range of product and service options. However, while we may be successful in developing the more sustainable products and waste management services, their uptake by the public is not assured. Various social and geographic factors may limit consumer uptake, as might individual drivers of behaviour. Nevertheless, a variety of approaches to the communication of both environmental issues and the availability of sustainable alternative products are available, and these may be applied in support of local, national and international waste management targets. Unfortunately, these communication and marketing methods are equally as available to the producers of less-sustainable products as they are to the producers of sustainable ones. Competing and apparently conflicting information may thus result in confusion on the part of the consumer, as well as variation in the acceptance of technical controls on plastic wastes, and compliance with regulations and laws. As a result, our ability to effectively prevent the global increase in plastic, microplastic and nanoplastic wastes remains a constant challenge.

References

Barnes, M., Chan-Halbrendt, C., Zhang, Q., & Abejon, N. (2011). Consumer preference and willingness to pay for non-plastic food containers in Honolulu, USA. *Journal of Environmental Protection*, *2*(9), 1264−1273. Available from https://doi.org/10.4236/jep.2011.29146.

Bharadwaj, B., Subedi, M. N., & Rai, R. K. (2023). Retailer's characteristics and compliance with the single-use plastic bag ban. *Sustainability Analytics and Modeling*, *3*, 100019. Available from https://doi.org/10.1016/j.samod.2023.100019.

Chen, K. J., Galinato, S. P., Marsh, T. L., Tozer, P. R., & Chouinard, H. H. (2020). Willingness to pay for attributes of biodegradable plastic mulches in the agricultural sector. *HortTechnology*, *30*(3), 437−447. Available from https://doi.org/10.21273/HORTTECH04518-20, https://journals.ashs.org/horttech/view/journals/horttech/30/3/article-p437.xml.

Civancik-Uslu, D., Puig, R., Hauschild, M., & Fullana-i-Palmer, P. (2019). Life cycle assessment of carrier bags and development of a littering indicator. *Science of the Total Environment*, *685*, 621−630. Available from https://doi.org/10.1016/j.scitotenv.2019.05.372, http://www.elsevier.com/locate/scitotenv.

Deeney, M., Green, R., Yan, X., Dooley, C., Yates, J., Rolker, H. B., & Kadiyala, S. (2023). Human health effects of recycling and reusing food sector consumer plastics: A systematic review and meta-analysis of life cycle assessments. *Journal of Cleaner Production*, *397*. Available from https://doi.org/10.1016/j.jclepro.2023.136567, https://www.journals.elsevier.com/journal-of-cleaner-production.

DEFRA. (2008). *A framework for proenvironmental behaviours*. London: Department for Environment, Food and Rural Affairs. https://assets.publishing.service.gov.uk/media/5a789f08ed915d04220640a4/pb13574-behaviours-report-080110.pdf.

De Tandt, E., Demuytere, C., Van Asbroeck, E., Moerman, H., Mys, N., Vyncke, G., Delva, L., Vermeulen, A., Ragaert, P., De Meester, S., & Ragaert, K. (2021). A recycler's perspective on the implications of REACH and food contact material (FCM) regulations for the mechanical recycling of FCM plastics. *Waste Management*, *119*, 315−329. Available from https://doi.org/10.1016/j.wasman.2020.10.012, http://www.elsevier.com/locate/wasman.

DeVetter, L. W., Goldberger, J. R., Miles, C., & Gomez, J. (2021). Grower acceptance of new end-of-life management strategies for plastic mulch in strawberry systems. *Acta Horticulturae*, *1309*, 659−662. Available from https://doi.org/10.17660/ActaHortic.2021.1309.95, https://www.actahort.org/members/showpdf?session = 18568.

Evans, J. S. B. (2011). Dual-process theories of reasoning: Contemporary issues and developmental applications. *Developmental review*, *31*(2−3), 86−102.

Evans, J. S. B., & Stanovich, K. E. (2013). Dual-process theories of higher cognition: Advancing the debate. *Perspectives on Psychological Science*, *8*(3), 223−241.

Fairhurst, G., & Sarr, R. (1996). *The art of framing*. Jossey-Bass.

Franks, B., Hanscomb, S., & Johnston, S. F. (2017). *Environmental ethics and behavioural change* (pp. 1−286). UK: Taylor and Francis. Available from https://www.routledge.com/products/9781315684611, https://doi.org/10.4324/9781315684611.

Hetherington, A. C., Borrion, A. L., Griffiths, O. G., & McManus, M. C. (2014). Use of LCA as a development tool within early research: Challenges and issues across different sectors. *The International Journal of Life Cycle Assessment*, 19, 130−143. Available from https://doi.org/10.1007/s11367-013-0627-8.

Hitt, C., Douglas, J., & Keoleian, G. (2023). Parametric life cycle assessment modeling of reusable and single-use restaurant food container systems. *Resources, Conservation and Recycling*, 190. Available from https://doi.org/10.1016/j.resconrec.2022.106862, http://www.elsevier.com/locate/resconrec.

Khan, M. M. H., Deviatkin, I., Havukainen, J., & Horttanainen, M. (2021). Environmental impacts of wooden, plastic, and wood-polymer composite pallet: a life cycle assessment approach. *The International Journal of Life Cycle Assessment*, 26, 1607−1622.

Liang, Y., Tan, Q., Song, Q., & Li, J. (2021). An analysis of the plastic waste trade and management in Asia. *Waste Management*, 119, 242−253. Available from https://doi.org/10.1016/j.wasman.2020.09.049, http://www.elsevier.com/locate/wasman.

Pani, S. K., & Pathak, A. A. (2021). Managing plastic packaging waste in emerging economies: The case of EPR in India. *Journal of Environmental Management*, 288. Available from https://doi.org/10.1016/j.jenvman.2021.112405, https://www.sciencedirect.com/journal/journal-of-environmental-management.

Paul, S., Sen, B., Das, S., Abbas, S. J., Pradhan, S. N., Sen, K., & Ali, S. I. (2023). Incarnation of bioplastics: Recuperation of plastic pollution. *International Journal of Environmental Analytical Chemistry*, 103(19), 8217−8240.

Polyportis, A., Magnier, L., & Mugge, R. (2023). Guidelines to foster consumer acceptance of products made from recycled plastics. *Circular Economy and Sustainability*, 3(2), 939−952. Available from https://doi.org/10.1007/s43615-022-00202-9, https://www.springer.com/journal/43615.

Rakib, M. A. N., Chang, H. J., & Jones, R. P. (2022). Effective sustainability messages triggering consumer emotion and action: An application of the social cognitive theory and the dual-process model. *Sustainability*, 14(5), 2505.

Sujitha, P., Swetha, N. B., & Gopalakrishnan, S. (2019). Awareness, acceptance and practice of plastic ban legislation among residents of an urban area in Kanchipuram district, Tamil Nadu: A cross sectional study. *International Journal Of Community Medicine And Public Health*, 7(1), 256. Available from https://doi.org/10.18203/2394-6040.ijcmph20195863.

Scherer, C., Emberger-Klein, A., & Menrad, K. (2017). Biogenic product alternatives for children: Consumer preferences for a set of sand toys made of bio-based plastic. *Sustainable Production and Consumption*, 10, 1−14. Available from https://doi.org/10.1016/j.spc.2016.11.001, http://www.journals.elsevier.com/sustainable-production-and-consumption/.

Van Asselt, J., Nian, Y., Soh, M., Morgan, S., & Gao, Z. (2022). Do plastic warning labels reduce consumers' willingness to pay for plastic egg packaging? − Evidence from a choice experiment. *Ecological Economics*, 198. Available from https://doi.org/10.1016/j.ecolecon.2022.107460, http://www.elsevier.com/inca/publications/store/5/0/3/3/0/5.

Vellini, M., & Savioli, M. (2009). Energy and environmental analysis of glass container production and recycling. *Energy*, 34(12), 2137−2143. Available from https://doi.org/10.1016/j.energy.2008.09.017, http://www.elsevier.com/inca/publications/store/4/8/3/.

Zanghelini, G. M., Cherubini, E., Dias, R., Kabe, Y. H. O., & Delgado, J. J. S. (2020). Comparative life cycle assessment of drinking straws in Brazil. *Journal of Cleaner Production*, 276. Available from https://doi.org/10.1016/j.jclepro.2020.123070, https://www.journals.elsevier.com/journal-of-cleaner-production.

Microplastic forecast

12

The production of plastic waste, its fate at end of life and the proliferation of microplastics are driven by a variety of socio-economic, technological and environmental factors. The rapid development of the plastics industry has subsequently been matched by the emergence of an increasing body of literature concerning the problems with both plastic and microplastic pollution, in addition to options for waste reduction and improved management. However, economic development and population growth sustain the demand for new plastics, and their subsequent disposal exceeds our efforts to limit their impacts. As a result, the standing stock of unmanaged wastes in the environment continues to increase. Nevertheless, there has been and will continue to be a great deal of nuance to the observed patterns in plastic transport, accumulation and impacts.

12.1 Lessons from the past

12.1.1 Durability and transport

In Chapter 1 we explored the role of historic events and existing environmental issues in driving the development of our first polymers. Had the individuals responsible for the synthesis of those early plastics estimated the future demand for their new products based on the annual use of scarce materials such as ivory and silk alone, they may have only predicted an output of a few hundred thousand tonnes per year rather than the hundreds of millions of tonnes produced today. The domination of plastic materials has been swift, expanding as new polymers and functional additives have been produced and novel applications or products have emerged. Who in the 1950s could have imagined the messaging and associated behavioural shift caused by plastics, or the spread of the personal electronics industries, the use of polymers in fast food and fast fashion? Apparently not the chemists. Nor could the inventor of Parkesine or polyamide have been thinking in terms of the decadal or 100-year longevity of this new family of materials, let alone the impact of this growth and lifespan on the cumulative mass of the resulting waste.

This is perhaps not surprising. Previously, the impacts of our overconsumption were primarily related to either intensive industrial processes or the act of resource extraction, for example mining, fishing and logging (or indeed, the hunting that led to the decline of elephant and rhinoceros populations which drove the quest for an

Microplastics. DOI: https://doi.org/10.1016/B978-0-443-13324-4.00023-6

alternative to ivory). These predominantly caused localised effects such as habitat loss or change, the mobilisation of harmful compounds or the release of pollutants. With modern plastics, while the impacts of resource extraction (whether of fossil fuels or renewable feedstocks) remain, the persistence of the resulting waste and its associated negative effects may impact environments globally.

Our apparent lack of awareness regarding the significant long-term impacts of pollutants of all kinds began to shift in the mid-20th century as a result of another set of chemical marvels now grouped under the term persistent organic pollutants. This period was already one of increasing environmental interest thanks to a few well-publicised accidental poisonings, occurrences of rivers catching fire and killer smogs. This same era heralded the publication of Rachel Carson's Silent Spring in 1962, which highlighted the impacts of long-lived pesticides and other chemicals on the environment. One such chemical is hexachlorobenzene (HCB), an orga-nochloride first synthesised at small scales in the 1890s and formerly used in agri-culture. The use of HCB became widespread over the same period as plastics were first making their dent in the market, beginning its use in agriculture in 1947. Its eventual industrial applications expanded from a pesticide to metal production, chemical industry applications and combustion in industrial and commercial set-tings until its impacts on both the environment and human health became apparent.

HCB has been seen to have carcinogenic properties and high toxicity in aquatic settings, where the potential for bioaccumulation is high. Indeed, in a notable example, accidental exposure as a result of consumption of bread made from treated seed resulted in 500 deaths including, where consumed by breast-feeding mothers, those of a number of infants. Although at less severe scales, the presence of HCB in foodstuffs has been identified as the most likely cause of widespread human uptake. The use of HCB was subsequently restricted in the late 60s and was one of the first substances covered under the Stockholm Convention. However, the presence of HCB has continued to be noted in ongoing screening, appearing in 76% of samples taken as part of the US National Human Adipose Tissue Survey (FY82), with EPA estimates made in the same period sug-gesting an average annual intake of 68, 22 and 5 μg for adults, toddlers and infants, respectively (Stanley et al., 1986).

Similar patterns are also noted with, Mirex, synthesised in 1946 and introduced as a pesticide in 1955; Heptachlor, the use of which as an insecticide began in the 40s, which also made an appearance in Silent Spring and was widely banned in the 1980s; and Toxaphene, which was widely used as a replacement for Dichlorodiphenyltrichloroethane (DDT), being first manufactured in the 1940s, peaking in use in 1974, before being reduced and subsequent widely banned in the late 80s and early 90s; along with a great many other compounds and substances (Fig. 12.1).

12.1.2 The impact of persistent organic pollutant control

Despite their widespread notoriety, the environmental half-lives of these persistent pollutants pale in relation to those of many plastics, with HCB persisting for up

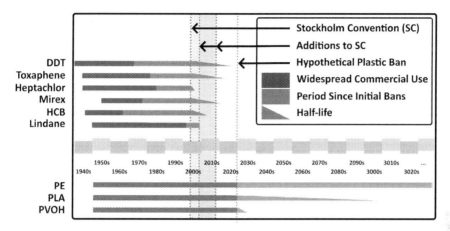

FIGURE 12.1 A comparison of the use and fate of persistent organic pollutants and plastics
A timeline of introduction, use and restriction of example pollutants controlled under the Stockholm Convention (SC). The apparent environmental half-lives of these contaminants are given following the introduction of the SC or addition to the SC's list of controlled substances to enable easier comparison. A comparative timeline is provided for a variety of plastic polymers, the apparent half-lives of which are based on the introduction of theoretical (and highly improbable) restrictions in 2030.

to 7.5 years. Estimates of the levels of HCB in the environment from 2005 suggested a range of between 10,000 and 26,000 tonnes, despite widespread restrictions on its agricultural use. This environmental distribution is spread across multiple environments including soils, sediments and oceans (Barber et al., 2005).

Similarly, DDT has an environmental half-life of 15 years, and while Heptachlor is susceptible to degradation by organisms and environmental factors and its breakdown in air may be as little as 6 hours, its apparent half-life in soils is between 6–9 months (Reed & Koshlukova, 2014). Additionally, Mirex has a half-life in soils of approximately 12 years and, once ingested, may stay in the body for over a year (Croom, 2012), and Toxaphene may have a soil half-life of between 1 and 14 years, easily entering waterbodies via both soil runoff and atmospheric transport (Wallace, 2014).

While the concentrations of these pollutants in the environment may markedly exceed recognised 'safe' levels, the length of the half-life and the impact of point sources result in very different environmental profiles to those of plastic pollution. High levels of POPs are associated with local sources or the transport of contaminated materials, and they may bind favourably to local sediments, resulting in semipredictable locally elevated concentrations as well as a steep surrounding concentration gradient. Conversely, while plastic pollution is also often elevated in relation to point sources, the existence of multiple diffuse sources as well as highly varied patterns of settlement, burial and remobilisation cause plastic litter to occur at high levels in unexpected places (Welden & Lusher, 2017).

12.1.3 **Effectiveness and impact of material replacement**

Of course, there are other key lessons that we might take from the management of historically significant environmental pollutants, particularly in relation to the unintended effects of regulatory and technological interventions. For example, there is the question of what to replace plastics with.

As we briefly touched upon in relation to Toxaphene and its role as a replacement for DDT, the 'cure' (the replacement product) may be nearly as environmentally damaging as the original 'disease'. We may observe similar unintended impacts in our management of airborne pollutants. Fluorinated gases (F-gas), such as hydrofluorocarbons, perfluorocarbons and sulphur hexafluoride, are the product of manufacturing, and are typically used as aerosol propellants as well as in insulation, fire protection and air conditioning (Sovacool et al., 2021). These substances were introduced in response to existing environmental concerns regarding chlorofluorocarbons (CFCs) and their effect on the ozone layer. As a result of their apparent negative impacts on ozone, limits on CFCs were set out in the Montreal Protocol (1985) and F-gas production rose to meet market needs.

However, F-gases are now known to contribute significantly to the greenhouse effect, indeed sulphur hexafluoride (SF_6) has the greatest climate forcing yet identified. The atmospheric concentrations of SF_6 alone have more than doubled over the past 20 years and, in 2021, it was estimated that the climate forcing arising from F-gas production was equivalent to approximately 6,300,000 mt of CO_2; although, for reference, CO_2 emissions in the same year were 37.9 Gt.

In recognition of the impact of one of the most widespread F-gases, hydrofluorocarbons, on both ozone and climate, the Kigali Amendment to the Montreal Protocol (ratified by 152 states as of writing) sets out plans for a staged reduction in their use as well as improvements in leak detection and prevention and correct disposal and handling at end of life. Additionally, there are a number of regional and national regulations, such as EU Regulation No 517/2014, which seek to ban the use of F-gases in new products and limit the volume that may be traded.

The impact of CFC control and the resulting climate impact of F-gases echoes the arguments made in earlier chapters, that the replacement of an environmentally damaging material with an *apparently* more sustainable alternative may result in unexpected secondary effects. While no single change in our use of plastics may represent a comparable problem to that outlined above, some issues have already become apparent, resulting in the rise and subsequent decline of options for plastic management. For example, the application of degradable fillers in macroplastics with the intention of reducing the lifespan of large plastic debris in the environment led to the unfortunate effect of increasing the rate of fragmentation and the production of microplastics, rather than the onset of degradation and eventual mineralisation (Abdelmoez et al., 2021). Similarly, the introduction of compostable and biobased plastics has been linked with the misdirection of wastes and the contamination of waste streams, reducing the effectiveness of existing waste management measures (Aldas et al., 2021). As we highlighted in our consideration of LCAs in the prevention of plastic wastes and our discussion of the influence of various behavioural factors on the reduction of plastic use, it is better to encourage sustainable action by curbing the use of all environmental resources than to fritter away our energies in trying to determine a technological fix that comes with its own environmental baggage.

12.1.4 **Effectiveness of reduced production**

While the impacts of macro-, micro- and nanoplastics remain less apparent than those associated with the contaminants considered above, this is no excuse for a lack of urgency on the part of policymakers and regulators concerned with their management. Nevertheless, there are a number of barriers to effective policymaking related to plastics, one of which we have not covered in previous chapters: the diversity of plastics themselves. Unlike the above pollutants, plastics are not a single compound or product group, resulting in varied physical and chemical properties, and applications which span multiple products and industries. Unlike the single pollutant approach, under which chemical compounds (or groups of compounds) are identified and restricted, plastic bans have typically focused on individual products (such as straws, bags, cotton buds and other single-use items), rather than specifically challenging polymers, despite these being more likely to become suspended in the water column or to persist for long period in the environment. Where enforceable bans do exist, the targets typically represent relatively low-hanging fruit for policymakers, products that may be excluded without significant effects on industry practice or consumer behaviour. Some of these plastics do not represent a significant contribution to plastic wastes and, subsequently, have a limited benefit to the environment.

Additionally, as indicated in Fig. 12.1, where bans are implemented, the longevity of plastics already in the environment may result in a significant period before the apparent benefits are felt. If we once again look at HCB we might observe that, despite widespread reductions in its application, environmental concentrations have been slow to reduce. Observations of HCB concentrations in air samples taken across the Great Lakes Region show little signs of diminishing over the past quarter century, with concentrations varying between 30 and 100 pg/m^3 (Hites et al., 2022). In the same region, elevated concentrations of HCB in samples of commercial fish species have been regularly recorded (Allen-Gil et al., 1997; Kucklick & Baker, 1998), with concentrations as high as 17 ppb in the sampled fish fillets (Newsome & Andrews, 1993; Zabik et al., 1995). These levels may be the result of ongoing bioconcentration, with the maximum HCB concentration factor recorded as up to 17 million times (Muir & Norstrom, 1994; Veith et al., 1979). Fortunately, the potential for the bioconcentration of plastic does not appear to be as marked or pervasive. Unfortunately, the environmental half-life of plastic is much longer than that of HCB and other POPs. As a result, the benefits of plastic bans and other management measures discussed in the previous chapters may take an even greater period either to be observable in the environment or to result in decreased negative effects on the community (Fig. 12.1).

12.2 Plastic predictions

12.2.1 Future plastic output under differing use and management scenarios

Our desire to minimise the production and mishandling of plastic wastes and inputs of microplastics to the environment has arisen as a result of our understanding of plastic impacts. Under business-as-usual scenarios, it is believed that plastic *production* will exceed 1,300 million tonnes by 2060 and that the estimated mass of plastic waste will be just in excess of 1,000 million tonnes (Global Plastics Outlook: Policy Scenarios to 2060, 2022), with some of the increased demand for plastic associated with projections of population growth, which indicate that there will be over 10 billion people by 2060. As a result, even with substantial action, the concentrations of plastics and microplastics in the environment (as well as their impacts on both abiotic conditions and biota) will continue to rise. However, the extent of this increase and the location at which these impacts will be felt remain uncertain despite a number of studies which have sought to project the production of plastic wastes into the coming decades.

Below is a Sankey diagram representing the estimated flow of plastic materials under current management measures. The estimates of plastic fate at end of life shown are taken from the OECD's recent report (Global Plastics Outlook: Economic Drivers, Environmental Impacts and Policy Options (2022)), which indicates that approximately 22% of plastic wastes are mismanaged or littered, with the rest either landfilled (46%), incinerated (17%) or collected for recycling (15%) (Fig. 12.2). Using 2019 figures for the production and management of plastics, this would equate to around 78 million tonnes of mismanaged wastes. Looking forward, if this relationship were to remain consistent, then the

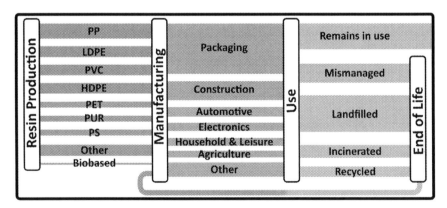

FIGURE 12.2 Material flows between stages of the plastic lifecycle

Proportions based on Plastics Europe (2022) and the OECD Global Plastics Outlook database.

proportion of mismanaged plastic produced in 2060 would equal around 223 million tonnes, approximately three times that today. Under this business-as-usual scenario, the mass of plastic reaching the oceans, estimated as 0.5% of the mass of plastic at end of life, would equate to 5.7 million tonnes in 2060.

However, it is not feasible to expect these patterns to remain the same, indeed, it is expected that plastic leakage to the environment will decrease. Fluctuation in consumer behaviour, the introduction of novel materials and improved waste management may all influence the creation and loss of plastic wastes. We can explore the potential impacts of these changes in the same way as we make predictions about climate change, by forecasting the effect of different scenarios on microplastic output and environmental conditions. Below we will consider the impacts of developing resource circularity, limiting product demand or reducing resource use according to the following 4 approaches: increased recycling, reduced consumption, alternative material use and a cumulative forecast.

Under the first scenario, let us consider the effects of either a 5%, 10% or 30% increase in the share of recycled plastics from our baseline. Applying these percentages to the 34.5 million tonnes of recycled material in 2021 would result in an increase of 1.7, 3.5, or 11.5 million tonnes of material, or a total recycled mass of 36.37, 38.1 or 46.17 million tonnes respectively. If we look at these weights as a proportion of plastic *production*, the increased availability of recycled material may reduce the requirement for virgin plastics by 0.5%, 1% or 3.3%.

Under our second scenario, we may attempt to reduce the consumption of plastics in either household goods or plastic packaging by 20%. These changes may reduce plastic demand by 1.4% and 8.8%, or 5.5 and 34.4 million tonnes, respectively. Finally, in the last of our individual scenarios, we may explore the impact of a 5% increase in the use of either glass bottles or compostable plastics in packaging. Currently approximately 690 billion glass bottles are produced per year, and a 5% increase would equate to a further 34.5 billion. Reducing PET output by the mass required to create the same number of bottles reduce plastic demand and waste by approximately 0.276 million tonnes (a result of the light weight of PET compared to glass), alternatively, a 5% increase in the use of bioplastic may result in a reduction in virgin PET use of 0.29 million tonnes. Under either scenario, the reduction in the mass of fossil-fuel derived virgin plastics is less than 0.1%.

If we combine the minimum increase in recycling, the changes in household plastic use, and the increase in the share of glass bottles from our 3 scenarios, we may see a reduction in virgin plastic production of 1.97 million tonnes (0.549%). Additionally, due to increased recycling of the resulting wastes, a reduction in the mass of mismanaged plastics of just above 1.3 million tonnes, a change of approximately 2%. While such a change is desirable, it must be recalled that some of the measures suggested, such as the 5% increase in the use of glass bottles, will increase water use, transport-associated CO_2 emissions and other factors in the resource chain.

12.2.2 The impact of reduced plastic pollution on microplastic concentrations

While the above reductions are notable, there remains uncertainity regarding the short-, medium- and long-term impacts of projected plastic waste output on the formation of secondary microplastics. Fortunately, we can use some more 'back-of-the-napkin' calculations to help us visualise these effects. Returning to our earlier baseline estimates of mismanaged plastic mass, we may attempt to convert this into microplastic abundance. The weight of mismanaged plastic waste generated in 2019, 78 million tonnes, may be converted to a volume based on polymer density. Here we have calculated an average polymer density of 1.02 g cm^{-3} from the specific gravities of polymers outlined in current production figures. This suggests a cube approximately 0.076 km^3 or 0.423 km long on every side (Fig. 12.3). If this cube was broken down into microplastic particles that were exactly 5 mm on every side, it would produce c.609 trillion (6.09E^{+14}) particles. However, every little 5 mm^3 particle contains 125 smaller 1 mm^3 particles, or 1000 tiny 0.5 mm^3 particles; thus, the abundance of microplastics may significantly increase in line with ongoing degradation. For example, in their recent study of microplastics in the upper ocean, Isobe et al. (2019) used a mode size of 1.2 mm to estimate plastic abundance. Were we to further fragmentation our plastic cube to this size, this would result in a microplastic output of 40 quadrillion particles.

Fortunately, not all of this misshandled waste ends up in the oceans, and this fragmentation of plastic to microplastic would not occur in one go. As already indicated, estimates of plastic waste entering the environment are approximately 0.5% of total waste production (not the larger 22% represented by the mismanaged proportion discussed above). This would be a cube of approximately 0.0017 km^3, or 0.1199 km on each side (Fig. 12.3), which would contain a potential 13.8 trillion 5 mm^3 microplastics. The degradation rate of plastics in the environment may be as low as a few percent of the total mass per year. For the purposes of this hypothetical example, we will use a low fragmentation rate of 0.1% of the total mass. Under this scenario, 2019 plastic waste may contribute 0.0138 trillion microplastics each year. While this lag time may be significant, the cumulative effect of the ongoing annual fragmentation of lost plastics from a single year would reach over 4 trillion microplastic particles by 2050!

The volume of our oceans is approximately 1.335 billion cubic kilometres. If we divide the potential 0.0138 trillion microplastics released from our 2019 cube by this ocean volume, we arrive at just over 10.3 microplastics per cubic kilometre in the first year of exposure. This figure appears to be[1] substantially lower than existing concentrations in surface waters alone. For example, Isobe and

---

[1] As indicated in earlier chapters, comparisons between studies made using square kilometres vs cubic kilometres present significant challenges. We can check our own calculations using some of the above studies, for example, if we take the upper estimate in Van Sebille's 2015 study (236,000 metric tonnes) and apply the above conversion steps to turn this mass into 5-mm microplastic particles, we return an estimated abundance of 46.2 trillion microplastics, which is within the range of particle concentrations reported in this paper.

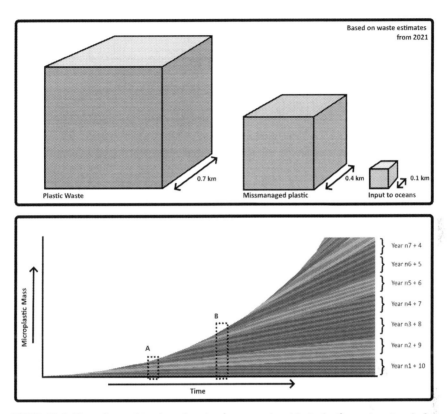

FIGURE 12.3 The volume of total produced, mismanaged and lost plastic wastes at end of life and the formation of microplastics in the environment

The top box utilises estimates of the mass of plastic waste generated in 2021 according to OECD estimates and the average density of its constituent polymers based on Plastics Europe production figures to provide an estimate plastic volume. The bottom box indicates the impact of ongoing macroplastic fragmentation on the standing stock of microplastics. For example, at time point A, the macroplastics introduced in year 1 (lower blue) have been releasing microplastics for four years, the macroplastics introduced in year 2 (tan) have been releasing microplastics for three years, the macroplastics introduced in year 3 have been releasing microplastics for 2 years, and the macroplastics introduced in year 4 have been releasing microplastics for one year. That is ten times the microplastic input in year one. Time point B shows a similar jump in the mass of microplastics released as a result of the ongoing fragmentation of the existing stock of environmental macroplastic debris.

colleagues' 2021 study estimating microplastic levels in the upper ocean (up to 3 m depth) alone returned a total of 24.4 trillion pieces, weighing between 8.2×10^4 and 57.8×10^4 tonnes (Isobe et al., 2021). Similarly, estimates of the mass of floating plastic in 2014 by van Sebille et al. (2015), suggest between 9.0×10^4 and 23.6×10^4 tonnes, 15−51 trillion particles. However, this difference may be explained by both the use of 5 mm as our base particle size (remember, a $5 \, \text{mm}^3$ cube of plastic contains 1000 $0.5 \, \text{mm}^3$ pieces), which results in a lower estimated plastic abundance, and the fact that the above value considers input from just one year's waste and does not account for the additive effect of the existing mass of marine litter. The cumulative effect of multiple years input of plastic litter on the mass of microplastic may be seen in Fig. 12.3.

The distribution of these future microplastics is also not expected to be even, driven by areas of increased plastic and microplastic input, rapid sinking and settlement (Zhang, 2017), as well as conditions at sites of aggregation. Indeed, much of the plastic introduced into the aquatic environment will be rapidly entrained into sediments. For example, estimates of the mass of microplastics on the ocean floor are as high as 14 million tonnes , a figure several orders of magnitude higher than that observed in the upper layers of the ocean (Barrett et al., 2020).

Looking forward again, if we return to the OECD figures of the projected mass of plastic wastes generated between 1950 and 2060, we may generate estimates of mismanaged waste for the period (Table 12.1). Decadal estimates of the total microplastic output arising from the roughly 7400 million tonnes of mishandled plastic suggest a minimum total mass of 956.63 million tonnes by 2060, or approximately 3788 trillion (3.789 E^{+15}) 5 mm microplastics. Such significant inputs increase the likelihood of many of the impacts of microplastics previously observed in laboratory studies. Subsequently, as the microplastics released continue to break down, further adding to the stock of small microplastics and nanoplastics, they may bring about more severe effects from the individual to the ecosystem level.

Table 12.1 Decadal estimates of the mass of both mismanaged plastic wastes and microplastic fragments based on OECD estimates of plastic waste production.

Decade	OECD estimates of waste plastic (Mt)	Decadal mass of mismanaged plastic (22% total Mt)	Hypothetical cumulative decadal contribution to microplastic (5 mm^3) mass (Mt) 1950–2060
1950–1959	17.82	3.92	2.06
1960–1969	94.00	20.68	9.85
1970–1979	318.66	70.11	30.06
1980–1989	630.75	138.77	52.73
1990–1999	1136.25	249.97	82.51
2000–2009	1990.02	437.80	122.80
2010–2019	3034.13	667.51	154.20
2020–2029	4050.83	891.18	161.69
2030–2039	5302.36	1166.52	153.24
2040–2049	6951.45	1529.32	124.52
2050–2060	9936.74	2186.08	62.98
Total	33,463.00	7361.86	956.63

In earlier chapters, it has been noted that the impacts of microplastic exposure on biological endpoints demonstrate either *threshold* or *hormesis*-type dose-response relationships (Agathokleous et al., 2021). Under the threshold scenario, effects are not observed until a 'tipping point' is reached, whereas a hormesis-type relationship may indicate an initial positive effect followed by negative impacts above a specific concentration. For example, adult *Daphnia magna* exposed to polystyrene microspheres (6-μm) at concentrations of 0, 5, 30 and 100 μg/L demonstrated increased growth in the 5 μg/L condition and marked dose-related decreased in growth thereafter (Eltemsah & Bøhn, 2019). As highlighted previously, the concentrations of microplastics applied in a high percentage of studies are far above that routinely reported in the environment (Bucci et al., 2020); however, while the increases in baseline plastic output indicated above are estimates only, the ongoing production of plastic wastes and secondary microplastics may be sufficient to manifest impacts formerly restricted to only the highest exposure groups in toxicity trials, moving these effects from the outlandish to the feasible.

Of course, these changes will not only be felt in marine systems, although they represent the main microplastic sink. Increased outputs of plastics will also be observed in terrestrial, airborne and freshwater environments. Although in the more transient of these settings, there may be only a limited period during which microplastics may form from macroplastic parent material. As a result, the increase in microplastic over current baselines may be much lower than that seen at long-term sinks. Subsequently, the unknown residence time across habitats prevents the prediction of increases in environmental concentrations at all but the largest spatial scales, such as global, regional or ecotype averages. Nevertheless, whatever the size of the change, whether our use and disposal of plastic continue to grow at their current rate or we implement significant changes, the outlook is clear: the mass or concentration of microplastics in the environment will continue to rise for decades to come.

12.3 Conclusion

Microplastic pollution is a problem born and perpetuated by our own ingenuity. Although similar to historic pollutants in many ways, microplastics and their parent plastic products are comparatively unmatched in their longevity. Through 75 years of iterative design, manufacture and disposal, we have created and mobilised a pollutant capable of outlasting its creators and subsequently increased the scale of its production by its inclusion in every aspect of the human environment. At some time or other, each of us will come into contact with it, ingest it, inhale it or perpetuate its spread. However, equally ingenious methods are being devised to enable its management. Methods that go beyond traditional approaches to the handling of solid wastes or chemical substances. These are paired with a global movement of concerned stakeholders people who seek to make changes to the way that they impact the environment, whether through technology or behavioural change. If consistently applied, these changes promise significant cumulative reductions in the sources of microplastics. Despite this, the multiple trade-offs in plastic management highlight that there is no clear path by which we might eliminate the introduction of microplastics into our environment, and it will be a long time until the effects of the standing stock of microplastics cease to be felt.

References

Abdelmoez, W., Dahab, I., Ragab, E. M., Abdelsalam, O. A., & Mustafa, A. (2021). Bio- and oxo-degradable plastics: Insights on facts and challenges. *Polymers for Advanced Technologies*, *32*(5), 1981–1996.

Agathokleous, E., Iavicoli, I., Barceló, D., & Calabrese, E. J. (2021). Micro/nanoplastics effects on organisms: A review focusing on 'dose. *Journal of Hazardous Materials*, *417*. Available from https://doi.org/10.1016/j.jhazmat.2021.126084, http://www.elsevier.com/locate/jhazmat.

Aldas, M., Pavon, C., De la Rosa-Ramirez, H., Ferri, J. M., Bertomeu, D., Samper, M. D., & Lopez-Martinez, J. (2021). The impact of biodegradable plastics in the properties of recycled polyethylene terephthalate. *Journal of Polymers and the Environment*, *29*(8), 2686–2700.

Allen-Gil, S. M., Gubala, C. P., Landers, R. W. , D. H., Wade, T. L., Sericano, J. L., & Curtis, L. R. (1997). Organochlorine pesticides and polychlorinated biphenyls (PCBs) in sediments and biota from four US Arctic lakes. *Archives of Environmental Contamination and Toxicology*, *33*(4), 378–387. Available from https://doi.org/10.1007/s002449900267.

Barber, J. L., Sweetman, A. J., Van Wijk, D., & Jones, K. C. (2005). Hexachlorobenzene in the global environment: Emissions, levels, distribution, trends and processes. *Science of the Total Environment*, *349*(1–3), 1–44. Available from https://doi.org/10.1016/j.scitotenv.2005.03.014.

Barrett, J., Chase, Z., Zhang, J., Banaszak Holl, M. M., Willis, K., Williams, A., Hardesty, B. D., & Wilcox, C. (2020). Microplastic pollution in deep-sea sediments from the Great Australian Bight. *Frontiers in Marine Science*, *7*, 808. Available from https://www.frontiersin.org/articles/10.3389/fmars.2020.576170.

Bucci, K., Tulio, M., & Rochman, C. M. (2020). What is known and unknown about the effects of plastic pollution: A meta-analysis and systematic review. *Ecological Applications*, *30*(2). Available from https://doi.org/10.1002/eap.2044, http://onlinelibrary.wiley.com/journal/10.1002/(ISSN)1939-5582.

Croom, E. (2012). Metabolism of xenobiotics of human environments. *Progress in Molecular Biology and Translational Science*, *112*. Available from https://doi.org/10.1016/B978-0-12-415813-9.00003-9, http://www.elsevier.com/books/book-series/progress-in-molecular-biology-and-translational-science#.

Eltemsah, Y. S., & Bøhn, T. (2019). Acute and chronic effects of polystyrene microplastics on juvenile and adult Daphnia magna. *Environmental Pollution*, *254*. Available from https://doi.org/10.1016/j.envpol.2019.07.087, https://www.journals.elsevier.com/environmental-pollution.

Global Plastics Outlook: Economic Drivers, Environmental Impacts and Policy Options. (2022). OECD. https://doi.org/10.1787/de747aef-en.

Global Plastics Outlook: Policy Scenarios to 2060. (2022). OECD. https://doi.org/10.1787/aa1edf33-en.

Hites, R. A., Bidleman, T. F., & Venier, M. (2022). Atmospheric concentrations of hexachlorobenzene and octachlorostyrene are uniform across the great lakes region and have not changed much in 25 years. *Environmental Science and Technology Letters*, 9 (8), 660−665. Available from https://doi.org/10.1021/acs.estlett.2c00444, http://pubs.acs.org/page/estlcu/about.html.

Isobe, A., Iwasaki, S., Uchida, K., & Tokai, T. (2019). Abundance of non-conservative microplastics in the upper ocean from 1957 to 2066. *Nature Communications*, 10(1). Available from https://doi.org/10.1038/s41467-019-08316-9, http://www.nature.com/ncomms/index.html.

Isobe, A., Azuma, T., Cordova, M. R., Cózar, A., Galgani, F., Hagita, R., Kanhai, L. D., Imai, K., Iwasaki, S., Kako, S. 'ichro, Kozlovskii, N., Lusher, A. L., Mason, S. A., Michida, Y., Mituhasi, T., Morii, Y., Mukai, T., Popova, A., Shimizu, K., & Zhang, W. (2021). A multilevel dataset of microplastic abundance in the world's upper ocean and the Laurentian Great Lakes. *Microplastics and Nanoplastics*, 1(1), 1−14. Available from https://doi.org/10.1186/s43591-021-00013-z.

Kucklick, J. R., & Baker, J. E. (1998). Organochlorines in Lake Superior's food web. *Environmental Science and Technology*, 32(9), 1192−1198. Available from https://doi.org/10.1021/es970794q, http://pubs.acs.org/journal/esthag.

Muir, D. C., & Norstrom, R. J. (1994). Persistent organic contaminants in Arctic marine and freshwater ecosystems. *Arctic Research of the United States*, 8, 136−146.

Newsome, W. H., & Andrews, P. (1993). Organochlorine pesticides and polychlorinated biphenyl congeners in commercial fish from the Great Lakes. *Journal of AOAC International*, 76(4), 707−710. Available from https://doi.org/10.1093/jaoac/76.4.707.

Reed, N. R., & Koshlukova, S. (2014). *Heptachlor. Encyclopedia of toxicology* (3rd edn, pp. 840−844). United States: Elsevier. Available from http://www.sciencedirect.com/science/book/9780123864550, https://doi.org/10.1016/B978-0-12-386454-3.00149-4.

Sovacool, B. K., Griffiths, S., Kim, J., & Bazilian, M. (2021). Climate change and industrial F-gases: A critical and systematic review of developments, sociotechnical systems and policy options for reducing synthetic greenhouse gas emissions. *Renewable and Sustainable Energy Reviews*, 141, 110759.

Stanley, J. S., Boggess, K. E., Onstot, J., Sack, T. M., Remmers, J. C., Breen, J., Kutz, F. W., Carra, J., Robinson, P., & Mack, G. A. (1986). PCDDs and PCDFs in human adipose tissue from the EPA FY82 NHATS repository. *Chlorinated Dioxins and Related Compounds 1985*, 15(9), 1605−1612. Available from https://doi.org/10.1016/0045-6535(86)90444-3, https://www.sciencedirect.com/science/article/pii/0045653586904443.

van Sebille, E., Wilcox, C., Lebreton, L., Maximenko, N., Hardesty, B. D., van Franeker, J. A., Eriksen, M., Siegel, D., Galgani, F., & Law, K. L. (2015). A global inventory of small floating plastic debris. *Environmental Research Letters*, 10(12), 124006. Available from https://doi.org/10.1088/1748-9326/10/12/124006, https://doi.org/10.1088/1748-9326/10/12/124006.

Veith, G. D., DeFoe, D. L., & Bergstedt, B. V. (1979). Measuring and estimating the bioconcentration factor of chemicals in fish. *Journal of the Fisheries Research Board of Canada*, 36(9), 1040−1048. Available from https://doi.org/10.1139/f79-146.

Wallace, D. R. (2014). *Toxaphene. Encyclopedia of toxicology: Third edition* (pp. 606–609). United States: Elsevier. Available from http://www.sciencedirect.com/science/book/9780123864550, 10.1016/B978-0-12-386454-3.00202-5.

Welden, N. A., & Lusher, A. L. (2017). Impacts of changing ocean circulation on the distribution of marine microplastic litter. *Integrated Environmental Assessment and Management, 13*(3), 483–487.

Zabik, M. E., Booren, A. M., Nettles, M., Song, J. H., Zabik, M. J., Welch, R., & Humphrey, H. (1995). Pesticides and total polychlorinated biphenyls in chinook salmon and carp harvested from the great lakes: Effects of skin-on and skin-off processing and selected cooking methods. *Journal of Agricultural and Food Chemistry, 43*(4), 993–1001. Available from https://doi.org/10.1021/jf00052a029.

Zhang, H. (2017). Transport of microplastics in coastal seas. *Estuarine, Coastal and Shelf Science, 199*, 74–86. Available from https://doi.org/10.1016/j.ecss.2017.09.032, http://www.elsevier.com/inca/publications/store/6/2/2/8/2/3/index.htt.

Glossary

Additives   May include antimicrobial compounds, antioxidants, antistatic agents, plasticisers, blowing agents, fillers, flame retardants, heat stabilisers, impact modifiers, lubricants, AV stabilisers, colourants (including pigments and soluble azocolorants), plasticisers and reinforcements.

Addition polymerisation   Step-growth polymer formation from monomers containing $C = C$ bonds.

Agriplastic   Plastic materials specific to agricultural settings, such as mulch films, irrigation systems and bale wrap.

Amorphous arrangement   A random arrangement of the polymer chain within the plastic, as a result of their atactic structure.

Anaerobic degradation   The microbial breakdown of materials in the absence of oxygen, resulting in the formation of a mixture of methane carbon dioxide and other gases as a by-product

Atactic arrangement   A polymer chain in which side groups are randomly distributed.

Bathymetry   The depth and three-dimensional structure of the bed of a water body.

Bioaccumulation   The build-up of a material or substance within the body of an organism. To be achieved, uptake must exceed elimination.

Bioconcentration factor   The relationship between the concentration of a substance observed in the environment and that of an organism.

Biofouling   The development of a colony or community of organism on the surface of a material.

Biodegradable plastics   Plastics that may be readily broken down by microbes and other animals able to produce suitable enzymes.

Biomagnification   Increasing levels of a material or substance observed as successive levels of a trophic chain or food web.

Bioplastic   Plastics synthesised from renewable feedstocks such as cellulose, starches, fats and oils.

Bulk sampling   The collection of a known volume of an environmental media.

Compostable plastic   A plastic that is biodegradable under the specific conditions found in either home or, predominantly, industrial composting.

Condensation polymerisation   The process of polymerisation in which two different monomers are added together, resulting in the release of water or other small molecule, as in the formation of polyesters.

Coriolis force/effect   The influence of the rotation of the Earth on the observed motion of water.

Dual process theory   The interaction between system 1 and system 2 thinking, or our fast instinctual response and more considered reasoning.

Extender producer responsibility   Policies under which the originator of a waste product is held responsible for its management.

End-of-waste status   Conversion of a waste product into a product or material for which there is market demand.

Extracellular polymeric substances   High-molecular-weight natural polymers synthesised by microorganisms which promote cell adhesion and biofilm formation.

Fetch   The distance over which the frictional forces of air currents may act over a water body to create waves and currents.

FTIR   Fourier Transformed Infrared Spectroscopy

Focal Plane Array (FPA)   An image sensor which utilises a grid of light detectors to generate infrared imagery.

Framing   The use of multiple communication measures to develop a context "frame" for the information which will influence the way in which it is received by the audience.

Free radical   A molecule which contains an unpaired electron, making them highly reactive and capable of influencing the bonds in other molecules.

Gasification   The heating of materials to high temperature ($>700°C$) at controlled oxygen levels, resulting in partial oxsidisation and the production of syngas.

Gyre   Basin scale circular water movement driven by wind patterns and Coriolis force.

Half-life   The time taken for a pollutant to reduce in concentration by 50%.

Hydrophobic   A lack of affinity for water exhibited by non-polar molecules.

Isotactic arrangement   A polymer chain in which side groups appear on one side alone.

Lentic   Still freshwater habitats, such as lakes.

Life cycle analysis   An assessment of the environmental footprint of a product, taking into account impact of the whole product lifespan on various environmental concerns, such as materials, water and energy footprint, land-use, transport and disposal.

Lotic   Moving freshwater environments, such as rivers

Limit of Detection   The lowest mass, volume or quantity at which a substance is reliably detected.

Limit of Quantitation   The lowest mass, volume or quantity at which a substance is reliably quantified.

Lithophyllic   A affinity for fats exhibited by non-polar molecules.

Macroplastics   Plastics above 10 mm (>5 mm when the mesoplastic category is not used)

Mesoplastic   Plastics with a size range between 5 and 10 mm.

Mesocosm   An enclosed environment, representative of the wider habitat or community, which enables experimenters to control key conditions of interest.

Microplastic   Plastic particles between 1 nm (using GESAMP definitions) and 5 mm; however, this upper limit has caused significant debate amongst the scientific community, with many authors highlighting its divergence from typical size ranges and preferring to use the 1 mm upper limit.

Monomer   Atoms or small molecules from which polymers are formed. Naturally occurring monomers include amino acids, fatty acids, and sugars. The monomers found in plastic include ethylene, phenol, propylene, styrene and vinyl chloride.

Nanoplastic   Plastic particles under either 1000 or 100 nm (definition variable depending on authority consulted).

Naphtha   A flammable fraction of crude oil.

Oxodegradable plastics   Plastics that may be fragmented by abiotic factors as a result of additives into the polymer.

Oxo-biodegradable   Plastics that are made more available to the action of biodegrading organisms as a result of changes to the polymer or the introduction of additives.

Packaging Export Recycling Note (PERN)/Packaging Recycling Note (PRN)   Evidence of recycling issued by an accredited reprocessor to a packaging producer to confirm that the legal requirement to achieve conversion to end-of-waste status has been met.

Parent plastic(s)   The macroplastic or mesoplastic from which secondary microplastics form.

Planned obsolescence   A design practice which limits the lifespan of a product through restricted durability or technological outmodedness.

Polyol   An organic compound containing multiple -OH groups

Polyolefin   Plastics derived from ethylene and propylene.

Polymer   Substances, both natural and synthetics, made of repeating monomer units.

Producer compliance schemes (PCS)   Associations of waste producers who take on the responsibility for ensuring the compliance of their members with relevant waste regulations.

Pyrolysis   The thermal breakdown of plastics under oxygen-free conditions to form fuels and other secondary chemicals.

Recalcitrance   Extreme resistance to change.

Reprocessor   Organisations responsible for the recycling or recovery of waste material to achieve end-of-waste status.

Sewage sludge   The solid residue arising as a result of wastewater treatment.

Semi-crystalline   A regular or semi-regular arrangement of the polymer chains within the plastic, driven by the presence of syndiotactic and isotactic chains.

Single action bias   The tendency of an individual to feel a reduced motivation to further action after changing a single behaviour.

Specific gravity   The relative density of a substance when compared to that of water at the same temperature.

System 1 thinking   Instinctive or automatic decision-making.

System 2 thinking   Conscious logical thought.

Syndiotactic arrangement   A polymer chain in which side groups appear on alternate sides.

Thermoset   Plastics which become irreversibly formed following curing. This permanence is the result of high levels of crosslinking between polymer chains.

Thermoplastics   Plastics which may be repeatedly heated and formed into new shapes.

Trophic transfer   The movement of a pollutant or other substance between levels of the food chain.

Upwelling   The bulk upward movement of water as a result of differential density, the action of wind and Coriolis force.

Willingness to accept   The amount of compensation or trade off that an individual is willing to take in exchange for giving up some good or service.

Willingness to pay   The maximum acceptable price for a product or service.

Windage   The impact of the wind on an objects course or movement.

Index

Note: Page numbers followed by "*b*," "*f*," and "*t*" refer to boxes, figures, and tables, respectively.

Printed in the United States
by Baker & Taylor Publisher Services